Engineering Surveying

problems and so~~lutions~~

Second Edition

F A Shepherd FRICS, C Eng, FI Min E, F Inst CES

Principal Lecturer in Engineering Surveying, Trent Polytechnic

Edward A

Based on material first published in 1968 as
Surveying Problems and Solutions

First edition published 1977
by Edward Arnold (Publishers) Ltd
41 Bedford Square, London WC1B 3DQ

Reprinted with corrections 1978

Second edition 1983

British Library Cataloguing in Publication Data

Shepherd, F. A.
   Engineering surveying problems and solutions.—
   2nd ed.
   1. Surveying
   I. Title
   526.9′02462   TA545

   ISBN   0-7131-3478-X

Preface to Second Edition

With the common use of the scientific calculator the presentation of the calculations are simplified and an opportunity has been taken to present some initial concepts of computing.

The development of electromagnetic distance measurement (EDM) as presented in the companion volume *Advanced Engineering Surveying* has to some extent diminished the need for the use of the steel tape as a measuring device except for short lengths and thus the topic is reduced.

Chapter 7, *Dip and fault problems*, and Chapter 10, *Areas and volumes*, are reintroduced as being essential for engineering surveying.

F A Shepherd
Nottingham 1982

Preface to First Edition

This book is based upon the previous work *Surveying Problems and Solutions* and deals with the basic mathematical aspects of Engineering Surveying, which is surveying as applied to construction and mining projects. With the acceptance of SI units as the standard system in virtually all educational institutions an attempt has been made both to revise and to fully metricate the 1968 edition. The title of the book has also been revised to emphasise the importance of survey in the industrial field.

As before, the main objective is to give guidance on the practical methods of solving typical problems posed in practice and in theory by various examination bodies. The general approach adopted is to give a theoretical analysis of each topic, followed by worked examples and finally by selected exercises for private study. It has also been decided to restructure the original work more conveniently so that this volume covers the fundamentals of measurement and those surveying techniques that are required by *all* students reading Land or Engineering Surveying as part of their studies. It is planned to include the remaining topics and to extend into more advanced surveying techniques in a separate volume which is in preparation.

The chapters have been arranged to follow a natural sequence, namely:

(a) Fundamental measurement:
    (i) linear measurement in the horizontal plane;
    (ii) angular measurement and its relationship to linear values, i.e. trigonometry;
    (iii) coordinates as a graphical and mathematical tool.
(b) Fundamental surveying techniques:
    (i) instrumentation;
    (ii) linear measurement in the vertical plane, i.e. levelling;
    (iii) traversing as a control system;
    (iv) stadia tacheometry as a detail system.

The book is suitable for all students in Universities, Polytechnics and Colleges of Further Education, as well as for supplementary postal tuition, in such courses as General Certificates of Education, Ordinary and Higher National Certificates, Diplomas and Degrees in Surveying, Construction, Architecture, Planning, Estate Management, Civil and Mining Engineering as well as for professional qualification for the Royal Institution of Chartered Surveyors.

I am greatly indebted to the Mining Qualifications Board (Ministry of Power) and the Controller of HM Stationery Office, who have given permission for the reproduction of examination questions. My thanks are also due to the Royal Institution of Chartered Surveyors, the Institution of Civil Engineers, to the Senates of the Universities of London and Nottingham, to the East Midland Educational Union and Trent Polytechnic, all of whom have allowed use of their examination questions and their subsequent conversion to SI units. The responsibility for the accuracy of these converted values is, of course, my own. My special thanks are due to my many colleagues at Nottingham who have offered help and advice, to the many correspondents who have written to me pointing out mistakes in the original script, and most especially to my wife for her consistent patience.

F A Shepherd
Nottingham 1977

Abbreviations used for examination papers

EMEU	East Midlands Educational Union
ICE	Institution of Civil Engineers
LU	University of London B Sc (Civil Engineering)
LU/E	University of London B Sc (Estate Management)
MQB/S	Mining Qualifications Board (Mining Surveyors)
MQB/M	Mining Qualifications Board (Colliery Managers)
MQB/UM	Mining Qualifications Board (Colliery Undermanagers)
TP	Trent Polytechnic
NU	University of Nottingham
RICS	Royal Institution of Chartered Surveyors (General)
RICS/M	Royal Institution of Chartered Surveyors (Mining)
RICS/ML	Royal Institution of Chartered Surveyors (Mining/Land)

Contents

1

Linear measurement

1.1 The basic principles of surveying

Fundamental rule 'Always work from the whole to the part.'
This implies 'precise control surveying' as the first consideration, followed by 'subsidiary detail surveying'.

A point C in a plane may be fixed relative to a given line AB in
one of the following ways:

1.1.1 Triangulation

See Fig. 1.1(a). Angular measurement from a fixed base line.
The length AB is known. The angles α and β are measured.

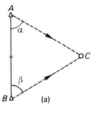

1.1.2 Trilateration

See Fig. 1.1(b). Linear measurement only. The lengths AC and
BC are measured or plotted. The position of C is always fixed
provided that $AC + BC > AB$.
Uses:
(a) Replacing triangulation with the use of microwave
    measuring equipment.
(b) Chain surveying.

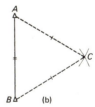

1.1.3 Polar coordinates

See Fig. 1.1(c). Linear and angular measurement.
Uses:
(a) Traversing.
(b) Setting out.
(c) Plotting by protractor.

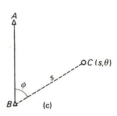

1.1.4 Rectangular coordinates

See Fig. 1.1(d). Linear measurement only at right angles.
Uses:
(a) Offsets.
(b) Setting out.
(c) Plotting.

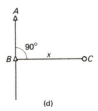

Fig. 1.1

1

1.2 General theory of measurement

The following points should be noted:

(1) There is no such thing as an exact measurement. All measurements contain some error, the magnitude of the error being dependent on the instruments used and the ability of the observer. Error may be defined as

$$e = L_m - L_t,$$

where L_m = measured value and L_t = true value.

(2) As the true value is never known, the true error is never determined.

(3) Estimated error or standard error based upon statistical data may be given as a fraction of the measured quantity and quoted as a relative accuracy or preferably as an error ratio, e.g. 100.00 m measured with an estimated error of ± 2 cm represents a relative accuracy or error ratio of 1/5000.

(4) Where readings are taken on a graduated scale to the nearest subdivision, the maximum error in estimation will be $\pm \frac{1}{2}$ division.

(5) Given a 50% probability a 'probable error' of $\pm \frac{1}{4}$ division may be assumed.

(6) Repeated measurement increases the accuracy by $\sqrt{n}$, where n is the number of repetitions. NB: this cannot be applied indefinitely.

(7) Agreement between repeated measurements does not imply accuracy but only consistency.

(8) Figures are rounded off to the required degree of precision, generally by increasing the last significant figure by 1 if the following figure is 5 or more. Thus

> 205.613 becomes 205.61 to 2 places;
> 205.615 becomes 205.62 to 2 places.

1.3 Significant figures in measurement and computation

In surveying, an indication of the accuracy attained may be shown by the number of significant figures used. These include digits of definite value plus the least accurate digit which is estimated and therefore subject to error, e.g. 123.45 contains 5 significant figures, with four certain and the last digit, 5, questionable. The error in the last digit may in this case be a maximum value of 0.005 or a probable error of ± 0.0025.

It is essential to recognise the difference between true errors, which have a known sign of $+$ or $-$, and random errors, which have indefinite signs $\pm$ indicating a probability that either sign may occur.

In the accumulation of errors of known sign, summation is

algebraic whilst the summation of random errors of $\pm$ value can only be computed by the root mean square value, i.e.

$$e_t = \sqrt{(\pm e_1^2 \pm e_2^2 \pm \cdots \pm e_n^2)}$$

Significant figures are shown as:

3 significant figures: 1.23, 12.3, 123, 0.123, 0.0123,
                         0.001 23, 1.23×10^{-2}
4 significant figures: 1.230, 12.30, 123.0, 1230, 1.230×10^4

Zeros at the end of the integers may cause some difficulties, as the 0 may or may not be significant and thus if any ambiguity exists the scientific notation should be used, e.g. 1.23×10^{-2} (where the significant figures are immediately obvious).

The number of significant figures required for calculations involving angles depends upon the relationship between degrees, minutes and seconds and their conversion into degrees and decimals, as follows:
As $x = s''/3600$ or $s'' = 3600\,x$

x	0.01	0.001	0.0001	0.000 01	0.000 28	0.002 78
s''	36"	3.6"	0.36"	0.04"	1"	10"

Thus in converting degrees, minutes and seconds into degrees the following number of places of decimals are recommended:

when working in minutes, 2 places of decimals;
when working in seconds, 4 places of decimals.

Many hand-held machines now have automatic conversion keys so that the problem is not significant as an intermediate result, but care must be taken in the final computed value by quoting the realistic number of significant figures.

The following points should be noted.

(1)   Field measurements should be consistent, thus dictating the number of significant figures in the derived or computed values.
(2)   In certain cases there may be an implied accuracy, e.g. a length of a building might be said to be 100 m, in which case the accuracy is not specified and can only be conjectured by reference to the conditions in which it is used; on measuring this length it might have been recorded as 100.004 m.
(3)   The accuracy of the angular and linear values should be compatible. If the effect of the angular and linear errors are to be equated (see Fig. 1.2) then

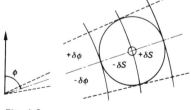

Fig. 1.2

$$\delta S = S.\delta\phi \quad \text{and} \quad \delta\phi'' = 206\,265\,\delta S/S$$

$\delta\phi''$	06' 53"	00' 41"	00' 21"	00' 08"	00' 04"	00' 01"
$\delta S/S$	1/500	1/5000	1/10 000	1/25 000	1/50 000	1/206 265

In general terms, angular values measured to 1" require distances to 1 mm.

Errors in computed results arise from:

(a) errors in measured or derived data; or
(b) errors in trigonometrical values used.

When these quantities are subjected to common arithmetical processes, the resultant values are frequently given false accuracies as illustrated in the following examples.

1.3.1 Addition

Considering maximum errors of known value,

if $\quad\quad s = x + y$

then $\quad s + \delta s = (x + \delta x) + (y + \delta y)$

where δs, δx and δy may be $+$ or $-$.

Considering probable errors of indefinite value,

$$s \pm e_s = (x \pm e_x) + (y \pm e_y)$$
$$= (x + y) \pm \sqrt{(e_x^2 + e_y^2)}$$
$$\pm e_s = \sqrt{(e_x^2 + e_y^2)} \tag{1.1}$$

Example 1.1 If $s = 4.92 + 5.137$ (measured quantities), the maximum errors will be 0.005 and 0.0005 and the probable errors will be ± 0.0025 and $\pm 0.000\,25$.

Then

$$s + \delta s = 10.057 \pm (0.005 + 0.0005)$$
$$= 10.057 \pm 0.0055$$
and $\quad s \pm e_s = 10.057 \pm \sqrt{(0.0025^2 + 0.000\,25^2)}$
$$= 10.057 \pm 0.002\,51$$

i.e. the maximum limits are 10.0625 and 10.0515; the most probable limits are 10.0595 and 10.0545.

It can be seen from the above that the sum may be expressed as 10.05 or 10.06 and thus the second decimal place is the most probable limit to which the derived quantity may be quoted, i.e. the accuracy of the sum must not exceed the least accurate figure used.

1.3.2 Subtraction

If $\quad\quad\quad\quad\quad d = x - y$

then

$$d + \delta d = (x - y) + (\delta x - \delta y)$$

the maximum error $= \delta d = (\delta x + \delta y)$

and $\quad\quad\quad\quad d \pm e_d = (x - y) \pm \sqrt{(e_x^2 + e_y^2)}$

the probable error $\pm e_d = \sqrt{(e_x^2 + e_y^2)}$ (as in addition)

Example 1.2   Let $d = 5.137 - 4.92$.

Then $\delta d = 0.0055$

and $\pm e_d = \pm 0.002\,51$

i.e. the maximum limits are 0.2225 and 0.2115; the most probable limits are 0.2195 and 0.2145.

The quoted value can only be 0.21 or 0.22 and the second decimal place is the most probable limit to which the derived quantity can be given, i.e. to the same accuracy as that of the least accurate figure used.

Example 1.3 $3.34 + 2.1 + 7.9634 \neq 13.4034$ but $= 13.4$.

Example 1.4 $4.021 - 1.75 + 6.0 \neq 8.271$ but $= 8.3$.

1.3.3   Multiplication

Let $p = xy$

Considering maximum values,
$$\delta p_x = y\delta x$$
$$\delta p_y = x\delta y$$
then the maximum error
$$\delta p = y\delta x + x\delta y$$
Considering probable errors,
$$e_p = \sqrt{(y^2 e_x^2 + x^2 e_y^2)} = \sqrt{[(e_x/x)^2 + (e_y/y)^2]}\,xy$$
the error ratio
$$e_p/p = \sqrt{[(e_x/x)^2 + (e_y/y)^2]} \qquad (1.2)$$

Example 1.5 $p = (3.82 + 0.005) \times (7.64 + 0.005) = 29.1848$ $\pm$ error.

For maximum error,
$$\delta p = (3.82 \times 0.005) + (7.64 \times 0.005)$$
$$= 0.0191 + 0.0382$$
$$= 0.0573, \text{ i.e. } 0.06$$

For most probable values,
$$p = (3.82 \pm 0.0025) \times (7.64 \pm 0.0025)$$
$$e_p = 29\sqrt{[(0.0025/3.82)^2 + (0.0025/7.64)^2]}$$
$$= \pm 0.02$$

The most probable limits are thus 29.20 and 29.16 and the value can only be given with certainty as 29.2.

For practical purposes, bearing in mind a rounding off process likely in more complex processes, the value may be given as 29.18, i.e. to the same accuracy as the least accurate figure used.

1.3.4   Division

Let $q = x/y$

Considering the maximum values,

$$\delta q_x = \delta x/y$$
$$\delta q_y = x\delta y/y^2$$

The maximum error

$$\delta q = \delta x/y + x\delta y/y^2$$

Considering the probable values,

$$e_q = \sqrt{[(e_x/y)^2 + (xe_y/y^2)^2]}$$
$$= \frac{x}{y}\sqrt{[(e_x/x)^2 + (e_y/y)^2]}$$

The error ratio

$$e_q/q = \sqrt{[(e_x/x)^2 + (e_y/y)^2]} \quad \text{(as in multiplication)} \quad (1.3)$$

Example 1.6 $q = 29.2/7.64 = 3.8220.$

For maximum error

$$\delta q = 0.05/7.64 + 29.2 \times 0.005/7.64^2$$
$$= 0.0065 + 0.0025 = 0.0090$$

For probable error

$$e_q = 3.8\sqrt{[(0.025/29.2)^2 + (0.0025/7.64)^2]}$$
$$= \pm 0.003$$

The most probable limits are thus 3.825 and 3.819. The value can only be given with certainty as 3.8.

Again for practical purposes involving rounding off, the value may be given as two places of decimals, i.e. 3.82, one more place of decimals than the least accurate value used and equivalent to the same number of significant figures.

1.3.5   Powers

If $r = x^n$

then $\delta r = nx^{n-1}\delta x$

the error ratio $\dfrac{\delta r}{r} = \dfrac{n\delta x}{x}$ (1.4)

Example 1.7 $r = 5.01^2 = 25.1001.$

For maximum value,

$$\delta r = 2 \times 5.01 \times 0.005 = 0.0501$$

For most probable value,

$$e_r = 2 \times 5.01 \times 0.0025 = 0.025$$

The most probable limits are thus 25.1251 or 25.0751, i.e. practically 25.10, the same accuracy as the original value.

Example 1.8 $r = \sqrt{25.10} = 5.0100$.

$\delta r = 0.0005$

$e_r = 0.000\,25$

The most probable limits are thus $5.010\,25$ and $5.009\,75$, i.e. practically 5.010, one more place of decimals than the original.

Example 1.9   A rectangle measures $3.82\,\text{m} \times 7.64\,\text{m}$ with errors of $\pm 5\,\text{mm}$. Express the area to the correct number of significant figures.

$$p = 3.82 \times 7.64 = 29.1848$$

The error ratios are

$$\frac{0.005}{3.82} \simeq \frac{1}{750}$$

and $\quad \dfrac{0.005}{7.64} \simeq \dfrac{1}{1500}$

$$\delta p = 29\left(\frac{1}{750} + \frac{1}{1500}\right) = \frac{29}{500} = \pm 0.06\,\text{m}$$

Therefore the area has limits of 29.12 to 29.24 and thus the answer can only be quoted as $29.18\,\text{m}^2$.

Example 1.10 $r = (6.04\,\text{m} \pm 0.005)^3$.

$\delta r = nx^{n-1}\delta x$

$= 3 \times 6.04^2 \times 0.005 = \pm 0.547$

The limits are therefore 220.349 ± 0.547 and the answer can thus be quoted as $220.3\,\text{m}^3$.

Example 1.11 $r = \sqrt{(25.10 \pm 0.01)}$.

$\delta r = \frac{1}{2} \times 25.10^{-1/2} \times 0.01 = \pm 0.001$

The limits are 5.01 ± 0.001.

Example 1.12   A rectangular building has sides approximately $480\,\text{m}$ and $300\,\text{m}$. If the area is to be determined to the nearest $10\,\text{m}^2$ what will be the maximum error permitted in each line, assuming equal precision ratios for each length? To what degree of accuracy should the lines be measured?

$A = 480 \times 300 = 144\,000\,\text{m}^2$

$\delta A = 10\,\text{m}^2$

therefore

$$\frac{\delta A}{A} = \frac{1}{14\,400} = \frac{\delta x}{x} + \frac{\delta y}{y}$$

but

$$\frac{\delta x}{x} = \frac{\delta y}{y}$$

therefore

$$\frac{\delta x}{x} + \frac{\delta y}{y} = \frac{2\delta x}{x}$$

therefore

$$\frac{\delta x}{x} = \frac{1}{2 \times 14\,400} = \frac{1}{28\,800}$$

i.e. the precision ratio of each line is $\dfrac{1}{28\,800}$,

This represents a maximum in 480 m of $\dfrac{480}{28\,800} = 0.0167\,\text{m}$

and in 300 m of $\dfrac{300}{28\,800} = 0.0104\,\text{m}$

If the number of significant figures in the area is 5, i.e. to the nearest $10\,\text{m}^2$, then each line also must be measured to at least 5 significant figures, i.e. 480.00 m and 300.00 m.

1.4   Chain surveying

The chain. There are three types:

(a)   Gunter's chain
       1 chain = 100 links = 66 ft = 20.12 m
       1 link = 0.66 ft = 7.92 in = 0.201 m
Its advantage lies in its relationship to the acre:
       10 sq chains = 100 000 sq links = 1 acre
(b)   Engineer's chain
       100 links = 100 ft
(c)   Metric chain
       100 links = 20 m
       1 link = 0.2 m

NB: The metric chain is thus only 0.12 m short of the Gunter chain.

Basic figures. There are many combinations of chain lines, all dependent on the linear dimensions forming trilateration, Fig. 1.3.

1.4.1   Corrections to the ground measurements

Standardisation. Where the length of the chain or tape does not agree with its nominal value, a correction must be made to the recorded value of a measured quantity.

The following rules apply:

(1)   If the tape is too long, the measurement will be too short— the correction will be positive.

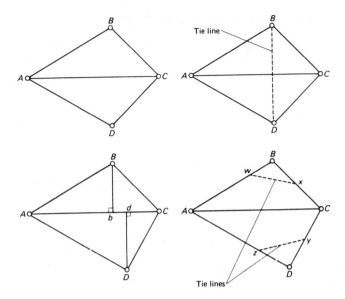

Fig. 1.3 Basic figures in chain surveying

(2) If the tape is too short, the measurement will be too long—the correction will be negative.

If a tape of nominal length l_n has a 'true' length l_t then

$$l_t = l_n + \delta l \tag{1.5}$$

and $\delta l / l_n$ = the error per unit length

If a line is measured as L_m and its 'true' length is L_t then

$$L_t = L_m + L_m \cdot \delta l / l_n$$
$$= L_m (1 + \delta l / l_n) \tag{1.6}$$

Alternatively,

$$L_t / L_m = l_t / l_n$$

and then

$$L_t = L_m \cdot l_t / l_n \tag{1.7}$$

Example 1.13 A chain of nominal length 20.00 m when compared with a standard measures 20.05 m. If this chain is used to measure a line AB and the recorded measurement is 131.35 m, what is the true length AB?

$$\delta l = l_t - l_n = 0.05$$

By equation 1.6,

$$L_t = 131.35(1 + 0.05/20.00) = 131.68 \text{ m}$$

By equation 1.7,

$$L_t = 131.35 \times 20.05/20.00 = 131.68 \text{ m}$$

Effect of standardisation on areas. Based on the principle of similar figures, true area (A_t) = apparent area $(A_m) \times (l_t/l_n)^2$

i.e. $\qquad A_t = A_m(l_t/l_n)^2$ (1.8)

or $\qquad A_t = A_m(1 + \delta l/l_n)^2$ (1.9)

Effect of standardisation on volumes. Based on the principle of similar volumes,

true volume $(V_t) = V_m(l_t/l_n)^3$ (1.10)

or $\qquad V_t = V_m(1 + \delta l/l_n)^3$ (1.11)

NB: Where the error in standardisation is small compared to the size of the area, the % error in area is approximately 2 × % error in length.

Example 1.14   A metric chain of nominal length 20.00 m is found to be 16 cm too long and on using it an area of 100.0 hectares is computed. (NB: 1 hectare = 10 000 m².)

The true area = $100.0 (20.16/20.00)^2$

$\qquad = 100.0 \times 1.016 = 101.6$ hectares

Alternatively,

linear error = 0.8 %

therefore

area error = $2 \times 0.8 \% = 1.6 \%$

true area = $100.0 + 100(1.6 \%) = 101.6$ hectares

This is derived from the binomial expansion of

$\qquad (1 + x)^2 = 1 + 2x + x^2$

i.e. if x is small, x^2 may be neglected, therefore

$\qquad (1 + x)^2 \simeq 1 + 2x$

Correction for slope (C_i).   This may be based upon

(1)   the angle of inclination, or
(2)   the difference in level between the ends of the line.

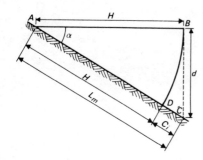

Fig. 1.4

In Fig. 1.4,

measured length $AC = L_m$
horizontal length $AB = H$
difference in level between A and $C = d$
angle of inclination $= \alpha$
correction to the measured line $= C_i$

(1)   *Given the angle of inclination* α,

$\qquad AB = AC \cos \alpha$

i.e. $\qquad H = L_m \cos \alpha$ (1.12)

$\qquad = L_m - C_i$

therefore $\qquad C_i = L_m(1 - \cos \alpha) = L_m$ versine α (1.13)

Example 1.15 In chaining, account should be taken of any significant effect of the slope of the ground on the accuracy of the horizontal length. Calculate the minimum angle of inclination that gives rise to relative accuracies of 1/1000 and 1/3000.

From equation 1.13,

$$C_i = L_m(1 - \cos \alpha)$$

therefore

$$\frac{C_i}{L_m} = 1/1000 = 1 - \cos \alpha$$

$$\alpha = \cos^{-1} 0.999\,00 = 2° 34' \quad \text{(i.e. 1 in 22)}$$

Similarly,

$$\cos \alpha = 1 - 1/3000$$

$$\alpha \simeq 1° 29' \quad \text{(i.e. 1 in 39)}$$

(2) *If the difference in level, d, is known,*

$$H = (L_m^2 - d^2)^{1/2} \tag{1.14}$$

or

$$L_m^2 = H^2 + d^2 = (L_m - C_i)^2 + d^2$$
$$= L_m^2 - 2L_m C_i + C_i^2 + d^2$$

therefore

$$C_i(C_i - 2L_m) = -d^2$$

$$C_i = \frac{-d^2}{C_i - 2L_m}$$

$$\simeq \frac{d^2}{2L_m} \quad \text{as } C_i \text{ is small compared with } 2L_m$$

$$\tag{1.15}$$

Rigorously, using the binomial expansion,

$$C_i = L_m - (L_m^2 - d^2)^{1/2}$$
$$= L_m - L_m(1 - d^2/L_m^2)^{1/2}$$
$$= L_m[1 - (1 - d^2/2L_m - d^4/8L_m^3 - \ldots)]$$
$$= d^2/2L_m + d^4/8L_m^3 \tag{1.16}$$

The use of the first term only gives the following relative accuracies.

Gradient	Error per 100 m	Relative accuracy
1 in 4	0.051 m	1/2000
1 in 8	0.0031 m	1/30 000
1 in 10	0.0013 m	1/80 000
1 in 20	0.0001 m	1/1000 000

Thus the approximation is acceptable for:

(1) Chain surveying under all general conditions.
(2) Traversing, gradients up to 1 in 10.
(3) Precise measurement (e.g. base lines), gradients up to 1 in 20.

Example 1.16   Let $AC = 126.300$ m; $\alpha = 9° 30'$; $d = 20.846$ m.

By equation 1.12,

$$H = 126.300 \cos 9° 30' \qquad\qquad = 124.568 \text{ m}$$

By equation 1.13,

$$C_i = 126.300 \, (1 - \cos 9°30')$$

$$= 1.732$$

$$H = 126.300 - 1.732 \qquad\qquad = 124.568$$

By equation 1.14,

$$H = (126.300^2 - 20.846^2)^{1/2} \qquad = 124.568$$

By equation 1.15,

$$C_i = \frac{20.846^2}{2 \times 126.300} \qquad\qquad = 1.720$$

By equation 1.16,

$$C_i = \frac{20.846^2}{2 \times 126.300} + \frac{20.846^4}{8 \times 126.300^3}$$

$$= 1.720 + 0.012 \qquad\qquad = 1.732$$

Then $H = 126.300 - 1.732 \qquad\qquad = 124.568$

For setting out purposes. Here the horizontal length (H) is given and the slope length (L_m) is required.

$$L_m = H \sec \alpha$$

$$C_i = H \sec \alpha - H$$

$$= H (\sec \alpha - 1) \tag{1.17}$$

Writing $\sec \alpha$ as a series $1 + \dfrac{\alpha^2}{2} + \dfrac{5\alpha^4}{24} + \ldots$, where α is in radians (see p. 61)

$$C_i = H(1 + \alpha^2/2 + 5\alpha^4/24 + \ldots - 1)$$

$$\simeq H\alpha^2/2 \tag{1.18}$$

$$\simeq \frac{H}{2}(0.017\,45\,\alpha)^2 \quad (\alpha \text{ in degrees})$$

$$\simeq 1.53H \times 10^{-4} \times \alpha^2 \quad (\alpha \text{ in degrees}) \tag{1.19}$$

$$\simeq 1.53 \times 10^{-2} \times \alpha^2 \text{ per } 100 \text{ m} \tag{1.20}$$

Example 1.17   Let $H = 100.00$ m; $\alpha = 5°$.

By equation 1.17,

$$C_i = 100.00(\sec 5° - 1) \qquad = 0.382 \text{ m}/100 \text{ m}$$

or by equation 1.19,

$$C_i = 1.53 \times 100 \times 10^{-4} \times 5^2 = 0.382 \text{ m}/100 \text{ m}$$

Correction per 100 m:

1°	0.015 m	6°	0.551 m
2°	0.061 m	7°	0.751 m
3°	0.137 m	8°	0.983 m
4°	0.244 m	9°	1.247 m
5°	0.382 m	10°	1.543 m

If the difference in level, d, is given, rigorously as before:

$$C_i = d^2/2H - d^4/8H^3 + \ldots \qquad (1.21)$$

NB: If the gradient of the ground is known as 1 vertical to n horizontal the angle of inclination (α) is given by

$$\alpha \simeq 57/n \quad (1 \text{ rad} \simeq 57.3°)$$

e.g.   1 and 10 gives $\simeq 57/10 = 5.7°$.

To find the horizontal length h given the gradient 1 in n and the measured length l (Fig. 1.5):

$$\frac{H}{L_m} = \frac{n}{\sqrt{(n^2 + 1)}}$$

therefore

$$H = L_m \cdot n / \sqrt{(n^2 + 1)} \qquad (1.22)$$

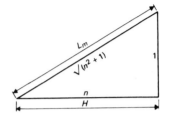

Fig. 1.5

Example 1.18   If a length of 300.00 m is measured on a slope of 1 in 3, find the horizontal length.

By equation 1.22,

$$H = 300.00 \times 3/\sqrt{10} = 284.61 \text{ m}$$

Example 1.19   Let $H = 300.00$ m and the gradient be 1 in 6.

By equation 1.22,

$$L_m = \frac{\sqrt{(n^2 + 1)}}{n}$$

$$L_m = (300.00\sqrt{37})/6 = 300.14 \text{ m}$$

1.4.2   The maximum length of offsets from chain lines

A point P is measured from a chain line ABC in such a way that B_1P is measured instead of BP, due to an error α in estimating the perpendicular.

On plotting, P_1 is fixed from B_1 (see Fig. 1.6). Thus the displacement on the plan due to the error in direction α is

$$PP_1 = B_1P\alpha \quad \text{(radians)}$$

$$= \frac{l\alpha''}{206\,265}$$

(NB: 1 radian = 206 265 seconds of arc.)

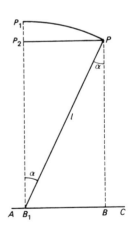

Fig. 1.6

If the maximum length PP_1 represents the minimum plottable point, i.e. say 0.2 mm which represents $0.0002\ \text{m} \times x$, where x is the representative fraction $1/x$, then

$$0.0002x = l\alpha/206\,265$$
$$l = 41.253x/\alpha''$$

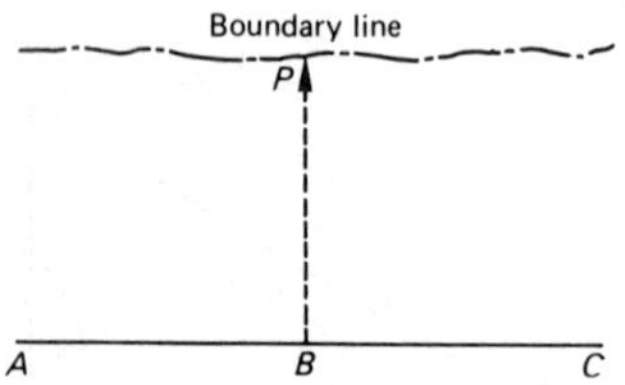

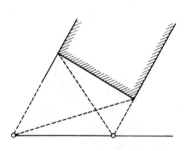

Boundary line

Fig. 1.7

Assuming the maximum error $\alpha = 4°$, i.e. $14\,400''$,

$$l = 41.253x/14\,400 \simeq 0.003x \qquad (1.23)$$

If the scale is $1/2500$, then $x = 2500$, and

$$l = 2500 \times 0.003 = 7.5\ \text{m}$$

If the point P lies on a fence approximately parallel to ABC, Fig. 1.7, then the plotted point will be in error by an amount $P_1 P_2 = l(1 - \cos\alpha)$, Fig. 1.6, therefore

$$l = 0.0002x/(1 - \cos\alpha) \qquad (1.24)$$

Example 1.20   Let $\alpha = 4°$.

By equation 1.24,

$$l = \frac{0.0002x}{1 - 0.9976}$$
$$= 0.083x \qquad (1.25)$$

Thus, if $x = 2500$,

$$l \simeq 208\ \text{m}$$

The error due to this source is almost negligible and the offset is only limited by practical considerations, e.g. the length of the tape.

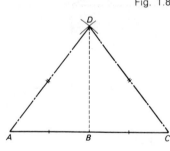

Fig. 1.8

It is thus apparent that in fixing the position of a point that is critical, e.g. the corner of a building, the length of a perpendicular offset is limited to $0.003x$ m, and beyond this length tie lines are required, the direction of the measurement being ignored, Fig. 1.8.

1.4.3   Setting out a right angle by chain

(a)   *From a point on the chain line* (Fig. 1.9),
  (i)   Measure off $BA = BC$
  (ii)   From A and C measure off $AD = CD$
(*Proof*: triangles ADB and DCB are congruent, thus $A\hat{B}D = D\hat{B}C = 90°$ as ABC is a straight line.)

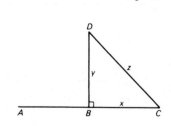

Fig. 1.9

(b)   *Using the principle of Pythagoras,*
$$z^2 = x^2 + y^2 \quad \text{(Fig. 1.10)}$$

By choosing suitable values the right angle may be set out. The basic relationship is

$$x : y : z :: 2n + 1 : 2n(n + 1) : 2n(n + 1) + 1 \qquad (1.26)$$

Fig. 1.10

If $n = 1$, $2n + 1 = 3$

$$2n(n+1) = 4$$

$$2n(n+1) + 1 = 5$$

Check: $[2n(n+1) + 1]^2 = (2n^2 + 2n + 1)^2$

$(2n+1)^2 + [2n(n+1)]^2 = 4n^2 + 4n + 1 + 4n^4 + 8n^3 + 4n^2$

$$= 4n^4 + 8n^3 + 8n^2 + 4n + 1$$

$$= (2n^2 + 2n + 1)^2$$

Check: $5^2 = 3^2 + 4^2$ or $25 = 9 + 16$

Similarly, if $n = 3/4$,

$$2n + 1 = \frac{6}{4} + 1 \quad = \frac{10}{4} \quad = \frac{40}{16}$$

$$2n(n+1) = \frac{6}{4}\left(\frac{3}{4} + 1\right) \quad = \frac{6}{4} \times \frac{7}{4} = \frac{42}{16}$$

$$2n(n+1) + 1 = \qquad \frac{42}{16} + \frac{16}{16} = \frac{58}{16}$$

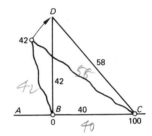

Thus the ratios become $40:42:58$ and this is probably the best combination for 100 unit measuring equipment; e.g. on the line ABC, Fig. 1.11, set out $BC = 40$ units. Then holding the ends of the chain at B and C the position of D is fixed by pulling taut at the 42/58 on the chain.

Fig. 1.11

Alternative values for n give the following:

$n = 2$ 5, 12, 13   (probably the best ratio for 30 m tapes)
$n = 3$ 7, 24, 25
$n = 4$ 9, 40, 41.

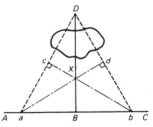

Fig. 1.12

1.4.4   To find the point on the chain line which produces a perpendicular from a point outside the line

(1)   *When the point is accessible* (Fig. 1.12). From the point D swing the chain of length $> DB$ to cut the chain line at a and b. The required position B is then the mid-point of ab.

(2)   *When the point is not accessible* (Fig. 1.13). From D set out lines Da and Db and, from these lines, perpendiculars ad and bc. The intersection of these lines at X gives the line DX which when produced gives B, the required point.

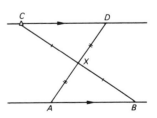

Fig. 1.13

To set out a line through a given point parallel to the given chain line (Fig. 1.14). Given the chain line AB and the given point C. From the given point C bisect the line CB at X. Measure AX and produce the line to D such that $AX = XD$. CD will then be parallel to AB.

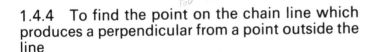

Fig. 1.14

1.4.5   Obstacles in chain surveying

(1)   *Obstacles to ranging*

   (a)   *Visibility from intermediates* (Fig. 1.15). Required to line
C and D on the line *AB*.

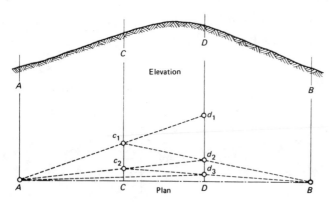

Fig. 1.15

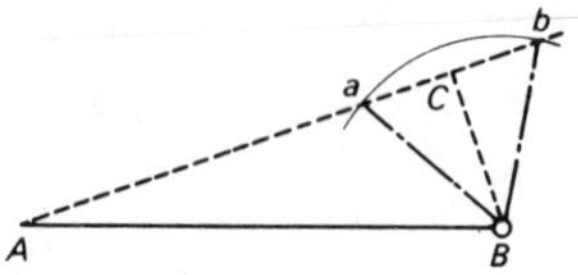

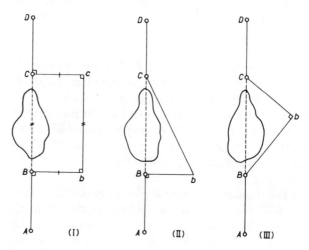

Fig. 1.16

   Place ranging pole at d_1 and line in c_1 on line Ad_1. From B
observe c_1 and move d_2 on to line Bc_1. Repetition will produce
c_2, c_3 and d_2, d_3, etc. until C and D lie on the line *AB*.

   (b)   *Non-visibility from intermediates* (Fig. 1.16). Required to
measure a long line *AB* in which A and B are not intervisible and
intermediates on these lines are not possible.

   Set out a 'random line' *AC* approximately on the line *AB*.

   From B find the perpendicular *BC* to line *AC* as above.
Measure *AC* and *BC*. Calculate *AB*.

(2)   *Obstacles to chaining*

(a)   *No obstacle to ranging*

   (i)   *Obstacle can be chained around.* There are many possible variations depending on whether a right angle is set out
(Fig. 1.17) or not (Fig. 1.18).

Fig. 1.17 By setting out right angles

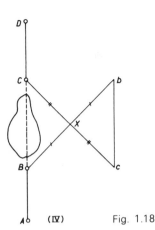
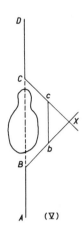
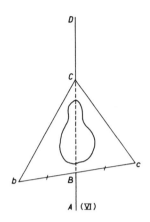

(IV)   Fig. 1.18   (V)   (VI)

I   Set out equal perpendiculars Bb and Cc; then $bc = BC$.

II   Set out Bb. Measure Bb and bC. Compute BC.

III   Set out line Bb. At b set out the right angle to give C on the chain line. Measure Bb and bC. Compute BC.

IV   and V. Set out parallel lines bc as described above to give similar figures, triangles BCX and bcX.

Then

$$BC = bc \times BX/bX \qquad (1.27)$$

VI   Set out line bc so that $bB = Bc$. Compute BC thus,

$$BC^2 = \frac{(bC)^2 Bc + (Cc)^2 bB}{bc} - bB \times Bc \qquad (1.28)$$

But $bB = Bc$, therefore

$$BC^2 = \frac{Bb(bC^2 - Cc^2)}{2Bb} - Bb \times Bb$$

$$= \tfrac{1}{2}(bC^2 + Cc^2) - Bb^2 \qquad (1.29)$$

Proof: In Fig. 1.19, using the cosine rule (assuming $\theta > 90°$; see p. 63),

$$p^2 = x^2 + d^2 + 2xd \cos \theta$$

and $$q^2 = y^2 + d^2 - 2yd \cos \theta$$

therefore

$$2d \cos \theta = (p^2 - x^2 - d^2)/x$$

$$= (y^2 + d^2 - q^2)/y$$

therefore

$$p^2 y - x^2 y - d^2 y = xy^2 + d^2 x - q^2 x$$

$$d^2(x + y) = q^2 x + p^2 y - xy(x + y)$$

$$d^2 = \frac{q^2 x + p^2 y}{x + y} - xy$$

$$d = \sqrt{\frac{q^2 x + p^2 y}{x + y} - xy} \qquad (1.30)$$

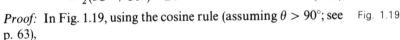

Fig. 1.19

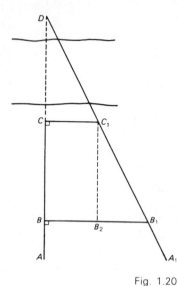

Fig. 1.20

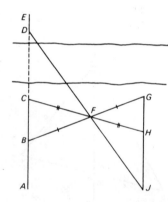

Fig. 1.21

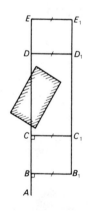

Fig. 1.22

If $x = y$,

$$d = \sqrt{\frac{(p^2 + q^2)}{2} - x^2} \qquad (1.31)$$

(ii) *Obstacle cannot be chained around.* A river or stream represents this type of obstacle. Again there are many variations depending on whether a right angle is set out or not.

By setting out right angles (Fig. 1.20). A random line DA_1 is set out and from perpendiculars at C and B points C_1 and B_1 are obtained.

By similar triangles DC_1C and $C_1B_1B_2$,

$$\frac{DC}{CB} = \frac{CC_1}{BB_1 - CC_1}$$

therefore

$$DC = \frac{CB \times CC_1}{BB_1 - CC_1}$$

Without setting out a right angle (Fig. 1.21). A point F is chosen. From points B and C on line AE, BF and CF are measured and produced to G and H.

$BF = FG$ and $CF = FH$. The intersection of DF and GH produce to intersect at J. Then $HJ = CD$.

(b) *Obstacles which obstruct ranging and chaining.* The obstruction, e.g. a building, prevents the line from being ranged and thus produced beyond the obstacle.

By setting out right angles (Fig. 1.22). On line ABC right angles are set out at B and C to produce B_1 and C_1, where $BB_1 = CC_1$.

B_1C_1 is now produced to give D_1 and E_1 where right angles are set out to give D and E, where $D_1D = E_1E = BB_1 = CC_1$. D and E are thus on the line ABC produced and $D_1C_1 = DC$.

Without setting out right angles (Fig. 1.23). On line ABC, CB is measured and G set out to form an equilateral triangle, i.e. $CB = CG = BG$. BG is produced to J.

An equilateral triangle HKJ sets out the line JE such that $JE = BJ$.

A further equilateral triangle ELD will restore the line ABC produced.

The missing length $BE = BJ = EJ$.

Exercises 1.1

1 The following measurements were made on inclined ground. Reduce the slope distances to the horizontal.

(a) 200.1 m   at a gradient of 1 in $2\frac{1}{2}$

(b) 125.42 m   at an inclination of $2° 30'$

(c) 142.365 m with a difference in level of 20.26 m

(Ans.   (a) 185.8 m; (b) 125.30 m; (c) 140.916 m)

2 Calculate the area in hectares represented by a plan area of 4 cm² on each of the plans drawn to scale 1/10 000, 1/2500, 1/500 respectively.
(NB: 1 hectare = 10 000 m².) (Ans. 4.00 ha, 0.25 ha, 0.01 ha)

3 A field was measured with a 20 m chain, 7 cm too long. The area thus found was 10 hectares. What is the true area?
(Ans. 10.07 ha)

4 A straight line *ABCD* was measured in three sections along a slope. The angles of inclination were measured with an Abney level and the following results obtained.

Line	Length (m)	Angle of slope
AB	102.50	+6° 30′
BC	57.95	+2° 00′
CD	48.20	−5° 00′

Determine the plan length of the line *AD*; thereafter calculate the gradient of a road to be made between *A* and *D*.
(Ans. *AD* 207.77 m; Δh 9.42 m; gradient +1 in 22.06)

5 A survey line was measured on sloping ground and recorded as 117.84 m. The difference in elevation between the ends was 5.88 m. The tape used was later found to be 20.03 m when compared with a standard 20 m tape. Calculate the horizontal length of the line. (Ans. 117.87 m)

6 An inclined line is measured as 53.637 m. Given that the angle of inclination is 4° 27′ 10″ and the difference in level between the ends of the line is 4.164 m, calculate the horizontal length using:
(a) $h = l \cos \alpha$;   (b) $c = l(1 - \cos \alpha)$
(c) $h = \sqrt{(l^2 - d^2)}$ (d) $c = d^2/2l$ (Ans. 53.475 m)

7 Find, without using tables, the horizontal length of a line recorded as 247.40 m when measured:
(a) on ground sloping at 1 in 4
(b) on ground sloping at 18° 26′ 00″ (tan 18° 26′ 00″ = 0.3333)
(Ans. (a) 240.01; (b) 234.73)

8 The distance between two stations *A* and *B* was measured with a metric chain and found to be 87.35 m. If the chain was subsequently found to be 20.04 m and the ground sloped regularly at 10° 51′, calculate the true horizontal distance *A* to *B*.
(Ans. 85.96 m)

9 Show that for small angles of slope the difference between the horizontal and slope length is $h^2/2l$ where *h* is the difference of vertical height of the two ends of the line of sloping length *l*.
If the errors in chaining are not to exceed 1 part in 1000 what is the greatest slope that can be ignored?
(LU Ans. 1 in 22.4)

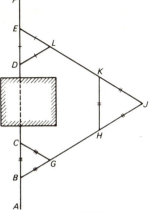

Fig. 1.23

Bibliography

ALLAN, A. L., HOLLWEY, J. R., and MAYNES, J. H. B., *Practical Field Surveying and Computations.* Heinemann (1968)

SMITH, J. R., *The Misuse of Figures in Elementary Calculations.* Unpublished paper, Portsmouth Polytechnic

2
Precise linear measurement

Where linear measurements are made with a steel tape accuracies of up to 1/30 000 may be attained provided that consideration is given to the following:

(a)   Standardisation
(b)   Malalignment
(c)   Reading or marking the tape
(d)   Slope
(e)   Temperature
(f)   Tension   (where applicable)
(g)   Sag due to the catenary curve   (where applicable)
(h)   Mean sea level
(i)   National Grid projection

NB: The last two are only necessary where grid lengths are required.

2.1   Standardisation

Steel field tapes are manufactured to a nominal length (l_n) under a specified tension (T_s) and at a standard temperature (t_s). After use the actual or 'true' length of the tape (l_t) may not be equal to the nominal length and the 'true' length can only be evaluated by comparison with a standard tape of length l_s or by use on a standard baseline of length L_s. The standard tape should be certified by the National Physical Laboratory (NPL) or its equivalent.

   To standardise the tape using a standard baseline the latter should be set out on a flat level surface with the index marks clearly defined and measured as L_m under a standard tension using the standard tape. As the temperature at the time of measurement (t_m) is unlikely to be at the standard (t_s) then a 'correction' for temperature must be applied (see Section 2.5). The true length of the line is then known as L_s.

   The same line is now measured with the field tape and its true length may then be quoted as:

(a)   the actual length of the field tape (l_f) at a specified tension and temperature, or
(b)   at a specified temperature calculated such that the actual length = the nominal length under the specified tension.

The latter is preferred as then the standardisation factor can be ignored.

If the field length (l_f) is known then the standardisation correction is

$$C_s = (l_f - l_n)/l_n \tag{2.1}$$

e.g. if the actual length of a field tape is 29.997 m, then

$$C_s = (29.997 - 30.000)/30.000 = -0.003 \text{ m}/30 \text{ m}$$
$$= -0.0001 \text{ m/m}$$

If the measured length of the standard baseline is given as L_m compared with its known length of L_s then

the standardisation error $= \delta L = L_m - L_s$

The length of the field tape under the measurement conditions on the baseline is given as

$$l_f = l_n(1 - \delta L/L_m) \tag{2.2}$$

2.2 Malalignment and deformation of the tape

Fig. 2.1 Malalignment of the tape

(a) *Malalignment* (Fig. 2.1). If the end of the tape is out of line by an amount d in a length L, the error will be

$$e_1 = d^2/2L \tag{2.3}$$

e.g. if $d = 100$ mm in a length 30.000 m then $e_1 = 0.000\,167$, i.e. $1/180\,000$.

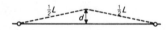

Fig. 2.2 Deformation of the tape

(b) *Deformation in the horizontal plane* (Fig. 2.2). If the tape is not pulled straight and the centre of the tape is out of line by a length d then

$$e_2 = d^2/L + d^2/L = 2d^2/L$$

e.g. if the deformation in the centre of the tape is 100 mm in a line of 30.000 m then

$$e_2 = 4e_1 = 0.000\,667, \text{ i.e. } 1/45\,000$$

(c) *Deformation in the vertical plane.* This is the same as (b) but is more difficult to detect. Any obvious change in gradient can be allowed for as part of the slope correction.

NB: In (a) and (b) alignment by eye is acceptable for all purposes except in extremely precise work.

2.3 Reading and marking the tape

Tapes graduated to 1 mm can be read by estimation to give a standard error of ± 0.2 mm and thus if both ends are read the standard error will probably be of the order of $\sqrt{2} \times 0.2$ mm, say ± 0.3 mm.

Setting the tape to a mark is usually considered less accurate.

2.4 Correction for slope (C_i)

As shown in Section 1.4.1 the reduction for slope may be based upon:

(a) the angle of inclination of the plane upon which the measurement is made, or

(b) the difference in level between the index marks.

(a) If the inclination of the plane (α) can be measured, then

$$H = L_m \cos \alpha \qquad \text{(equation 1.11)}$$

$$C_i = L_m (1 - \cos \alpha) \qquad \text{(equation 1.12)}$$

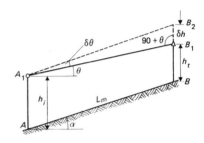

As the precise value of α can rarely be measured directly the following modification should be noted.

Let h_i = height of the instrument (e.g. the theodolite)

h_t = height of the target (e.g. a levelling staff)

θ = measured vertical angle

L_m = measured line on the plane of inclination (α)

In Fig. 2.3,

$$\alpha = \theta + \delta\theta$$

In the triangle $A_1 B_2 B_1$, by the sine rule (see p. 64),

$$\sin \delta\theta = (h_i - h_t) \sin(90 + \theta)/L_m$$

$$= \delta h \cos \theta / L_m \qquad (2.5)$$

As $\delta\theta$ is small then:

$$\delta\theta'' = 206\,265\, \delta h \cos\theta / L_m \qquad (2.6)$$

or $\quad \delta\theta° = 57.2958\, \delta h \cos\theta / L_m \qquad (2.7)$

NB: The sign of $\delta\theta$ is the same as δh (Fig. 2.4).

Fig. 2.3

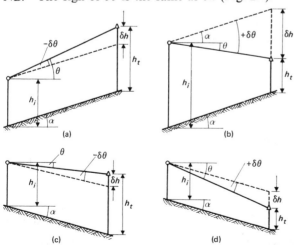

(a)

(b)

(c)

(d)

Fig. 2.4

Example 2.1   Given $h_i = 1.370$ m, $h_t = 1.675$ m, $\theta = 4° 30' 00''$, $L_m = 106.680$ m.

By equation 2.7,

$$\delta\theta° = 57.2958 (1.370 - 1.675) \cos 4.5/106.680$$

$$= -0.1633$$

$$\alpha = 4.5000 - 0.1633 = 4.3367$$

Then $H = 106.680 \cos 4.3367 = 106.375$ m

If the effect had been ignored, then

$$H' = 106.680 \cos 4.5 = 106.351 \text{ m}$$

$$\text{Error} = \quad 0.024 \text{ m}$$

Example 2.2   A length AD is measured in three bays with ground distances recorded as follows:

$AB = 42.361$ m, $BC = 25.734$ m, and $CD = 52.114$ m.

From a theodolite station at A, instrument height 1.46 m, staff readings were taken with a mean vertical angle of $-3° 24' 30''$ to B 1.58, to C 0.96 and D 0.96. Calculate the horizontal length of the line AD. (TP)

Let $d_1 = 1.46, d_2 = 1.58, d_3 = 0.96, d_4 = 0.96$. It is necessary to reduce each bay separately (see Fig. 2.5):

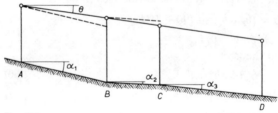

Fig. 2.5

Line AB $h_i = d_1 = 1.46, h_t = d_2 = 1.58$

then $\delta\theta° = 57.2958 (1.46 - 1.58) \cos - 3.4083/42.361 = -0.1620$

$\alpha_1 = -3.4083 - 0.1620 = -3.5703$

$H_1 = 42.361 \cos - 3.5703$ $= 42.2788$ m

Line BC $h_i = d_2 = 1.58, h_t = d_3 = 0.96$

then $\delta\theta° = 57.1945 (1.58 - 0.96)/25.734 = 1.3780$

$\alpha_2 = -3.4083 + 1.3780 = -2.0303$

$H_2 = 25.734 \cos - 2.0303$ $= 25.7178$ m

Line CD $h_i = d_3 = 0.96 = h_t = d_4$

$\alpha_3 = \theta$

$H_3 = 52.114 \cos - 3.4083$ $= 52.0218$ m

Total length H $= 120.0184$

say 120.018 m

(b)   If the difference in level of the index marks is measured as d then by equations 1.14 and 1.16

$$H = \sqrt{(L_m^2 - d^2)}$$

and $C_i = d^2/2L_m + d^4/8L_m^3 + \ldots$

2.4.1   Errors in the reduction

(a)   Given the value $\alpha \pm \delta\alpha$ and

$$H = L_m \cos \alpha$$

By differentiating with respect to α,

$$\delta H = -L_m \sin \alpha . \delta\alpha \qquad (2.8)$$
$$= -H \tan \alpha . \delta\alpha \qquad (2.9)$$

If $L_m = 100.000$ m and $\alpha = 3° 00' 00'' \pm 30''$ then

$$H = 100.000 \cos 3.0000 = 99.8630 \text{ m}$$
$$\delta H_\alpha = 99.8630 \tan 3.0000 \times 30/206\,265 = 0.0008 \text{ m}$$
$$\delta H/H = 0.0008/99.8630 \simeq 1/130\,000$$

To limit the error to $1/30\,000$ then

$$\delta\alpha'' = 206\,265/(30\,000 \tan 3.0000°)$$
$$= 131'' \quad \text{i.e.} \quad 2' 11''$$

Thus the vertical angle should be measured with normal care.

(b)   Given the level difference between the index marks as $d + \delta d$ and

$$H^2 = L_m^2 - d^2$$

then $2H.\delta H_d = -2d.\delta d$

$$\delta H_d = -d.\delta d/H \qquad (2.10)$$

If $L_m = 100.000$ m and $d = 5.240 \pm 0.010$ m then

$$H = \sqrt{(100^2 - 5.240^2)} = 99.863 \text{ m}$$
$$\delta H_d = 5.240 \times 0.010/99.863 = \pm 0.0005 \text{ m}$$

i.e. $\delta H/H = 0.0005/99.863 \simeq 1/190\,000$

To limit the error to $1/30\,000$,

$$\delta d = 99.863^2/(30\,000 \times 5.240) = \pm 0.063 \text{ m}$$

It can be seen that this is easily attained.

2.5   Correction for temperature (C_t)

The measuring band is standardised at a given temperature t_s. If in the field the temperature of the band is recorded as t_m then the band will expand or contract and a correction to the measured length is given as

$$C_t = L_m . \alpha (t_m - t_s) \qquad (2.11)$$

where L_m is the measured length and α is the coefficient of linear expansion of the metal.

The coefficient of linear expansion (α) of a solid is defined as 'the increase in length per unit length of the solid when its temperature changes by one degree'. For steel the average value of α is given as

$$\alpha = 11.2 \times 10^{-6} \text{ per } °C$$

The range of linear coefficients α is thus given as:

	per 1°C
Steel	10.6 to 12.2 ($\times 10^{-6}$)
Invar	5.4 to 7.2 ($\times 10^{-7}$)

With reference to the standardisation of the field tape: The true length of the field tape under the manufacturer's specifications is given as

$$l_t = l_f[1 - \alpha(t_m - t_s)] \tag{2.12}$$

To find the temperature at which the true length = the nominal length then

$$t'_s = t_m + (l_n - l_f)/(l_n . \alpha) \tag{2.13}$$

Example 2.3 On a test line the following data was recorded:
Standard 'invar' tape   30.0000 m at 20° C and 50 N
measured length   28.8853 m at 10° C and 50 N
Steel field tape   nominal length '30' m at 20° C and 50 N
measured length   28.8895 m at 10° C and 50 N
(Assume α values: invar $5 \times 10^{-7}/°$ C; steel $11 \times 10^{-6}/°$ C.)

To derive the true length of the test line using the invar tape,
$$L_s = L_m[1 + \alpha(t_m - t_s)]$$
$$= 28.8853[1 + 5 \times 10^{-7}(10 - 20)] \qquad = 28.8852 \text{ m}$$
The comparative length measured with the field tape = 28.8895 m

$$\delta L = \quad 0.0043 \text{ m}$$

The length of the field tape under the measurement conditions is given by equation 2.2:
$$l_f = l_n(1 - \delta L/L_m)$$
$$= 30.0000(1 - 0.0043/28.8895) \qquad = 29.9955 \text{ m}$$
The length of the field tape under the manufacturer's specifications is given by equation 2.12:
$$l_t = l_f[1 - \alpha(t_m - t_s)]$$
$$= 29.9955[1 - 11 \times 10^{-6}(10 - 20)] \qquad = 29.9988$$

The standardisation 'correction' is then $-0.0012 \text{ m}/30 \text{ m}$.
As the tape is too short, the measurement will be too long; then

$$C_s = -0.0012 \times 28.8895/30 \qquad = -0.00116$$
$$C_t = 28.8895 \times 11 \times 10^{-6}(10 - 20) = -0.00318$$
$$C_{\text{total}} = -0.00434$$

The true length of the line is then

$$28.8895 - 0.0043 = 28.8852\,\text{m}$$

Alternatively, to find the temperature at which the true length equals the nominal length, by equation 2.13:

$$t'_s = t_m + (l_n - l_f)/(l_n \cdot \alpha)$$
$$= 10.0 + 0.0045/30 \times 11 \times 10^{-6} \qquad = 23.6°\,\text{C}$$

Then $\quad C_t = 28.8895 \times 11 \times 10^{-6}(10 - 23.6) \ = \ -0.0043\,\text{m}$

$\qquad L_s = 28.8895 - 0.0043 = 28.8852\,\text{m}$

It can be seen that the latter is simpler to apply.

Example 2.4   A traverse line is 152.400 m long. If the tape used in the field is 50.000 m when standardised at 17° C, what correction must be applied if the temperature at the time of measurement is 23° C? (Assume $\alpha = 11.2 \times 10^{-6}$ per °C.)

By equation 2.11,

$$C_t = 152.400 \times 11.2 \times 10^{-6} \times (23 - 17)$$
$$= +0.010\,\text{m}$$

Example 2.5   If a field tape when standardised at 17.2° C measured 100.0052 m, at what temperature will it be exactly its nominal value? (Assume $\alpha = 11.2 \times 10^{-6}$ per °C.)

$$\delta l = 0.0052\,\text{m}$$

By equation 2.13,

$$t'_s = 17.2 - \frac{0.0052}{100 \times 11.2 \times 10^{-6}}$$
$$= 17.2 - 4.6 = 12.6°\,\text{C}$$

2.5.1   Errors due to the wrongly applied temperature

As $\qquad L_t = L_m[1 + \alpha(\Delta t)]$

then $\quad \delta L_t = L_m \cdot \alpha \cdot \delta(\Delta t)$

If $\delta L/L_m$ is limited to $1/30\,000$, then

$$\delta(\Delta t) = 1/(30\,000 \cdot \alpha)$$
$$= 1/(30\,000 \times 11 \times 10^{-6}) = \pm 3.0°\,\text{C}$$

It has been suggested from practical observations in recording that the actual temperatures for ground and catenary measurement are $\pm 3°\,\text{C}$ and $\pm 1.5°\,\text{C}$ respectively.

2.6   Correction for tension (C_T)

The measuring band is standardised at a given tension T_s. If in the field the applied tension is T_m then the tape will, due to its own elasticity, expand or contract in accordance with Hooke's Law.

A correction factor is thus given as

$$C_T = L_m(T_m - T_s)/A.E \qquad (2.14)$$

where L_m = the measured length (the value of C_T is in the same unit as L)

A = cross-sectional area of the tape

E = Young's modulus of elasticity, i.e. stress/strain.

Weight is the gravitational force attracting a body to the centre of the earth. Gravitational attraction varies over the earth's surface and thus the weight of the body varies.

Mass of a body is dependent only on the quantity of the material in it and thus is constant.

As force = mass × acceleration

weight = mass × acceleration due to gravitational attraction

i.e. $w = m \times g$

'g' varies with the location of the mass, but for most purposes the international standard value for g is $9.806\,65\,\text{ms}^{-2}$.

In SI units a mass of 1 kg has a weight of $1 \times 9.807\,\text{kg\,ms}^{-2}$, i.e. 9.807 newtons.

In practice, when a body is 'weighed' it will invariably erroneously be given units of kilograms instead of newtons, but the engineer/surveyor must clearly distinguish between mass and weight for computational purposes. British Standard BS 4484: Part 1: 1969 says, 'Tension is normally measured in newtons but as spring balances are normally used for checking tension the figure may be given in kilograms.' Considering this, it has been decided here to use the term 'mass'; weight (w) per unit length is replaced by mass (m) per unit length.

The units used for T, A and E must be compatible, e.g.

$$T_2(\text{N}) \quad A_2(\text{m}^2) \quad E_2(\text{N\,m}^{-2}) \quad (\text{SI units})$$

Based on the use of the International System of Units (SI units) the unit of force is the newton (N), i.e. the force required to accelerate a mass of 1 kg by 1 metre per second.

For steel, $E \simeq 19.3$ to $20.7 \times 10^{10}\,\text{N\,m}^{-2}$

For invar, $E \simeq 13.8$ to $15.2 \times 10^{10}\,\text{N\,m}^{-2}$

Notes

(1)   If $T_m = T_s$ no correction is necessary.

(2)   It is generally considered good practice to over-tension to minimise deformation of the tape, the amount of tension being strictly recorded and the correction applied.

(3)   The cross-sectional area (A) of the tape may be physically measured using a mechanical micrometer, or it may be computed from the total mass M of the tape of length l and a value ρ for the density of the material:

$$A = Mg/L\rho \qquad (2.15)$$

Example 2.6   A tape is 30.000 m at a standard tension of 100 N and measures in cross-section 6.0 mm × 0.2 mm. If the applied tension is 78 N and $E = 1.93 \times 10^5 \, \text{N} \, \text{mm}^{-2}$, calculate the correction to be applied.

By equation 2.14,

$$C_T = \frac{30.000(78 - 100)}{6.0 \times 0.2 \times 1.93 \times 10^5} = -0.003 \, \text{m}$$

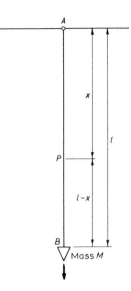

Fig. 2.6

Measurement in the vertical plane. When a metal tape AB of length l is fixed at one end A and then freely suspended, it will lengthen by a value s due to the tension T applied by the earth's gravitational pull g on the mass ml of the tape.

The tension at the bottom of the tape is zero and at the fixed point a maximum value (mgl). It is thus not uniform, and the stress varies along its length.

If a mass M is added at the lower end B (Fig. 2.6) then the tension at any point P along the tape is given as

$$T = Mg + mg(l - x) \qquad (1)$$

where M = the attached mass

m = the mass of the tape per unit length

g = gravitational acceleration

l = the total length of the tape

x = the length from the fixed point to P

i.e. T = the force due to the gravitational effect on the mass below the point P.

If the cross-sectional area of the tape is A and Young's modulus is E, then the strain at P is given as

$$\delta s / \delta x = T/AE$$

$$\delta s = \delta x \, T/AE \quad \text{(compare this with equation 2.14)}$$

and $T = AE \, \delta s / \delta x \qquad (2)$

NB: δs is a small increment of elongation; δx is a small increment of the length x.

Equating (1) and (2) gives

$$AE \delta s / \delta x = Mg + mgl - mgx$$

Integrating,

$$AEs = Mgx + mglx - (mgx^2/2) + d$$

When $x = 0$ and $s = 0$ then $d = 0$, therefore

$$s = \frac{gx}{AE} \left[M + \tfrac{1}{2} m(2l - x) \right] \qquad (2.16)$$

If $x = l$,

$$s = \frac{gl}{AE} \left(M + \frac{ml}{2} \right) \qquad (2.17)$$

and, when $M = 0$,

$$s = mgl^2 / 2AE \qquad (2.18)$$

Taking into account the standardisation tension factor, a negative extension must be allowed for initially as the tape is not tensioned up to standard. That is,

$$s_1 = -xT_s/AE$$

assuming $x = l, T_s = $ standard tension.

Thus the general equation for precise measurement will be

$$s = \frac{gx}{AE}\left(M + \tfrac{1}{2}m(2l - x) - \frac{T_s}{g}\right) \qquad (2.19)$$

Example 2.7 Calculate the elongation at (1) 300 m and (2) 1000 m of a 1000 m mine shaft measuring tape hanging vertically due to its own mass. The modulus of elasticity is $2 \times 10^5 \, \text{N mm}^{-2}$, the mass of the tape is $0.07 \, \text{kg m}^{-1}$ and the cross-sectional area of the tape is $9.7 \, \text{mm}^2$.

By equation 2.16,

$$s = \frac{gx}{AE}\left[M + \tfrac{1}{2}m(2l - x)\right]$$

As $M = 0$,

$$s = \frac{mgx}{2AE}(2l - x)$$

For (1): $s = \dfrac{0.07 \times 9.81 \times 300\,(2000 - 300)}{2 \times 9.7 \times 2 \times 10^5}$

$$= 1.7698 \times 10^{-7} \times 300 \times 1700 = 0.090 \text{ m}$$

For (2): $x = l$

$$s = mgl^2/2AE$$

$$= 1.7698 \times 10^{-7} \times 1000^2 = 0.177 \text{ m}$$

Example 2.8 If the same tape is standardised as 1000.000 m at 180 N tension, what is the true length of the shaft recorded as 999.532 m?

Elongation due to its own mass, by equation 2.16,

$$s_1 = \frac{gx}{AE}\left[M + \frac{m}{2}(2l - x)\right]$$

$$= \frac{9.807 \times 999.532}{9.7 \times 2 \times 10^5}\left[0 + \frac{0.07}{2}(2000.000 - 999.532)\right]$$

$$= 0.1770 \text{ m}$$

Elongation due to standard tension,

$$s_2 = -lT_s/AE$$

$$= -\frac{999.532 \times -180}{9.7 \times 2 \times 10^5}$$

$$= 0.0927 \text{ m}$$

Total elongation

$$s = 0.1770 - 0.0927$$

$$= 0.0843 \text{ m}\quad \text{correction to measurement}$$

True length

$$= 999.532 + 0.084 = 999.616 \text{ m}$$

Alternatively, by equation 2.18,

$$s = (mgl)l/2AE \quad \text{where } mgl/2 = \text{applied tension}$$

$$T_m = mgl/2 = 0.5 \times 0.07 \times 9.807 \times 999.532 = 343 \text{ N}$$

Then the tension correction

$$= l(T_m - T_s)/AE$$

$$= 999.532\,(343 - 180)/9.7 \times 2 \times 10^5$$

$$= 0.084 \text{ as above.}$$

Alternatively, by equation 2.19,

$$s = \frac{gx}{AE}\left[M + \frac{m}{2}(2l - x) - \frac{T_s}{g} \right]$$

$$= \frac{9.807 \times 999.532}{9.7 \times 2 \times 10^5}\left[0 + \frac{0.07}{2}(2000.000 - 999.532) - \frac{180}{9.807} \right]$$

$$= 0.084 \text{ m}$$

2.6.1   Errors due to variation from the recorded value of tension

As

$$L_t = L_m(1 + \Delta T/A.E) \tag{2.20}$$

$$\delta L = L_m.\delta(\Delta T)/A.E \tag{2.21}$$

If the error ratio $\delta L/L_m$ is not to exceed $1/30\,000$ then

$$\delta(\Delta T) = A.E/30\,000$$

If $A = 2 \text{ mm}^2$ and $E = 2.2 \times 10^5 \text{ N mm}^{-2}$ then

$$\delta(\Delta T) = 2 \times 2.2 \times 10^5/30\,000 = \pm 14.7 \text{ N}$$

2.7   Correction for sag (C_c)

The measuring band may be standardised in two ways, (1) on the flat or (2) in catenary.

If the band is used in a manner contrary to the standard conditions some correction is necessary.

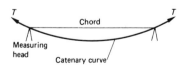

Fig. 2.7 Measurement in catenary

(1) *If standardised on the flat and used in catenary* (Fig. 2.7) the general equation for correction is applied, viz.

$$C_c = -(mg)^2\, L_m^3/24T^2$$

$$= -\frac{L_m^3}{24}\left(\frac{mg}{T}\right)^2 \tag{2.22}$$

$$= -\frac{L_m}{24}\left(\frac{Mg}{T}\right)^2 \qquad (2.23)$$

where L_m = tape length recorded
m = mass of tape per unit length (kg m^{-1})
g = gravitational acceleration
M = total mass of tape in use = mL_m(kg)
T = applied tension (N).

(2)   *If standardised in catenary*:
(a) The length of the chord may be given relative to the length of the tape, or
(b) the length of the tape in catenary may be given.

(i)   If the tape is used on the flat a positive sag correction must be applied.
(ii)   If the tape is used in catenary at a tension T_m which is different from the standard tension T_s, the correction will be the difference between the two relative corrections, i.e.

$$C_c = -\frac{(Mg)^2\, L_m}{24}\left(\frac{1}{T_m^2}-\frac{1}{T_s^2}\right) \qquad (2.24)$$

If $T_m > T_s$ the correction will be positive.

(iii)   If standardised in catenary using a length L_s and then applied in the field at a different length L_m the corrected length may be computed as follows.
    The standardised chord length is converted into a standardised arc length. A comparison with the nominal length of the tape thus gives the standardisation error.
    The recorded arc length may now be converted into a standardised arc length and thereafter corrected for sag to give a standardised chord length.

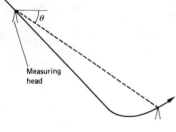

Fig. 2.8

    The sag correction is an acceptable approximation based on the assumption that the measuring heads are at the same level. If the heads are at considerably different levels (Fig. 2.8) the correction should be

$$C_c = -\frac{L_m^3}{24}\left(\frac{mg\cos\theta}{T}\right)^2\left(1\pm\frac{mg\,L_m\sin\theta}{T}\right) \qquad (2.25)$$

the sign depending on whether the tension is applied at the upper or lower end of the tape.
    For general purposes,

$$C_c = c_1\cos^2\theta$$

$$= -\frac{L_m^3}{24}\left(\frac{mg\cos\theta}{T}\right)^2 \qquad (2.26)$$

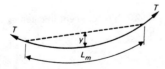

Fig. 2.9 Mass of the tape determined in the field

The mass of the tape determined in the field. The catenary sag of the tape can be used to determine the mass of the tape, Fig. 2.9.

If y is the measured sag at the mid-point, then the mass per unit length is given as

$$m = 8Ty/g L_m^2 \qquad (2.27)$$

or the amount of sag

$$y = mg\, L_m^2/8T \qquad (2.28)$$

where m = mass/unit length
$\quad\quad\;\; T$ = applied tension
$\quad\quad\;\; y$ = vertical sag at the mid-point
$\quad\quad\;\; L_m$ = length of tape between supports.

Example 2.9 Calculate the horizontal length between two supports approximately level, if the recorded length is 30.5522 m, the tape has a mass for this length of 0.425 kg and the applied tension is 90 N.

By equation 2.23,

$$C_c = -\frac{L_m}{24}\left(\frac{Mg}{T}\right)^2$$

$$= -\frac{(0.425 \times 9.807)^2 \times 30.5522}{24 \times 90^2} = -0.0027\,\text{m}$$

Horizontal length $= 30.5522 - 0.0027 = 30.5495\,\text{m}$

Example 2.10 A '30 m' tape standardised in catenary as 29.9850 m at 110 N is used in the field with a tension of 90 N. Calculate the sag correction if the mass of the tape is $0.0312\,\text{kg m}^{-1}$.

Standardised chord length with 110 N $\qquad = \quad 29.9850$

Sag correction $= (0.0312 \times 9.81/110)^2\, 30^3/24 \qquad = + \quad 0.0087$

Standardised arc length $\qquad\qquad\qquad\qquad = \quad 29.9937$

Sag correction in the field $= (0.0312 \times 9.81/90)^2 30^3/24 = - \quad 0.0130$

Reduced field length $\qquad\qquad\qquad\qquad\quad = \quad 29.9807$

Alternatively, by equation 2.24,

$$C_c = 30^3 (0.0312 \times 9.81)^2 (1/90^2 - 1/110^2)/24 = -0.0043$$

Therefore the reduced field length $= 29.9850 - 0.0043 \quad = \quad 29.9807$

Example 2.11 A 30 m tape suspended in catenary with a tension of 130 N is used to record a length of 29.3655 m between the supports. At the mid-point the sag is measured as 0.168 m. Calculate the mass of the tape per unit length.

By equation 2.27,

$$m = \frac{8Ty}{gl^2} = \frac{8 \times 130 \times 0.168}{9.807 \times 29.3655^2} = 0.021\,\text{kg m}^{-1}$$

Example 2.12   A tape of nominal length '30 m' is standardised in catenary at 45 N tension and found to be 29.9850 m. If the mass of the tape is 0.015 kg m^{-1} calculate the horizontal length of a span recorded as 15.3645 m.

Standardised chord length	=	29.9850 m
Sag correction $c_1 = (0.015 \times 9.81/45)^2\,30^3/24$	= +	0.0120
Standardised arc length	=	29.9970
Standardisation error per 30 m	=	− 0.0030
Recorded arc length	=	15.3645
Standardisation error = $15.3645 \times 0.0030/30$	= −	0.0015
Standardised arc length	=	15.3630
Sag correction = $c_1 \times (15.3645/30)^3$	=	− 0.0016
Standardised chord length	=	15.3614

Example 2.13   A copper transmission line 12 mm in diameter is stretched between two points 300 m apart at the same level, with a tension of 5×10^3 N, when the temperature is 32° C. It is necessary to define its limiting positions when the temperature varies. Making use of the corrections for sag, temperature, and elasticity normally applied to base line measurements by tape in catenary, find the tension at a temperature of − 12° C and the sag in the two cases. For copper, Young's modulus = 7×10^4 N mm^{-2}, density = 9×10^3 kg m^{-3} and coefficient of linear expansion = 1.7×10^{-5} per °C.   (LU)

The mass per unit length is

$$m = r^2\,\rho\pi$$
$$= 0.006^2 \times 9 \times 10^3 \times 3.1416 = 1.018\,\text{kg m}^{-1}$$

The length of wire needed at 32°C = 300.000 + δl

where $\delta l_1 = (mg)^2\,l^3/24T^2$

$$= \frac{(1.018 \times 9.807)^2 \times 300.000^3}{24 \times 5000^2}$$

$$= 4.485\,\text{m}$$

The approximate length of the wire is thus 304.49 m and this value may be used in the above expression to give a better approximation. That is,

$$\delta l_2 = \frac{(1.018 \times 9.807)^2 \times 304.49^3}{24 \times 5000^2} = 4.690\,\text{m}$$

The better approximation for the sag length is 304.69 m.

Amount of sag = $y = (mg)l^2/8T$

$$= \frac{1.018 \times 9.807 \times 304.69^2}{8 \times 5000} = 23.17\,\text{m}$$

When the temperature falls to $-12°\,C$,

    contraction of wire $= l\alpha\Delta t$

$$= 304.69 \times 1.7 \times 10^{-5} \times 44 = 0.23\,\text{m}$$

Adjusted length of wire $= 304.69 - 0.23 = 304.46\,\text{m}$.

    As $y \propto l^2$,

$$y_2 = y_1\,(l_1/l_2)^2$$

$$= 23.17 \times (304.46/304.69)^2 = 23.14\,\text{m}$$

Similarly, as $y \propto 1/T$,

$$T_2 = T_1\,(y_1/y_2)$$

$$= 5000\,(23.17/23.14) = 5006\,\text{N}$$

NB: A change of $6\,\text{N}$ in tension will expand the tape by

$$\delta l = l\Delta T/AE$$

2.7.1   Errors in the catenary correction

Where the tape has been standardised on the flat and is then used in catenary with the measuring heads at different levels, the approximation formula is given as

$$C_c = -\frac{L_m^3}{24}\left(\frac{mg\cos\theta}{T}\right)^2 \qquad \text{(Equation 2.26)}$$

where θ is the angle of inclination of the chord between measuring heads. The value of $\cos^2\theta$ becomes negligible when θ is small.

    The sources of error are derived from:

(a)   an error in the mass of the tape per unit length, m,
(b)   an error in the angular value, θ,
(c)   an error in the tension applied, T.

By successive differentiation,

$$\delta C_m = \frac{-2L_m^2 mg^2\cos^2\theta\,\delta m}{24T^2} \qquad (2.29)$$

$$= 2C\,\delta m/m \qquad (2.30)$$

i.e. $\delta C_m/C = 2\delta m/m$ $\qquad\qquad\qquad\qquad$ (2.31)

This may be due to an error in the measurement of the weight of the tape or due to foreign matter on the tape, e.g. dirt.

$$\delta C_\theta = \frac{-l^3(mg)^2}{24T^2}\sin 2\theta\,\delta\theta \qquad (2.32)$$

$$= 2C\tan\theta\delta\theta \qquad (2.33)$$

$\delta C_\theta/C = 2\tan\theta\delta\theta$ $\qquad\qquad\qquad\qquad$ (2.34)

Variation due to error in tension $T, \delta T$

$$\delta C_T = \frac{2l^3(mg)^2\cos^2\theta\,\delta T}{24T^3} \qquad (2.35)$$

$$= \frac{-2C\,\delta T}{T} \tag{2.36}$$

$$\delta C_T/C = -2\delta T/T \tag{2.37}$$

The compounded effect of a variation in tension gives

$$\frac{2l^3(mg)^2\cos^2\theta\,\delta T}{24T^2} + \frac{l\delta T}{AE} = \frac{l\delta T}{12}\left[\left(\frac{lmg\cos\theta}{T}\right)^2 + \frac{12}{AE}\right] \tag{2.38}$$

Example 2.14   If $l = 30.000$ m, $m = 0.014 \pm 0.0014$ kg m^{-1}, $\theta = 2°\,00'\,00'' \pm 10''$ and $T = 45 \pm 4.5$ N, then

$$C_c = \frac{(mg)^2\,l^3\cos^2\theta}{24T^2} = \frac{(0.014 \times 9.807)^2\,30^3\cos^2 2°}{24 \times 45^2}$$

$$= 0.0105\,\text{m} \quad (0.011\,\text{m})$$

The effect of an error in the measurement of mass of the tape (δm) gives

$$\delta C_m = \frac{2C\,\delta m}{m} = \frac{2 \times 0.0105 \times 0.0014}{0.014}$$

$$= 0.0021\,\text{m}$$

i.e. 10% error in the measurement of the mass of the tape produces an error of $1/14\,000$.

The effect of an error in the value of $\theta, \delta\theta$ gives

$$\delta C_\theta = \frac{2C\tan\theta\,\delta\theta''}{206\,265} = 3.6 \times 10^{-8}\,\text{m}$$

(this is obviously negligible)

The effect of an error in the value of $T, \delta T$, gives

$$\delta C_T = 2C\,\delta T/T = 2 \times 0.0105 \times 4.5/45 = 0.0021\,\text{m}$$

i.e. 10% error in tension produces an error of $1/14\,000$.

Example 2.15   A base line is measured and subsequent calculations show that its total length is 1391.662 m. It is later discovered that the tension was recorded incorrectly, the proper figure being 45 N less than that stated in the field book, extracts from which are given below. Assuming that the base line was measured in 46 bays of nominal length 30 m and one bay of nominal length 12 m calculate the error incurred.

Extract from field notes:

standard temperature	10° C
standard tension	90 N
measured temperature	7.2 ° C
measured tension	178 N
Young's modulus	2.2×10^5 N mm^{-2}
cross-sectional area of tape	3.2 mm × 1.3 mm
density of steel	7.75×10^3 kg m^{-3} (TP)

Mass of steel tape per 30 m

$$= 0.0032 \times 0.0013 \times 7.75 \times 10^3 \times 30 = 0.967 \, \text{kg}$$

From equation 2.14, $C_T = \dfrac{L(T_m - T_s)}{AE}$

Then the error due to wrongly applied tension

$$= C_T - C'_T$$

$$= \frac{L(T_m - T_s)}{AE} - \frac{L(T'_m - T_s)}{AE}$$

$$= \frac{L}{AE}(T_m - T'_m)$$

where T_m = true applied tension
T'_m = assumed applied tension.

Therefore

$$\text{error} = \frac{1392 \times 45}{3.2 \times 1.3 \times 2.2 \times 10^5} = -0.0684 \, \text{m}$$

From equation 2.23, correction for sag

$$C_c = \frac{(Mg)^2 l}{24T^2}$$

Therefore error due to wrongly applied tension

$$= C_c - C'_c$$

$$= -\frac{(Mg)^2 l}{24}\left(\frac{1}{T^2} - \frac{1}{T_1^2}\right)$$

Error for 30 m bay

$$= \frac{-30(0.967 \times 9.81)^2}{24}\left(\frac{1}{133^2} - \frac{1}{178^2}\right)$$

$$= -30 \times 3.749 \times 2.497 \times 10^{-5}$$

$$= -0.002\,81 \, \text{m}$$

Error for 46 bays $= -0.1292 \, \text{m}$

Error for 12 m bay $= \dfrac{-12^3(0.967 \times 9.81)^2}{30^2} \times 2.497 \times 10^{-5}$

$$= -0.000\,18 \, \text{m}$$

Therefore

Total sag error $= -0.1294 \, \text{m}$
Total tension error $= -0.0684 \, \text{m}$
Total error $= -0.1978 \, \text{m}$

i.e. apparent reduced length is 0.1978 m too large.

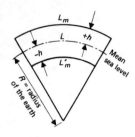

Fig. 2.10 Reduction to mean sea level

2.8   Reduction to mean sea level

If the length at mean sea level is L and h = height of line above or below mean sea level (Fig. 2.10), then

$$L/L_m = R/(R+h)$$

Therefore

$$L = L_m R/(R+h) \tag{2.39}$$

As $C_m = L_m - L = L_m[1 - R/(R+h)]$

then $C_m = L_m h/(R+h) \tag{2.40}$

Notes

(1)   If h is $-$ve, i.e. below MSL, then C_m is $-$ve.

(2)   If $R \simeq 6380\,\text{km}$, $C_m \simeq 6.38\,h \times 10^{-6}$ per m. $\tag{2.41}$

(3) $\delta L \propto \delta h$.

2.9   Reduction of ground length (L_m) to grid length (L_g)

$$L_g = L_m \times F \quad \text{(the local scale factor)} \tag{2.42}$$

The local scale factor depends on the properties of the projection.

Here we will consider only the Modified Transverse Mercator projection as adopted by the Ordnance Survey in the British Isles.

Local scale factor (F) for short lines (say $< 10\,\text{km}$)

$$F = F_0\left(1 + \frac{Y^2}{2\rho v}\right) \tag{2.43}$$

where F_0 = the local scale factor at the central meridian
       Y = the Easting in metres from the true origin
       ρ = the radius of curvature to the meridian
       v = the radius of curvature at right angles to the meridian.

Assuming $\rho \simeq v = R$, then

$$F = F_0\left(1 + \frac{Y^2}{2R^2}\right) \tag{2.44}$$

and if R is assumed to be $6380\,\text{km}$ then

$$1/2R^2 = 1 \cdot 228 \times 10^{-8}$$

For practical purposes,

$$F \simeq F_0(1 + 1.228\,Y^2 \times 10^{-8}) \tag{2.45}$$

$$\simeq 0.999\,6013\,(1 + 1.228\,Y^2 \times 10^{-8}) \tag{2.46}$$

(see Fig. 2.11).

NB: Y = Eastings $- 400\,\text{km}$.

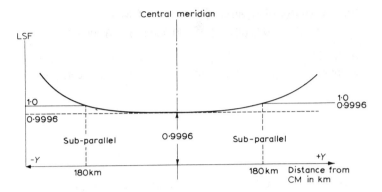

Fig. 2.11

Example 2.16 Calculate (a) the local scale factors for each corner of the grid square TA54 (i.e. grid coordinates of SW corner), (b) the local scale factor at the centre of the square.

See Chapter 4, page 123. Refer to Fig. 2.12.

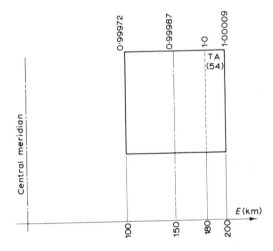

Fig. 2.12

(a) (i) At the SW corner, coordinates are 500 km E, i.e. 100 km E of central meridian. Therefore, from equation 2.46,

$$LSF = 0.999\,6013\,(1 + 100^2 \times 1.23 \times 10^{-8})$$
$$= 0.999\,601 + 0.000\,123 = 0.999\,724$$

(ii) At SE corner, coordinates are 600 km E, i.e. 200 km E of CM. Therefore

$$LSF = 0.999\,6013\,(1 + 200^2 \times 1.23 \times 10^{-8})$$
$$= 0.999\,601 + 2^2 \times 0.000\,123$$
$$= 0.999\,601 + 0.000\,492 = 1.000\,093$$

(b)   At centre of square, 150 km from CM,

$$LSF = 0.999\,601 + 1.5^2 \times 0.000\,123 = 0.999\,878$$

As the local scale factor is applicable only to distances measured at mean sea level a compounded factor (F') may be applied as follows. As

$$L_g = L_m \times F$$

then $$F' = F\left(\frac{R}{R+h}\right) \qquad\qquad (2.47)$$

Example 2.17   Calculate the local scale factors applicable to a place 415 km E and to coal seams at depths of 150 m, 300 m, and 600 m respectively. Assume the local radius of the earth to be 6380 km.

$$LSF = 0.999\,6013\,(1 + 1.228\,Y^2 \times 10^{-8})$$

(Equation 2.46)

Correction for mean sea level

$$F' = F\left(\frac{R}{R-h}\right) = F.K$$

At mean sea level,

$$LSF_0 = 0.999\,6013(1 + 1.228 \times 15^2 \times 10^{-8})$$
$$= 0.999\,6041$$

At 150 m below mean sea level,
   Scale factor for depth,

$$K_1 = 6380/6379.850 = 1.000\,0235$$

LSF at depth,

$$F'_1 = 0.999\,6041 \times 1.000\,0235 = \underline{0.999\,6276}$$

At 300 m below mean sea level,

$$K_2 = 6380/6379.7 = 1.000\,0470$$

LSF at depth

$$F'_2 = 1.000\,0470\,F \qquad\qquad = \underline{0.999\,6510}$$

At 600 m below mean sea level,

$$K_3 = 6380/6379.4 = 1.000\,0940$$

LSF at depth

$$F'_3 = 1.000\,940\,F \qquad\qquad = \underline{0.999\,6981}$$

The variation of local scale factor with position. The effect of LSF varies with the position of the measured line on the National Grid and depends on the distance of the line from the central meridian.

   By equation 2.45,

$$F = F_0(1 + 1.228\,Y^2 \times 10^{-8})$$

Thus the value varies with Y^2, where $Y = $ Easting of line

-400 km, and it is necessary to determine the range of Eastings for which a given LSF may be applied to retain a desired accuracy. If

$$F = F_0(1+KY^2) \quad \text{where } K = 1.228 \times 10^{-8}$$

then
$$F + \Delta F = F_0[1+K(Y+\Delta Y)^2]$$
$$= F_0[1+KY^2+K(2Y\Delta Y+\Delta Y^2)]$$
$$\Delta F = F_0 K(2Y\Delta Y+\Delta Y^2)$$

The error ratio is given as

$$\frac{\Delta F}{F} = \frac{K(2Y\Delta Y+\Delta Y^2)}{1+KY^2} = \frac{1}{X}$$

(the permitted relative accuracy)

$$2Y\Delta Y+\Delta Y^2 = \frac{1+KY^2}{KX} = M, \text{ say}$$

then
$$\Delta Y^2 + 2Y\Delta Y - M = 0$$

This is a quadratic equation in ΔY; then

$$\Delta Y = \frac{-2Y \pm \sqrt{(4Y^2+4M)}}{2}$$
$$\Delta Y = -Y \pm \sqrt{(Y^2+M)} \tag{2.48}$$

Example 2.18   Given that $1/X = \pm 1/30\,000$ and $Y = \pm 57.0$ km (i.e. Easting of line is either 457.0 km or 343.0 km).

By equation 2.48,
$$\Delta Y = -Y \pm \sqrt{(Y^2+M)}$$

where
$$M = \frac{1+KY^2}{KX} = \frac{1+1.228 \times 10^{-8} \times 57.0^2}{1.228 \times 10^{-8} \times 30\,000}$$

$$= 2714.5$$
$$\Delta Y = \mp Y \pm \sqrt{(Y^2 \mp M)}$$
$$= \mp 57.0 \pm \sqrt{(57.0^2 - 2714.5)} \qquad \text{(a)}$$
$$= \mp 57.0 \pm 23.1 = \pm 80.1 \quad \text{or} \quad \pm 33.9$$

or
$$\Delta Y = \pm 57.0 \pm \sqrt{(57.0^2 + 2714.5)} \qquad \text{(b)}$$
$$= \pm 57.0 \pm 77.2 = \pm 134.2 \quad \text{or} \quad \pm 20.2$$

Thus the coordinate range for 457.0 km E is

$$457.0 + 20.2 = 477.2 \text{ km} \quad \text{E}$$

or
$$457.0 - 33.9 = 423.1 \text{ km} \quad \text{E}$$

whilst the coordinate range for 343.0 km E is

$$343.0 + 33.9 = 376.9 \text{ km} \quad \text{E}$$
$$343.0 - 20.2 = 322.8 \text{ km} \quad \text{E}$$

This may be shown diagrammatically, Fig. 2.13.

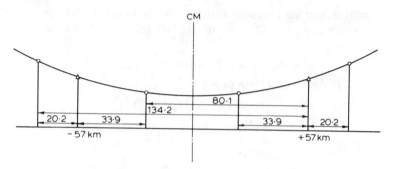

Fig. 2.13

An alternative solution may be found by varying the LSF by the permitted error ratio, i.e.

$$\text{LSF (at } Y = 57.0) = 0.999\,6013(1 + 1.228 \times 57.0^2 \times 10^{-8})$$
$$= 0.999\,6412$$
$$1/30\,000 \text{ of LSF} = 0.999\,6412/30\,000 = \pm 0.000\,0333$$

thus the limits of LSF

for $Y_1 = 0.999\,6745$

and   for $Y_2 = 0.999\,6079$

Substituting these values into equation 2.45,

$$0.999\,6745 = 0.999\,6013(1 + K\,Y^2)$$

$$Y^2 = \left(\frac{0.999\,6745}{0.999\,6013} - 1 \right) \div K$$

$$Y_1 = 77.2$$
$$\Delta Y = 77.2 - 57.0 = 20.2$$

Similarly $Y_2 = 23.1$

$$\Delta Y = 23.1 - 57.0 = 33.9$$

It can be seen that Y_1 and Y_2 above are shown in the previous equation as $(Y^2 - M)$.

Example 2.19   An invar reference tape was compared with standard on the flat at the National Physical Laboratory at 20° C and 100 N and found to be 30.0073 m in length.

The first bay of a colliery triangulation base line was measured in catenary using the reference tape and then with the invar field tape at a temperature of 15° C and with a tension of 100 N. The means of these measurements were 29.8189 m and 29.8332 m respectively. The second bay of the base line was measured in catenary using the field tape at 13° C and 100 N and the resulting mean measurement was 29.9349 m.

Given:

(a) the coefficient of expansion of invar = 5.4×10^{-7} per °C
(b) the mass of the tape/metre run   = 0.0123 kg m^{-1}

(c) the inclination of the second bay $= 3°15'00''$
(d) the mean height of the second bay 250 m AOD
Assuming the radius of the earth to be 6367.3 km, calculate the horizontal length of the second bay reduced to Ordnance Datum.

To find the 'true' length of the test line,

$L_m = 29.8189$

$C_t = 29.8189 \times 5.4 \times 10^{-7} \times (15 - 20) = -0.0001$

$C_s = 0.0073 \, \text{m}/30 \, \text{m} \qquad\qquad = +0.0073$

$C_c = \dfrac{29.8189^3}{24} \left(\dfrac{0.0123 \times 9.81}{100} \right)^2 \qquad = -0.0016$

$$C_{\text{total}} = \quad 0.0056$$

therefore

$L_t = 29.8189 + 0.0056 = 29.8245$

The field tape applied under the same conditions when corrected for sag gives

$29.8332 - 0.0016 = 29.8316$

The difference represents the standardisation correction

$$= -0.0071$$

Tabulating,

	+	−
Standardisation		
$\quad C_s$		0.0071
Sag		
$\quad C_c =$ as before		0.0016
Temperature		
$\quad C_t = L\alpha(t_m - t_s)$		
$\qquad = 29.9349 \times 5.4 \times 10^{-7} \times (15 - 20)$		
$\qquad = -0.0001$		0.0001
Slope		
$\quad C_i = -l(1 - \cos\theta)$		
$\qquad = -29.9349\,(0.001\,61) = -0.0482$		0.0482
Sea level		
$\quad C_m = lh/R$		
$\qquad = 29.9349 \times 250/6367.3 \times 10^3 = 0.0012$		0.0012
	$C_{\text{total}} =$	0.0582

Horizontal length reduced to mean sea level

$= 29.9349 - 0.0582 = 29.8767 \, \text{m}$

Example 2.20   The details given below refer to the measurement of the first '30 m' bay of a base line. Determine the correct length of the bay reduced to mean sea level.

With the tape hanging in catenary at a tension of 100 N and at a mean temperature 14° C the recorded length was 29.9824 m. The difference in height between the ends was 0.46 m and the site was 490 m above msl.

The tape had previously been standardised in catenary at a tension of 75 N and at a temperature of 16° C and the distance between the zeros was 29.9970 m. $R = 6367.27$ km; mass of the tape $= 0.0193$ kg m^{-1}; sectional area of tape $= 3.61$ mm^2; $E = 2.2 \times 10^5$ N mm^{-2}; temperature coefficient of expansion of tape $= 12 \times 10^{-6}$ per °C.

Corrections:

	+	−
Standardisation		
Tape is 29.9970 at 75 N and 14° C		
$\quad C_s = (29.9970 - 30.000)$ per 30 m		
$\quad\quad = 0.003/30$ m per m		0.0030
Temperature		
$\quad C_t = L \propto (t_m - t_s)$		
$\quad\quad = 30 \times 12 \times 10^{-6}(14 - 16) =$		0.0007
Tension		
$\quad C_T = L(T_m - T_s)/AE$		
$\quad\quad = \dfrac{30(100 - 75)}{3.61 \times 2.2 \times 10^5}$	0.0009	
Slope		
$\quad C_i = -d^2/2L$		
$\quad\quad = -0.46^2/2 \times 30$		0.0035
Sag		
$\quad C_c =$ difference between the corrections for field and standard tensions		
$\quad\quad = \dfrac{-(mg)^2 L^3}{24}\left[\dfrac{1}{T_m^2} - \dfrac{1}{T_s^2}\right]$		
$\quad\quad = -\dfrac{(0.0193 \times 9.81)^2 30^3}{24}\left[\dfrac{1}{100^2} - \dfrac{1}{75^2}\right]$	0.0031	
Height		
$\quad C_h = hl/R$		
$\quad\quad = \dfrac{490 \times 30}{6\,367\,270}$		−0.0023
	+0.0040	−0.0095
	0.0040	
		−0.0055

Reduced length $= 29.9824 - 0.0055 = 29.9769$ m.

Example 2.21 A steel tape has the following specifications:
Mass 0.5 kg
Cross-sectional area 2 mm²
Young's modulus 20×10^{10} N m⁻²
Length at 20° C and 50 N is 30.005 m
Coefficient of linear expansion 11×10^{-6} per °C
It is to be used in catenary but in order to reduce the number of corrections to be applied to the measured lengths it is suggested that:
(a) the standard temperature be adjusted so that the actual length is equal to the nominal length of 30.000 m; and
(b) the tape be used at a tension such that the effects of sag and tension will be compensating.
Calculate the necessary temperature and tension. (NB: the acceptable tension will be in the region of 100 N.) (TP)

(a) By equation 2.11,

$$C_t = L_m \alpha (t_m - t_s)$$

therefore

$$\Delta T = C_t / L_m \alpha$$
$$= 0.005/30 \times 11 \times 10^{-6} = 15.2°$$

Therefore to contract the tape by 0.005 m the temperature would need to be reduced by 15.2° C, i.e. $20.0 - 15.2 = 4.8° C$ (new standard temperature).

(b) Correction for sag

$$C_s = \frac{(mg)^2 L_m^3}{24 T_m^2}$$

Correction for tension

$$C_T = \frac{L_m (T_m - T_s)}{AE}$$

In order that they equate,

$$C_s = C_T$$

$$\frac{(mg L_m)^2 L_m}{24 T_m^2} = \frac{L_m (T_m - T_s)}{AE}$$

therefore

$$\frac{AE(mg L_m)^2}{24} = T_m^3 - T_m^2 T_s$$

then

$$T_m^3 - 50 T_m^2 - \frac{(0.5 \times 9.81)^2 \times 2 \times 2 \times 10^5}{24} = 0$$

i.e.

$$T_m^3 - 50 T_m^2 - 400\,984 = 0$$

Solving this cubic equation by Newton's approximation

method,

$$f(T_m) = T_m^3 - 50T_m^2 - 400\,984$$
$$f'(T_m) = 3T_m^2 - 100T_m$$

Assuming $T_m \simeq 100\,\text{N}$

$$f(100) = 10^6 - 5 \times 10^5 - 400\,984 = 99\,016$$
$$f'(100) = 3 \times 10^4 - 10^4 = 20\,000$$

$$\delta T_m = \frac{f(100)}{f'(100)} = 4.95 \quad \text{therefore } c = -5$$

$$T_m' = 100 - 5 = 95$$
$$f(95) = 95^3 - 50 \times 95^2 - 400\,984 = \; 5141$$
$$f'(95) = 3 \times 95^2 - 100 \times 95 = 17\,575$$
$$\delta T_m = 5141/17\,575 = 0.29 \quad \text{therefore } c = -0.3$$
$$T_m'' = 95.0 - 0.3 = 94.7, \text{ say } 95\,\text{N}.$$

Example 2.22

(a) For the establishment of a precise datum via a vertical shaft, a '150 m' steel tape was used, being freely suspended with a plumbob of mass 20 kg attached. The tape was standardised on the flat as 149.9985 m under a tension of 100 N at a temperature of 20° C, has a mass of 2.5 kg and a cross-sectional area of 35 mm². From the following recorded data, calculate the level of the underground inset mark.

Surface bench mark 73.420 m AOD.
Staff reading at bench mark 1.326 m.
Staff reading at tape index mark 2.763 m.
Recorded length at the underground inset mark 99.934 m.
Recorded mean temperature in the shaft 4° C.
Coefficient of linear expansion 10.6×10^{-6} per ° C.
Young's modulus of elasticity $20 \times 10^{10}\,\text{N}\,\text{m}^{-2}$.

(b) What mass should be attached to the tape in order that the recorded length is affected only by the temperature change.

(RICS/M)

By equation 2.19,

$$s = \frac{gx}{AE}\left(M + \tfrac{1}{2}m(2l - x) - \frac{T_s}{g}\right)$$

$$= \frac{9.81 \times 99.934}{35 \times 2 \times 10^5}\left(20 + \frac{0.5 \times 2.5}{150}(300 - 99.934) - \frac{100}{9.81}\right)$$

$$= +0.0016\,\text{m}$$

Temperature correction

$$C_t = L_m\alpha(t_m - t_s)$$
$$= 99.934 \times 10.6 \times 10^{-6} \times (4 - 20) = -0.0169$$

Standardisation

$$C_s = 0.0015 \times 99.934/149.9985$$

$$= -0.0010\,\text{m}$$

Total correction $= -0.0163\,\text{m}$
Measured length $= 99.934\,\text{m}$
Adjusted length $= 99.918\,\text{m}$ -99.918

Levels at the surface

BS	FS	Fall	RL		
1.326			73.420		
	2.763	1.437	71.983		$+71.983$
1.326	Level at inset mark				-27.935 AOD
2.763					

-1.437

Alternative solution (1)
Tension affecting tape

Tension at $0\,\text{m}$ $= Mg + mgl$

Tension at $x\,\text{m}$ $= Mg + mg(l-x)$

Mean tension $= \frac{1}{2}[(Mg + mgl) + (Mg + mg(l-x))]$

$= Mg + \frac{1}{2}mg(2l - x)$

$= g[M + \frac{1}{2}m(2l - x)]$

$= 212.6\,\text{N}$

Correction for tension

$$C_T = L_m \frac{(T_m - T_s)}{AE}$$

$$= \frac{99.934\,(212.6 - 100)}{35 \times 2 \times 10^5}$$

$$= +0.0016 \text{ as before}$$

Temperature correction and standardisation as before.

Alternative solution (2)
At $0\,\text{N}$ tension

$$C_T = L_m\,(T_m - T_s)/AE = 150(0 - 100)/(35 \times 2 \times 10^5)$$

$$= -0.0021\,\text{m}$$

Standard length at $0\,\text{N}$ $= 149.9985 - 0.0021$

$= 149.9964$

Standardisation $= -0.0036\,\text{m}/150\,\text{m}$

Standardisation correction $= -0.0036 \times 100/150 = -0.0024$

Elongation $c = \dfrac{gx}{AE}[Mx + \frac{1}{2}m(2l - x)] = +0.0030$

Temperature correction as before	-0.0169
Total correction	-0.0163

The total corrections are -0.0163

Temperature correction	-0.0169
Difference	-0.0006

Therefore adjustment to tension correction $= -0.0006$
Therefore by equation 2.19,

$$\Delta s = \frac{gx}{AE}\left[\Delta M + \tfrac{1}{2}m(2l-x) - \frac{T_s}{g}\right]$$

$$\Delta M = \frac{AE\,\Delta s}{gx} - \tfrac{1}{2}m(2l-x) + \frac{T_s}{g}$$

$$= \frac{35 \times 2 \times 10^5 \times (-0.0006)}{9.81 \times 99.934} - \frac{2.5}{300}(300 - 99.934) + \frac{100}{9.81}$$

$$= 4.2\,\text{kg}$$

New $M = \underline{15.8\,\text{kg}}$

$$Check \quad s = \frac{9.81 \times 99.934}{35 \times 2 \times 10^5}\left(15.8 + \frac{2.5}{300}(300-99.934) - \frac{100}{9.81}\right)$$

$$= +0.0010$$

Standardisation $= \underline{-0.0010}$

Exercises 2.1

1 Calculate the horizontal length of the line measured as follows:

Bay	Inclined length (m)	Difference in level between ends of bay (m)
1	29.634	1.78
2	29.443	0.98
3	29.556	0.43
4	18.612	0.54

(Ans.   107.164 m)

2 Calculate the horizontal length of the line measured as follows:

Bay	Inclined length (m)	Angle of inclination
1	11.588	4° 00′ 00″
2	27.455	6° 10′ 00″
3	28.275	5° 40′ 00″
4	24.295	3° 12′ 00″

(Ans.   91.250 m)

3 Calculate the horizontal length of the line measured as follows:

Bay	Inclined length (m)	Vertical angle	Target height	Instrument height
1	27.632	4° 36′ 10″	0.28	1.21
2	58.422	4° 36′ 10″	1.48	
3	17.198	4° 36′ 10″	1.48	

(Ans.   102.913 m)

4 A tape is found to have a length of 29.997 m when standardised at a temperature of 15° C. Find the temperature at which it will be its nominal length.
(Assume $\alpha = 11 \times 10^{-6}$ per ° C.)   (Ans.   24.1° C)

5 A '30 m' tape has been standardised at 27° C and its true length at this temperature is 30.0023 m. A line is measured at 24° C and recorded as 348.694 m. Find its true length assuming the coefficient of linear expansion is 11.2×10^{-6} per ° C.
(Ans.   348.708 m)

6 A '30 m' tape has a total mass of 0.450 kg and has a density of $8890 \, \text{kg m}^{-3}$. If it is of nominal length at a tension of 50 N, calculate the correction to be applied if it is used at a tension of 110 N to measure a line of 487.107 m.
(Assume $E = 2 \times 10^5 \, \text{N mm}^{-2}$.)   (Ans.   +0.087 m)

7 Calculate the elongation of a '100 m' tape suspended under its own weight at points on the tape at (a) 30.480 m and (b) 91.440 m, given that $E = 20.7 \times 10^{10} \, \text{N m}^{-2}$, the mass of the tape is $0.0744 \, \text{kg m}^{-1}$ and the cross-sectional area $9.6 \times 10^{-6} \, \text{m}^2$.   (Ans.   (a) 0.0009 m; (b) 0.0018 m)

8 Calculate the horizontal length between supports approximately level if the recorded length is 30.5522 m, the tape has a mass of 0.425 kg for 30 m, and the applied tension in the form of a suspended mass is 9.072 kg.   (Ans.   30.5494 m)

9 It is necessary to find the mass of the tape by measuring its sag when suspended in catenary with both ends level. If this tape is 30.000 m long and the amount of sag is recorded as 238 mm at its midpoint under a tension of 90 N, what is the mass of the tape per metre?   (Ans. $0.0194 \, \text{kg m}^{-1}$)

10 A tape standardised as 29.995 m in catenary at 110 N is used in the field with a tension of 90 N. Calculate the horizontal length if the recorded length is 29.984 m and the mass of the tape is $0.0312 \, \text{kg m}^{-1}$.   (Ans.   29.975 m)

11 Calculate the grid length of a line measured as 246.384 m at a position with approximate grid coordinates 325 210 m E, 414 640 m N.   (Ans.   246.303 m)

Exercises 2.2 (General)

1 A nominal distance of '30 m' was set out with a 30 m tape from a mark on the top of one peg to a mark on the top of another, the tape being in catenary under a pull of 90 N and at a mean temperature of 21.1° C. The top of one peg was 0.370 m below the top of the other. The tape had been standardised in catenary under a pull of 110 N and at a temperature of 16.7° C.

Calculate the horizontal distance between the marks on the two pegs and reduce it to mean sea level. The top of the higher peg was 250 m above mean sea level. (Assume radius of the earth 6370 km, density of steel 7.75×10^3 kg m^{-3}, section of tape 3.13 mm × 1.20 mm, Young's modulus 2×10^5 N mm^{-2}, coefficient of linear expansion 11.2×10^{-6} per °C.

(Ans.   29.995 m)

2 A steel tape is found to be 29.9956 m long at 14° C under a tension of 50 N. The tape has the following specifications:

> width   10 mm
> thickness   0.45 mm
> Young's modulus of elasticity 2×10^5 N mm^{-2}
> coefficient of thermal expansion 11.2×10^{-6} per °C

Determine the tension to be applied to the tape to give a length of precisely 30 m at a temperature of 20° C.   (Ans.   122 N)

3 (a)   Calculate the sag correction for a '100 m' tape used in catenary in three equal spans if the tape has a mass of 1.36 kg/100 m and it is used under a tension of 90 N.
(b)   It is desired to find the mass of the tape by measuring its sag when suspended in catenary with both ends level. If the distance between the supports is 30.480 m and the sag amounts to 238 mm at its midspan under a tension of 90 N, what is its mass per unit length?   (Ans.   (a) 0.010 m; (b) 0.0188 kg m^{-1})

4 In the preliminary standardisation tests on a black carbon-steel tape, the general specifications were recorded as:
> nominal length '30 m'
> cross-sectional area 6.28 mm × 0.31 mm

(a)   Compute Young's modulus of elasticity of the tape from the given data.

The tape was freely suspended, variable loads were applied whilst the variable lengths were observed using a level fitted with a parallel-plate micrometer.

Tape reading at the fixed end	0.3500 m	
Tape reading at the free end	29.6680 m	applied load 10 kg
	29.6612 m	20 kg
	29.6536 m	30 kg
	29.6458 m	40 kg
	29.6535 m	30 kg
	29.6615 m	20 kg
	29.6682 m	10 kg

(b)   The same tape was then suspended in catenary with its maximum length between supports. The amount of sag at its central point was measured as 0.168 m whilst the applied tension was 100 N. Compute the mass of the tape per unit length.

(RICS   Ans.   (a) $2 \times 10^5 \, \text{N mm}^{-2}$; (b) $0.0149 \, \text{kg m}^{-1}$)

5   A base line was measured with a '30 m' invar tape which had been standardised on the flat under a tension of 67 N and at a temperature of 15.6° C and shown to be 30.0046 m. From the data given below, determine the horizontal grid length of the line.

Bay	Mean recorded length (m)	Difference in level between supports (m)	Air temperature (° C)
1	29.8682	0.65	10.1
2	29.8530	0.49	10.1
3	29.8742	1.19	10.2
4	29.8540	1.30	10.3
5	15.4670	0.27	10.3

Modulus of elasticity for invar $15 \times 10^{10} \, \text{N m}^{-2}$
Coefficient of linear expansion 6×10^{-7} per ° C
Applied tension 110 N
Mass per metre run of tape $0.015 \, \text{kg m}^{-1}$
Cross-sectional area of tape $2.58 \, \text{mm}^2$
Average reduced level of site 230 m AOD
Approximate coordinates of site 314 000 m E, 597 000 m N
Radius of the earth 6350 km

(RICS   Ans.   134.8316 m)

Bibliography

BANNISTER, A., and RAYMOND, S., *Surveying*, 4th edition. Pitman (1977)
CLARK, D., *Plane and Geodetic Surveying for Engineers*, 6th edition, **1** (1972), **2** (1973). Constable
COOPER, M. A. R., *Fundamentals of Survey Measurement and Analysis*. Crosby Lockwood Staples (1974)
HOLLAND, J. L., WARDELL, K., and WEBSTER, A. G., *Surveying*, **1**, **2**. *Coal Mining Series*. Virtue
IRVINE, W., *Surveying for Construction*, 2nd edition. McGraw-Hill (1980)
ORDNANCE SURVEY, *Constants, Formulae and Methods used in Transverse Mercator*. HMSO
SCHOFIELD, W., *Engineering Surveying*, 3rd edition. Butterworth (1978)

3

Surveying trigonometry

'Who conquers the triangle half conquers his subject'
M H Haddock

Of all the branches of mathematics, trigonometry is the most important to the surveyor, forming the essential basis of all calculations and computation processes. It is therefore essential that a thorough working knowledge is acquired and this chapter summarises the basic requirements.

3.1 Angular measurement

There are two ways of dividing the circle:

(a) the degree system
(b) the continental 'grade' system.

The latter divides the circle into 4 quadrants of 100 grades each and thereafter subdivides on a decimal system. It has little to commend it apart from its decimalisation which could be applied equally to the degree system. It has found little favour and will not be considered here.

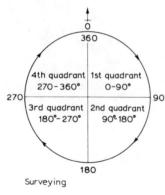

Surveying

3.1.1 The degree system

The circle is divided into 360 equal parts or degrees, each degree into 60 minutes, and each minute into 60 seconds. The following symbols are used:

degrees($°$) minutes($'$) seconds($''$)

so that 47 degrees 26 minutes 6 seconds is written as

47° 26' 06''

NB: The use of 06'' is preferred in surveying to 6'' so as to remove any doubts in recorded or computed values.

In *mathematics* the angle is assumed to rotate *anticlockwise* whilst in *surveying* the direction of rotation is assumed *clockwise*.

This variance in no way alters the subsequent calculations but is merely a different notation.

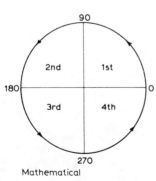

Mathematical

Fig. 3.1 Comparison of notations

3.1.2 Measurement of angles

The theodolite is the surveyor's most important instrument and is used for measuring horizontal and vertical angles.

Angles measured in the horizontal plane (Fig. 3.2(a)) are derived from two pointings of the telescope. Thus

angle θ = reading B − reading A

$$= \phi_B - \phi_A$$

ϕ_A and ϕ_B may be considered as arbitrary bearings (Fig. 3.2(b)).

A bearing is a clockwise angle measured from an origin known as a meridian (see p. 86).

Angles measured in the vertical plane are derived from single pointings, the angle being measured relative to the horizon known as a vertical angle (α) which may be + (an elevation; Fig. 3.2(c)) or − (a depression; Fig. 3.2(d)). In modern instruments the origin is the zenith and the circle is then graduated clockwise. The latter is less ambiguous.

$$Z = 90 - \alpha$$

where α may be + or − .

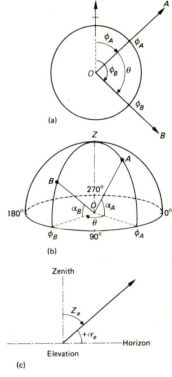

(a)

(b)

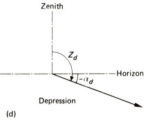

Zenith

(c) Elevation

Zenith

(d) Depression

Fig. 3.2

3.2 Trigonometrical ratios

Consider Fig. 3.3. Assume radius = 1.

sine (abbreviated sin) angle $\theta \quad = \dfrac{AB}{OA} = \dfrac{AB}{1} = AB = GO$

cosine (abbreviated cos) angle $\theta \quad = \dfrac{OB}{OA} = \dfrac{OB}{1} = OB = GA$

tangent (abbreviated tan) angle $\theta = \dfrac{AB}{OB} = \dfrac{\sin \theta}{\cos \theta} = \dfrac{DC}{OC} = \dfrac{DC}{1}$

cotangent θ (cot θ) $\quad = \dfrac{1}{\tan \theta} = \dfrac{\cos \theta}{\sin \theta} = \dfrac{OB}{AB} = \dfrac{FE}{FO} = \dfrac{FE}{1}$

cosecant θ (cosec θ) $\quad = \dfrac{1}{\sin \theta} = \dfrac{OA}{AB} = \dfrac{OE}{OF} = \dfrac{OE}{1}$

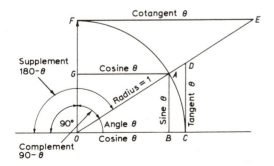

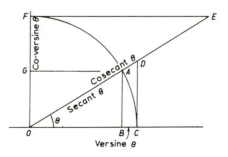

Fig. 3.3

$$\text{secant } \theta \text{ (sec } \theta) \qquad = \frac{1}{\cos \theta} = \frac{OA}{OB} = \frac{OD}{OC} = \frac{OD}{1}$$

$$\text{versine } \theta \text{ (vers } \theta) \qquad = 1 - \cos \theta = OC - OB$$

$$\text{coversine } \theta \text{ (covers } \theta) \quad = 1 - \sin \theta = OF - OG$$

NB: In mathematical shorthand $\sin^{-1} x$ means the angle x whose sine is . . . .

If $OA = $ radius $= 1$ then, by Pythagoras,

$$\sin^2 \theta + \cos^2 \theta = 1 \tag{3.1}$$

$$\sin^2 \theta = 1 - \cos^2 \theta \tag{3.2}$$

$$\cos^2 \theta = 1 - \sin^2 \theta \tag{3.3}$$

Dividing equation 3.1 by $\cos^2 \theta$,

$$\frac{\sin^2 \theta}{\cos^2 \theta} + \frac{\cos^2 \theta}{\cos^2 \theta} = \frac{1}{\cos^2 \theta}$$

i.e. $\qquad \tan^2 \theta + 1 = \sec^2 \theta$

$$\tan^2 \theta = \sec^2 \theta - 1 \tag{3.4}$$

Dividing equation 3.1 by $\sin^2 \theta$,

$$\frac{\sin^2 \theta}{\sin^2 \theta} + \frac{\cos^2 \theta}{\sin^2 \theta} = \frac{1}{\sin^2 \theta}$$

i.e. $\qquad 1 + \cot^2 \theta = \operatorname{cosec}^2 \theta \tag{3.5}$

$$\sin \theta = \sqrt{(1 - \cos^2 \theta)} \tag{3.6}$$

$$\cos \theta = \sqrt{(1 - \sin^2 \theta)} \tag{3.7}$$

$$\tan \theta = \frac{\sin \theta}{\cos \theta} = \frac{\sin \theta}{\sqrt{(1 - \sin^2 \theta)}} \tag{3.8}$$

or

$$= \frac{\sqrt{(1 - \cos^2 \theta)}}{\cos \theta} \tag{3.9}$$

$$\cos \theta = \frac{1}{\sec \theta} = \frac{1}{\sqrt{(1 + \tan^2 \theta)}} \tag{3.10}$$

$$\sin \theta = \frac{1}{\operatorname{cosec} \theta} = \frac{1}{\sqrt{(1 + \cot^2 \theta)}} \tag{3.11}$$

The above shows that, by manipulating the equations, any function can be expressed in terms of any other function.

3.2.1   Complementary angles

The complement of an acute angle is the difference between the angle and $90°$, i.e.

$$\text{if angle } A = 30°$$

$$\text{its complement} = 90° - 30° = 60°$$

the sine of an angle   = cosine of its complement
the cosine of an angle   = sine of its complement
the tangent of an angle   = cotangent of its complement
the secant of an angle   = cosecant of its complement
the cosecant of an angle   = secant of its complement

3.2.2 Supplementary angles

The supplement of an angle is the difference between the angle
and $180°$, i.e.

$$\text{if angle } A = 30°$$

$$\text{its supplement} = 180° - 30° = 150°$$

the sine of an angle = the sine of its supplement
cosine of an angle = the cosine of its supplement (but a
                      negative value)
tangent of an angle = the tangent of its supplement (but a
                      negative value)

These relationships are best illustrated by graphs.

Sine graph (Fig. 3.4). Let the line OA of length 1 rotate
anticlockwise. Then the height above the horizontal axis
represents the sine of the angle of rotation.

At $90°$ it reaches a maximum = 1
At $180°$ it returns to the axis
At $270°$ it reaches a minimum $= -1$

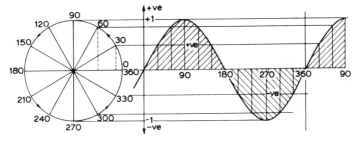

Fig. 3.4 The sine graph

It can be seen from the graph that

$$\sin 30° = \sin (180 - 30), \text{ i.e. } \sin 150°$$
$$= -\sin (180 + 30), \text{ i.e. } -\sin 210°$$
$$= -\sin (360 - 30), \text{ i.e. } -\sin 330°$$

Thus   the sine of all angles 0–$180°$ are $+ve$ (positive)
        the sine of all angles $180°$–$360°$ are $-ve$ (negative)

Cosine graph (Fig. 3.5). This is the same as the sine graph but
displaced by $90°$.

$$\cos 30° = -\cos (180 - 30) = -\cos 150°$$
$$= -\cos (180 + 30) = -\cos 210°$$
$$= +\cos (360 - 30) = +\cos 330°$$

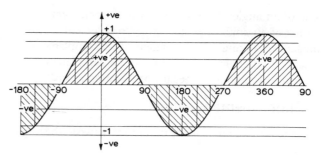

Fig. 3.5 The cosine graph

Thus   the cosine of all angles 0–90° and 270°–360° are $+ve$
the cosine of all angles 90°–270° are $-ve$

Tangent graph (Fig. 3.6). This is discontinuous as shown.

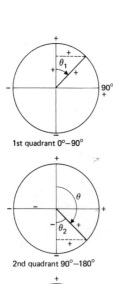

1st quadrant 0°–90°

2nd quadrant 90°–180°

3rd quadrant 180°–270°

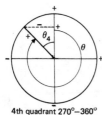

4th quadrant 270°–360°

Fig. 3.7

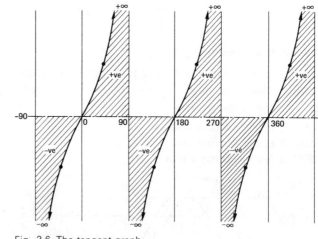

Fig. 3.6 The tangent graph

$$\tan 30° = \tan (180 + 30), \text{ i.e. } \tan 210°$$
$$= -\tan (180 - 30), \text{ i.e. } -\tan 150°$$
$$= -\tan (360 - 30), \text{ i.e. } -\tan 330°$$

Thus the tangents of all angles 0–90° and 180°–270° are $+ve$
the tangents of all angles 90°–180° and 270°–360° are $-ve$

Comparing these values based on the clockwise notation, the sign of the function can be seen from Fig. 3.7.

Let the rotating arm be $+ve$ $\theta°$

1st quadrant $(\theta_1 = \theta)$

$$\sin \theta_1 = \frac{+}{+}, \text{ i.e. } +$$

$$\cos \theta_1 = \frac{+}{+}, \text{ i.e. } +$$

$$\tan \theta_1 = \frac{+}{+}, \text{ i.e. } +$$

2nd quadrant $\theta_2 = (180 - \theta)$

$$\sin \theta_2 = \frac{+}{+}, \text{ i.e. } +$$

$$\cos \theta_2 = \frac{-}{+}, \text{ i.e. } -$$

$$\tan \theta_2 = \frac{-}{+}, \text{ i.e. } -$$

3rd quadrant $\theta_3 = (\theta - 180)$

$$\sin \theta_3 = \frac{-}{+}, \text{ i.e. } -$$

$$\cos \theta_3 = \frac{-}{+}, \text{ i.e. } -$$

$$\tan \theta_3 = \frac{-}{-}, \text{ i.e. } +$$

4th quadrant $\theta_4 = (360 - \theta)$

$$\sin \theta_4 = \frac{-}{+}, \text{ i.e. } -$$

$$\cos \theta_4 = \frac{+}{+}, \text{ i.e. } +$$

$$\tan \theta_4 = \frac{-}{+}, \text{ i.e. } -$$

3.2.3 Basis of tables of trigonometrical functions

Trigonometrical tables may be prepared, based on the following series:

$$\sin \theta = \theta - \frac{\theta^3}{3!} + \frac{\theta^5}{5!} - \frac{\theta^7}{7!} + \ldots \qquad (3.12)$$

where θ is expressed as radians (see p. 61) and
     3! is factorial 3, i.e. $3 \times 2 \times 1$
     5! is factorial 5, i.e. $5 \times 4 \times 3 \times 2 \times 1$

$$\cos \theta = 1 - \frac{\theta^2}{2!} + \frac{\theta^4}{4!} - \frac{\theta^6}{6!} + \ldots \qquad (3.13)$$

This information is readily available in many varied forms and to the number of places of decimals required for the particular problem in hand.

The following number of places of decimals are recommended:

    for degrees only, 4 places of decimals;
    for degrees and minutes, 5 places of decimals;
    for degrees, minutes and seconds, 6 places of decimals.

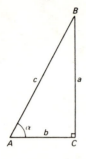

Fig. 3.8

3.3   Easy problems based on the solution of the right angled triangle

In any triangle there are six parts: three sides and three angles. The usual notation is that the side opposite angle A is side a, as shown in Fig. 3.8. The following facts are known about the given right angled triangle ABC.

$$\text{angle } C = 90°$$
$$\text{angle } A + \text{angle } B = 90°$$
$$c^2 = a^2 + b^2 \quad \text{(by Pythagoras)}$$

For an acute angle α the trigonometrical ratios are defined as:

$$\sin \alpha \; = a/c \quad \cos \alpha = b/c \quad \tan \alpha = a/b$$
$$\operatorname{cosec} \alpha = c/a \quad \sec \alpha = c/b \quad \cot \alpha = b/a$$

If $BAC = \hat{A} = \alpha$ then $\hat{B} = 90 - \alpha$ and so

$$\sin \alpha = \cos(90 - \alpha) \quad \tan \alpha = \cot(90 - \alpha) \quad \sec \alpha = \operatorname{cosec}(90 - \alpha)$$
$$\cos \alpha = \sin(90 - \alpha) \quad \cot \alpha = \tan(90 - \alpha) \quad \operatorname{cosec} \alpha = \sec(90 - \alpha)$$

α	$\sin \alpha$	$\cos \alpha$	$\tan \alpha$
0	0	1	0
90	1	0	∞
180	0	-1	0
270	-1	0	$-\infty$

Note

$$\sin(-\alpha) = -\sin \alpha \qquad \sin(180 - \alpha) = \sin \alpha$$
$$\cos(-\alpha) = +\cos \alpha \qquad \cos(180 - \alpha) = -\cos \alpha$$
$$\tan(-\alpha) = -\tan \alpha \qquad \tan(180 - \alpha) = -\tan \alpha$$

To find the remaining parts of the triangle it is necessary to know 3 *parts* (in the case of the right angled triangle, one angle $= 90°$ and therefore only 2 other facts are required).

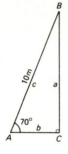

Fig. 3.9

Example 3.1   In a right angled triangle ABC, the hypotenuse AB is 10.000 metres long, whilst angle A is 70° 00′. Calculate the remaining parts of the triangle.

As the hypotenuse is AB (c) the right angle is at C (Fig. 3.9). Then

$$a/c = \sin 70° \, 00'$$

therefore

$$a = c \sin A$$
$$= 10.000 \sin 70° \, 00' = 9.3969 \text{ (say 9.397 m)}$$

therefore

$$b = c \cos A$$
$$= 10 \times 0.342\,02 \qquad = 3.4202 \text{ (say 3.420 m)}$$
$$\text{angle } B = 90° - 70° = 20°$$

Check

$$B = \tan^{-1} (b/a) = \tan^{-1} (3.4202/9.3969)$$
$$= 20°$$

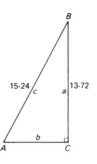

Fig. 3.10

Example 3.2   It is necessary to climb a vertical wall 13.72 m high with a ladder 15.24 m long; Fig. 3.10. Find:
(a)   how far from the foot of the wall the ladder must be placed;
(b)   the inclination of the ladder.

$$A = \sin^{-1} (13.72/15.24) = 64.193° = 64° \; 11' \; 30''$$
$$B = 90° - A \qquad\quad = 25.807° = 25° \; 48' \; 30''$$
$$b = c \cos A = 15.24 \cos 64.193 = 6.635, \text{ say } 6.64$$

Answer: (a)   6.64 m; (b)   64° 11′ 30″ from the horizontal.

Example 3.3   A ship sails 48.28 km on a bearing 030° 00′. It then changes course and sails a further 80.47 km 315° 00′. Find:
(a)   the bearing back to its starting point;
(b)   the distance back to its starting point.
NB: See Chapter 4 on bearings.

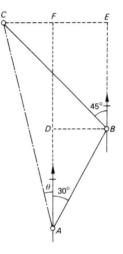

To solve this problem two triangles, *ADB* and *BCE*, are joined to form a resultant third *ACF* (Fig. 3.11).

$$AD = AB \cos 30° \; 00' = 48.28 \cos 30.000° = 41.812 \, \text{km}$$
$$DB = AB \sin 30° \; 00' = 48.28 \sin 30.000° = 24.140 \, \text{km}$$
$$BE = BC \; \cos 45° \; 00' = 80.47 \cos 45.000° = 56.901 \, \text{km} \; (= CE)$$

Therefore

Fig. 3.11

$$CF = CE - DB = 56.901 - 24.140 \qquad = 32.761 \, \text{km}$$
$$AF = AD + BE = 41.812 + 56.901 \qquad = 98.713 \, \text{km}$$
$$\theta = \tan^{-1} (CF/AF) = \tan^{-1} (32.761/98.713) = 18.360°$$
$$AC = AF/\cos \theta = 98.713/\cos 18.360 \qquad = 104.007 \, \text{km}$$

Answer: (a)   Bearing $AC = (360 - \theta) = 361.640° = 341° \; 38' \; 30''$;
(b) length $AC = 104.01$ km.
NB: This problem can be solved more satisfactorily by using coordinates; see Chapter 4.

Example 3.4   An angle of elevation of 45° 00′ was observed to the top of a tower. 42.00 metres nearer to the tower a further angle of elevation of 60° 00′ was observed. Find
(a)   the height of the tower,
(b)   the distance the observer is from the foot of the tower.

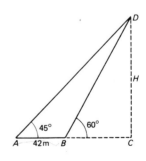

In Fig. 3.12,

Fig. 3.12

$$AC = H \cot A$$
$$BC = H \cot B$$
$$AB = AC - BC = H \; (\cot A - \cot B)$$

Therefore

$$H = \frac{AB}{\cot A - \cot B} = \frac{42.00}{\cot 45 - \cot 60} = 99.373$$

$$BC = H \cot B \quad = 99.373 \cot 60 \qquad = 57.373$$

Check $AC = DC \qquad = 99.373$

$$BC = AC - AB = 99.373 - 42.000 \qquad = 57.373$$

Answer: (a)   Height of tower = 99.37 m;
(b)   Distance of observer from foot of tower = 57.37 m.

Exercises 3.1

1   A flagstaff 27.43 m high is held up by ropes, each being attached to the top of the flagstaff and to a peg in the ground and inclined at 30° 00′ to the vertical; find the lengths of the ropes and the distances of the pegs from the foot of the flagstaff.
(Ans.   31.67 m; 15.84 m)

2   From the top of a mast of a ship 22.86 m high the angle of depression of an object is 20° 00′. Find the distance of the object from the ship.   (Ans.   62.81 m)

3   A tower has an elevation 60° 00′ from a point due north of it and 45° 00′ from a point due south. If the two points are 200.0 metres apart, find the height of the tower and its distance from each point of observation.   (Ans.   126.8 m; 73.2 m; 126.8 m)

4   A boat is 457 m from the foot of a vertical cliff. To the top of the cliff and the top of a building standing on the edge of the cliff, angles of elevation were observed as 30° and 33° respectively. Find the height of the building to the nearest metre.
(Ans.   33 m)

5   A vertical stick 3.00 m long casts a shadow from the sun of 1.75 m. What is the elevation of the sun?   (Ans.   59° 45′)

6   *X* and *Y* start walking in directions N 17° 00′ W and N 73° 00′ E; find their distance apart after three hours and the direction of the line joining them. *X* walks at 3 km an hour and *Y* at 4 km an hour.   (Ans.   15 km S 70° 08′ E)

7   *A*, *B* and *C* are three places. *B* is 30 km N 67½° E of *A*, and *C* is 40 km S 22½° E of *B*. Find the distance and bearing of *C* from *A*.   (Ans.   50 km; S 59° 22′ E)

3.4   Circular measure

The circumference of a circle = $2\pi r$ where $\pi = 3.1416$ approx.

3.4.1 The radian

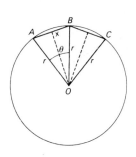

Consider an arc AB of a circle, centre O, radius r, which subtends an angle θ at O (Fig. 3.13). The length of the arc AB is proportional to the angle θ. As the circumference $= 2\pi r$ for an angle 360°,

$$AB = 2\pi r \times \theta/360 = (\pi/180)r\theta$$

This factor $(\pi/180)$ is the length of an arc for 1° and thus results the circular unit of measurement known as the *radian*, i.e.

$$\theta \text{ rad} = (\pi/180)\theta°$$

Fig. 3.13

Then $2\pi \text{ rad} = 360°$

$$1 \text{ rad} = 360/2\pi = 57.295\,779\ldots° \simeq 206\,264.8''$$

$$1° = 0.017\,4532\ldots \text{ rad}$$

Example 3.5

$$64° 11' 33'' = 64.1925° = 64.1925/57.2958$$
$$= 1.120\,37 \text{ rad}$$

or $64° 11' 33'' = 231\,093'' = 231\,093/206\,265$
$$= 1.120\,37 \text{ rad}$$

It now follows that the length of an arc of a circle of radius r and subtending θ radians at the centre of the circle can be written as

$$\text{arc} = r \cdot \theta \text{ rad} \qquad (3.14)$$

This is generally superior to the use of the formula

$$\text{arc} = 2\pi r \times \frac{\theta°}{360} \qquad (3.15)$$

NB: When θ is written it implies θ radians.

To find the area of a circle. A regular polygon $ABC \ldots A$ is drawn inside a circle; see Fig. 3.13.

Draw OX perpendicular to AB.

Then area of polygon

$$= \tfrac{1}{2}OX(AB + BC + \ldots) = \tfrac{1}{2}OX(\text{perimeter of polygon})$$

When the number of sides of the polygon is increased to infinity (∞), OX becomes the radius, the perimeter becomes the circumference, and the polygon becomes the circle; therefore

$$\text{area of circle} = \tfrac{1}{2} \cdot r \cdot 2\pi r = \pi r^2$$

The area of the sector OAB.

$$\frac{\text{area of sector}}{\text{area of circle}} = \frac{\theta}{2\pi}$$

therefore

$$\text{area of sector} = \frac{\pi r^2 \theta}{2\pi} = \tfrac{1}{2}r^2\theta \qquad (3.16)$$

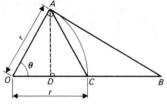

Fig. 3.14

3.4.2 Small angles and approximations

For any angle $\theta < 90°$ (i.e. $< \pi/2$ radians) $\tan \theta > \theta > \sin \theta$.

Let angle $AOC = \theta$, $OA = OC = r$, and let AB be a tangent to the arc AC at A to cut OC produced at B. Draw AD perpendicular ($\perp$) to OB (Fig. 3.14). Then

$$\text{area of triangle } OAB = \tfrac{1}{2}OA \cdot AB$$
$$= \tfrac{1}{2}r \cdot r \tan \theta = \tfrac{1}{2}r^2 \tan \theta$$
$$\text{area of sector } OAC \qquad = \tfrac{1}{2}r^2 \theta$$
$$\text{area of triangle } OAC = \tfrac{1}{2}OC \cdot AD$$
$$= \tfrac{1}{2}r \cdot r \sin \theta = \tfrac{1}{2}r^2 \sin \theta$$

Now triangle $OAB >$ sector $OAC >$ triangle OAC. Therefore

$$\tfrac{1}{2}r^2 \tan \theta > \tfrac{1}{2}r^2 \theta > \tfrac{1}{2}r^2 \sin \theta$$

therefore

$$\tan \theta > \theta > \sin \theta$$

This is obviously true for all values of $\theta < \pi/2$.

Take θ to be very small. Divide each term by $\sin \theta$; then

$$\frac{1}{\cos \theta} > \frac{\theta}{\sin \theta} > 1$$

It is known that as $\theta \to 0$ then $\cos \theta \to 1$. Thus

$$\cos \theta \simeq 1 \text{ when } \theta \text{ is small}$$

therefore

$$\frac{\theta}{\sin \theta} \text{ must also be nearly 1.}$$

The results shows that $\sin \theta$ may be replaced by θ.

Similarly, dividing each term by $\tan \theta$,

$$1 > \frac{\theta}{\tan \theta} > \frac{\sin \theta}{\tan \theta}$$

i.e. $$1 > \frac{\theta}{\tan \theta} > \cos \theta$$

Tan θ may also be replaced by θ.

It can thus be shown that for very small angles

$$\tan \theta \simeq \theta \simeq \sin \theta$$

Angle	Tangent	Radian	Sine
1° 00′ 00″	0.017 46	0.017 45	0.017 45
1° 30′ 00″	0.026 19	0.026 18	0.026 18
2° 00′ 00″	0.034 92	0.034 91	0.034 90
2° 30′ 00″	0.043 66	0.043 63	0.043 62
3° 00′ 00″	0.052 41	0.052 36	0.052 34
3° 30′ 00″	0.061 16	0.061 09	0.061 05
4° 00′ 00″	0.069 93	0.069 81	0.069 76
4° 30′ 00″	0.078 70	0.078 54	0.078 46
5° 00′ 00″	0.087 49	0.087 27	0.087 16

From these it can be seen that θ may be substituted for $\sin \theta$ or $\tan \theta$ to 5 figures up to $2°$, whilst θ may be substituted for $\sin \theta$ up to $5°$ and for $\tan \theta$ up to $4°$ to 4 figures, thus allowing approximations to be made when angles are less than $4°$.

3.5 Trigonometrical ratios of compound angles

These may be tabulated as follows:

$$\sin (A \pm B) = \sin A \cos B \pm \cos A \sin B \qquad (3.17)$$

$$\cos (A \pm B) = \cos A \cos B \mp \sin A \sin B \qquad (3.18)$$

$$\tan (A \pm B) = (\tan A \pm \tan B)/(1 \mp \tan A \tan B) \quad (3.19)$$

If $A = B$:

$$\sin 2A = 2 \sin A \cos A \qquad (3.20)$$

$$\cos 2A = \cos^2 A - \sin^2 A = 1 - 2 \sin^2 A = 2 \cos^2 A - 1 \qquad (3.21)$$

$$\tan 2A = (2 \tan A)/(1 - \tan^2 A) \qquad (3.22)$$

3.6 Transformation of products and sums

These may be tabulated as follows:

$$\sin (A + B) + \sin (A - B) = 2 \sin A \cos B \qquad (3.23)$$

$$\sin (A + B) - \sin (A - B) = 2 \cos A \sin B \qquad (3.24)$$

$$\cos (A + B) + \cos (A - B) = 2 \cos A \cos B \qquad (3.25)$$

$$\cos (A + B) - \cos (A - B) = -2 \sin A \sin B \qquad (3.26)$$

$$\sin A + \sin B = 2 \sin \frac{(A + B)}{2} \cos \frac{(A - B)}{2} \qquad (3.27)$$

$$\sin A - \sin B = 2 \sin \frac{(A - B)}{2} \cos \frac{(A + B)}{2} \qquad (3.28)$$

$$\cos A + \cos B = 2 \cos \frac{(A + B)}{2} \cos \frac{(A - B)}{2} \qquad (3.29)$$

$$\cos A - \cos B = -2 \sin \frac{(A + B)}{2} \sin \frac{(A - B)}{2} \qquad (3.30)$$

3.7 The solution of triangles

The following important formulae are now proved:

Sine rule $\quad \dfrac{a}{\sin A} = \dfrac{b}{\sin B} = \dfrac{c}{\sin C} = 2R \qquad (3.31)$

Cosine rule $\quad c^2 = a^2 + b^2 - 2ab \cos C \qquad (3.32)$

$$\sin C = \frac{2}{ab} \sqrt{s(s - a)(s - b)(s - c)} \qquad (3.33)$$

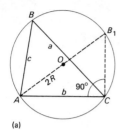

(a)

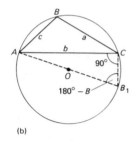

(b)

Fig. 3.15 The sine rule

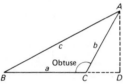

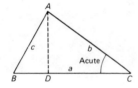

Fig. 3.16 The cosine rule

$$\text{Area of triangle} = \tfrac{1}{2}ab\sin C \tag{3.34}$$

$$= \sqrt{s(s-a)(s-b)(s-c)} \tag{3.35}$$

$$\text{Half-angle formulae} \quad \sin\frac{A}{2} = \sqrt{\frac{(s-b)(s-c)}{bc}} \tag{3.36}$$

$$\cos\frac{A}{2} = \sqrt{\frac{s(s-a)}{bc}} \tag{3.37}$$

$$\tan\frac{A}{2} = \sqrt{\frac{(s-b)(s-c)}{s(s-a)}} \tag{3.38}$$

$$\text{Napier's tangent rule} \quad \tan\frac{B-C}{2} = \frac{b-c}{b+c}\tan\frac{B+C}{2} \tag{3.39}$$

3.7.1   Sine rule

Let triangle ABC be drawn with circumscribing circle.
   Let AB_1 be a diameter through A (angle ABC = angle AB_1C).

In Fig. 3.15(a), $AC/AB_1 = \sin B$

In Fig. 3.15(b), $AC/AB_1 = \sin(180-B) = \sin B$

therefore $b/2R = \sin B$

therefore $b/\sin B = 2R$

Similarly

$$\frac{a}{\sin A} = \frac{b}{\sin B} = \frac{c}{\sin C} = 2R \tag{3.40}$$

3.7.2   Cosine rule

In Fig. 3.16,

$$\begin{aligned}
AB^2 &= AD^2 + BD^2 \text{ (Pythagoras)} \\
&= AD^2 + (BC-CD)^2 \\
&= b^2\sin^2 C + (BC - b\cos C)^2
\end{aligned} \Bigg\} \text{ with } C \text{ acute}$$

$$\begin{aligned}
&= AD^2 + (BC+CD)^2 \\
&= b^2\sin^2(180-C) + \{BC + b\cos(180-C)\}^2 \\
&= b^2\sin^2 C + (BC - b\cos C)^2
\end{aligned} \Bigg\} \text{ with } C \text{ obtuse}$$

therefore

$$AB^2 = b^2\sin^2 C + (BC - b\cos C)^2 \quad \text{in either case}$$

therefore

$$\begin{aligned}
c^2 &= b^2\sin^2 C + a^2 - 2ab\cos C + b^2\cos^2 C \\
&= a^2 + b^2(\sin^2 C + \cos^2 C) - 2ab\cos C \\
c^2 &= a^2 + b^2 - 2ab\cos C
\end{aligned} \tag{3.41}$$

3.7.3   Area of a triangle

From $c^2 = a^2 + b^2 - 2ab \cos C$

$$\cos C = \frac{a^2 + b^2 - c^2}{2ab}$$

therefore

$$\sin^2 C = 1 - \cos^2 C = 1 - \left(\frac{a^2 + b^2 - c^2}{2ab}\right)^2$$

$$= \left(1 + \frac{a^2 + b^2 - c^2}{2ab}\right)\left(1 - \frac{a^2 + b^2 - c^2}{2ab}\right)$$

$$= \frac{(a+b)^2 - c^2}{2ab} \times \frac{c^2 - (a-b)^2}{2ab}$$

$$= \frac{(a+b+c)(-c+a+b)(c-a+b)(c+a-b)}{(2ab)^2}$$

$$= \frac{4s(s-a)(s-b)(s-c)}{a^2 b^2}$$

where $2s = a + b + c$. Thus

$$\sin C = \frac{2}{ab}\sqrt{s(s-a)(s-b)(s-c)} \qquad (3.42)$$

In Fig. 3.16,

$$\text{area of triangle} = \tfrac{1}{2} AD \cdot BC$$

$$= \tfrac{1}{2} ab \sin C \qquad (3.43)$$

$$= \tfrac{1}{2} ab \frac{2}{ab}\sqrt{s(s-a)(s-b)(s-c)}$$

$$= \sqrt{s(s-a)(s-b)(s-c)} \qquad (3.44)$$

3.7.4   Half-angle formulae

From equation 3.21,

$$\sin^2 \tfrac{1}{2} A = \tfrac{1}{2}(1 - \cos A) = \tfrac{1}{2}\left(1 - \frac{b^2 + c^2 - a^2}{2bc}\right)$$

$$= \frac{a^2 - (b-c)^2}{4bc} = \frac{(a-b+c)(a+b-c)}{4bc}$$

$$= \frac{(s-b)(s-c)}{bc}$$

therefore

$$\sin \frac{A}{2} = \sqrt{\frac{(s-b)(s-c)}{bc}} \qquad (3.45)$$

Similarly,

$$\cos^2 \tfrac{1}{2} A = \tfrac{1}{2}(1 + \cos A) = \tfrac{1}{2}\left(1 + \frac{b^2 + c^2 - a^2}{2bc}\right)$$

$$= \frac{(b+c)^2 - a^2}{4bc}$$

$$= \frac{(b+c+a)(b+c-a)}{4bc}$$

$$= \frac{s(s-a)}{bc}$$

$$\cos \frac{A}{2} = \sqrt{\frac{s(s-a)}{bc}} \tag{3.46}$$

Finally,

$$\tan \frac{A}{2} = \frac{\sin \dfrac{A}{2}}{\cos \dfrac{A}{2}} = \sqrt{\frac{(s-b)(s-c)}{s(s-a)}} \tag{3.47}$$

The last formula is preferred as $(s-b)+(s-c)+(s-a) = s$, which provides an arithmetical check.

3.7.5   Napier's tangent rule

From the sine rule,

$$\frac{b}{c} = \frac{\sin B}{\sin C}$$

then

$$\frac{b-c}{b+c} = \frac{\sin B - \sin C}{\sin B + \sin C}$$

$$= \frac{2\cos \dfrac{B+C}{2} \sin \dfrac{B-C}{2}}{2\sin \dfrac{B+C}{2} \cos \dfrac{B-C}{2}}$$

(by equations 3.27 and 3.28)

$$= \frac{\tan \dfrac{B-C}{2}}{\tan \dfrac{B+C}{2}}$$

therefore

$$\tan \frac{B-C}{2} = \frac{b-c}{b+c} \tan \frac{B+C}{2} \tag{3.48}$$

3.7.6 Problems involving the solution of triangles

All problems come within the following four cases:

(1)   *Given two sides and one angle (not included)* to find the other angles.
Solution: sine rule solution ambiguous as illustrated in Fig. 3.17. Given AB and AC with angle B, AC may cut line BC at C_1 or C_2.

(2)   *Given all the angles and one side* to find all the other sides.
Solution: sine rule.

(3)   *Given two sides and the included angle.*
Solution: either cos rule to find remaining side, or Napier's tangent rule.

(4)   *Given the three sides.*
Solution: either cos rule or half-angle formula.

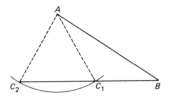

Fig. 3.17 The ambiguous case of the sine rule

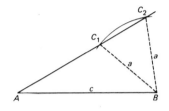

Fig. 3.18

Example 3.6   *Problem 1*
Let $c = 466.00 \text{ m}$
   $a = 190.50 \text{ m}$
   $\hat{A} = 22° \, 15' \, 00''$
Find angles C.
Using the sine rule,

$$C_2 = \sin^{-1} 466.00 \sin 22.2500/190.50 = 67.8573°$$

or $C_1 = \sin^{-1}(180 - C_2)$ $= 112.1427°$

Problem 2: to find side b,

$$b = a \sin B / \sin A$$

$$b_1 = 190.50 \sin (67.8573 + 22.2500)/\sin 22.2500$$

$$= 503.104$$

or $b_2 = 190.50 \sin (112.1427 + 22.2500)/\sin 22.2500$

$$= 359.500$$

therefore

   $C_1 = 112° \, 08' \, 30''$; $C_2 = 67° \, 51' \, 30''$
   $b_1 = 503.10 \text{ m}$; $b_2 = 359.50 \text{ m}$
   $B_1 = 45° \, 36' \, 30''$; $B_2 = 89° \, 53' \, 30''$

Example 3.7   *Problem 3*
Let $a = 636.00 \text{ m}$
   $c = 818.00 \text{ m}$
   $B = 97° \, 30' \, 00''$
Find b, A and C.

By the cosine rule,

$$b^2 = a^2 + c^2 - 2ac \cos B$$

therefore
$$b = \sqrt{(636^2 + 818^2 - 2 \times 636 \times 818 \times \cos 97.500)}$$
$$= 1099.742 \quad (\text{say } 1099.74\,\text{m})$$

By the sine rule,
$$A = \sin^{-1}(a \sin B/b)$$
$$= \sin^{-1}(636.00 \sin 97.500/1099.742) = 34° 59' 10''$$
$$B = 180 - (97° 30' 00'' + 34° 59' 10'') \quad = 47° 30' 50''$$

Alternatively, using Napier's tangent rule,

$$\tfrac{1}{2}(C - A) = \tan^{-1}\left[\frac{818 - 636}{818 + 636} \tan\left(\frac{180 - 97.500}{2}\right)\right]$$
$$= 6.2643°$$
$$\tfrac{1}{2}(C + A) = 41.2500$$
$$C = 47.5143 \quad (\text{say } 47° 30' 50'')$$
$$A = 34.9857 \quad (\text{say } 34° 59' 10'')$$

then
$$b = a \sin B/\sin A$$
$$= 636.00 \sin 97.500/\sin 34.9857$$
$$= 1099.74$$

Example 3.8 *Problem 4*
Let $a = 381$
    $b = 719$
    $c = 932$
Find the angles.

From
$$a^2 = b^2 + c^2 - 2bc \cos A$$

$$A = \cos^{-1}\left(\frac{b^2 + c^2 - a^2}{2bc}\right)$$

$$= \cos^{-1}\left(\frac{719^2 + 932^2 - 381^2}{2 \times 719 \times 932}\right) = 22.2500 \quad \text{say } 22° 15' 00''$$

$$B = \cos^{-1}\left(\frac{a^2 + c^2 - b^2}{2ac}\right)$$

$$= \cos^{-1}\left(\frac{381^2 + 932^2 - 719^2}{2 \times 381 \times 932}\right) = 45.6074 \quad \text{say } 45° 36' 30''$$

$$C = \cos^{-1}\left(\frac{a^2 + b^2 - c^2}{2ab}\right)$$

$$= \cos^{-1}\left(\frac{381^2 + 719^2 - 932^2}{2 \times 381 \times 719}\right) = 112.1426 \quad \text{say } 112° 08' 30''$$

Check
$$\sum = 180.0000 \qquad 180° 00' 00''$$

Alternatively, by half-angle formula,

$$\tan\frac{A}{2} = \sqrt{\frac{(s-b)(s-c)}{s(s-a)}}$$

$$\begin{array}{llll}
a = & 381 & s-a = & 635 \\
b = & 719 & s-b = & 297 \\
c = & \underline{932} & s-c = & 84 \\
2s = & \underline{2032} & & \\
s = & 1016 & s = & \overline{1016}
\end{array}$$

$$A = 2\tan^{-1}\sqrt{[(297 \times 84)/(1016 \times 635)]} = 22.2500$$
$$B = 2\tan^{-1}\sqrt{[(635 \times 84)/(1016 \times 297)]} = 45.6074$$
$$C = 2\tan^{-1}\sqrt{[(635 \times 297)/(1016 \times 84)]} = 112.1425$$

Example 3.9   The sides of a triangle ABC measure as follows: $AB = 36.036$ m, $AC = 30.146$ m and $BC = 6.021$ m.
(a)   Calculate to the nearest 20 seconds, the angle BAC.
(b)   Assuming that the standard error in measuring any of the sides is ± 0.003 m, give an estimate of the standard error in the angle A.

(a)   By the cosine rule (Fig. 3.19),

$$A = \cos^{-1}(b^2 + c^2 - a^2)/2bc = 2.1716°$$
$$B = \cos^{-1}(a^2 + c^2 - b^2)/2ac = 10.9364°$$
$$C = \cos^{-1}(a^2 + b^2 - c^2)/2ab = \underline{166.8921°}$$

Check $\qquad\qquad\qquad\qquad \sum = 180.0001°$

Fig. 3.19

(b)   The standard error is 0.003 m.
   The effect on the angle A of varying the three sides is best calculated by varying each of the sides in turn whilst the remaining two sides are held constant. To carry out this process, the equation must be successively differentiated and a better equation for this purpose is the cosine rule. Thus

$$\cos A = (b^2 + c^2 - a^2)/2bc$$

Partially differentiating with respect to a,

$$-\sin A\,\delta A_a = -2a\delta a/2bc$$

therefore

$$\delta A_a = a\delta a/bc \sin A$$

Partially differentiating with respect to b,

$$-\sin A\,\delta A_b = \left(\frac{(2bc \times 2b) - (b^2 + c^2 - a^2)\,(2c)}{4b^2 c^2}\right)\delta b$$

$$= \left(\frac{a^2 + b^2 - c^2}{2b^2 c}\right)\delta b$$

but $\qquad \cos C = \dfrac{a^2 + b^2 - c^2}{2ab}$

therefore

$$\delta A_b = -\frac{a \cos C}{bc \sin A} \delta b = -\delta A_a \cos C \quad \text{(as } \delta a = \delta b\text{)}$$

Similarly, from the symmetry of the function:

$$\delta A_c = -\frac{a \cos B}{bc \sin A} \delta c = -\delta A_a \cos B \quad \text{(as } \delta a = \delta c\text{)}$$

Substituting values into the equations gives:

$$\delta A_a = \frac{6.021 \times \pm 0.003 \times 206\,265}{30.146 \times 36.036 \times \sin 2° \, 10' \, 20''} = \pm 90.49''$$

$$\delta A_b = \delta A_a \cos 166° \, 53' \, 30'' \qquad = \pm 88.13''$$
$$\delta A_c = \delta A_a \cos 10° \, 56' \, 10'' \qquad = \pm 88.85''$$

Therefore total standard error $= \sqrt{\delta A_a^2 + \delta A_b^2 + \delta A_c^2}$
$$= \sqrt{90.49^2 + 88.13^2 + 88.85^2}$$
$$= \pm 154''$$

The angle A is then $2° \, 10' \, 18'' \pm 02' \, 34''$ and thus the angle can only be quoted as $2° \, 10' \pm 02'$.

Exercises 3.2

1 Given $BC = 156.00$, $AC = 140.00$, angle $ACB = 34° \, 54'$, calculate the missing parts.

(Ans. $AB = 90.07$; $B = 62° \, 47'$; $A = 82° \, 19'$)

2 Given $c = 41.5$, $b = 52.4$, $a = 71.6$, calculate the largest angle:
(a) by the cosine rule and
(b) by the half-angle tangent rule.   (Ans. $A = 98° \, 42'$)

3 Given $a = 56.80$, $A = 48° \, 37'$, $B = 83° \, 15'$, calculate the missing parts.   (Ans. $b = 75.18$; $c = 56.38$; $C = 48° \, 08'$)

4 Fig. 3.20 represents a roof truss with dimensions as shown. Calculate the length AE. $BD = 27.00$ m.

(Ans. $AE = 19.41$ m)

Fig. 3.20

3.8 Heights and distances

3.8.1 To find the height of an object having a vertical face

In Fig. 3.21, given the height of the observer $= h_i$
the measured vertical angle $= \theta$
the slope of the ground $= \alpha$
the slope length $= D$
the horizontal length $= H$
the vertical difference $= V$

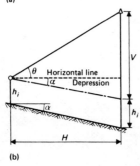

(a)

(b)

Fig. 3.21

In the general case:

$$V = H \tan \theta - H \tan \alpha = H (\tan \theta - \tan \alpha) \qquad (3.49)$$
$$= D \cos \alpha (\tan \theta - \tan \alpha) \qquad (3.50)$$
$$= D (\cos \alpha \tan \theta - \sin \alpha) \qquad (3.51)$$

When $\alpha = 0$ then $D = H$ (Fig. 3.22) and

$$V = H \tan \theta \qquad (3.52)$$

The height of the object above the ground $= V + h_i$.

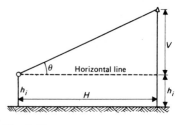

Fig. 3.22

3.8.2 To find the height of an object when the base is inaccessible

A base line AB is established and the vertical angles are measured from its extremities as θ_a and θ_b.

(a) *Base line AB in line with the object* (Fig. 3.23)

Angles measured at A: vertical angle to object θ_a
at B: vertical angle to object θ_b

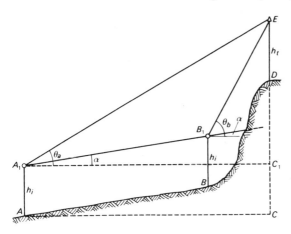

Fig. 3.23

Slope of the line $= \alpha$

In triangle $A_1 E B_1$,

$$A_1 = \theta_a - \alpha$$
$$B_1 = 180 - (\theta_b - \alpha)$$
$$E = \theta_b - \theta_a$$

then $A_1 E = A_1 B_1 \sin B_1 / \sin E$

$$= AB \sin (\theta_b - \alpha)/\sin (\theta_b - \theta_a) \qquad (3.53)$$

The height of the object at E above the ground at A is

$$EC = EC_1 + h_i$$
$$= A_1 E \sin \theta_a + h_i$$
$$= AB \sin (\theta_b - \alpha) \sin \theta_a \operatorname{cosec} (\theta_b - \theta_a) + h_i \qquad (3.54)$$

The height of the ground at D above the ground at A is

$$DC = EC_1 + h_i - h_t$$
$$= EC + h_t$$

If the ground of the base line is level, then $\alpha = 0$ (Fig. 3.24) and

$$A_1 E = AB \sin \theta_b \operatorname{cosec} (\theta_b - \theta_a)$$

$$EC = \frac{AB \sin \theta_b \sin \theta_a}{\sin \theta_b \cos \theta_a - \cos \theta_b \sin \theta_a} + h_i$$

$$= \frac{AB}{\cot \theta_a - \cot \theta_b} + h_i \qquad (3.55)$$

$$DC = EC - h_t$$

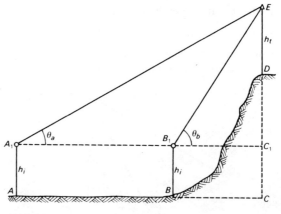

Fig. 3.24

(b)   *Base line AB is not in line with the object* (Fig. 3.25)

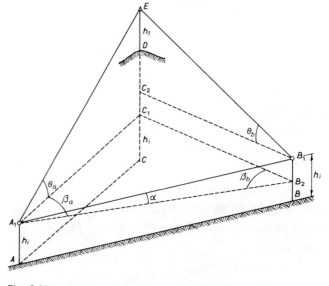

Fig. 3.25

Angles measured at A: horizontal β_a;   vertical θ_a
at B: horizontal β_b;   vertical θ_b

Slope of the ground $= \alpha$;   horizontal plane $A_1 B_2 C_1$

$$A_1 B_2 = AB \cos \alpha$$

$$A_1 C_1 = A_1 B_2 \sin \beta_b \, \text{cosec} \, (\beta_a + \beta_b)$$

$$EC_1 = A_1 C_1 \tan \theta_a$$

$$= AB \cos \alpha \sin \beta_b \, \text{cosec} \, (\beta_a + \beta_b) \tan \theta_a \qquad (3.56)$$

The height of the object E above the ground at A is

$$EC = EC_1 + h_i$$

Also, as

$$B_1 C_2 = B_2 C_1 = A_1 B_2 \sin \beta_a \, \text{cosec} \, (\beta_a + \beta_b)$$

$$EC_2 = B_1 C_2 \tan \theta_b$$

$$EC = EC_2 + B_1 B_2 + h_i$$

$$= AB[\cos \alpha \sin \beta_a \, \text{cosec} \, (\beta_a + \beta_b) \tan \theta_b + \sin \alpha] + h_i$$
$$(3.57)$$

The height of the ground at D above the ground at A is

$$DC = EC_1 + h_t$$

$$= EC_2 + h_t + AB \sin \alpha$$

If the ground is level $\alpha = 0$ (Fig. 3.26); then

$$EC_1 = AB \sin \beta_b \, \text{cosec} \, (\beta_a + \beta_b) \tan \theta_a \qquad (3.58)$$

or $\qquad\quad = AB \sin \beta_a \, \text{cosec} \, (\beta_a + \beta_b) \tan \theta_b \qquad (3.59)$

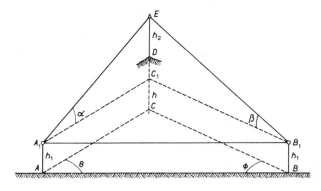

Fig. 3.26

Example 3.10   The following observations were made from a base line AB measured as 127.552 m.

At A the theodolite height was 1.65 m and at B 1.46 m whilst the target height was 1.50 m.

Angles measured at A: horizontal angle $EAB = 80°\, 20'\, 42''$
vertical angle $\qquad\quad = 12°\, 15'\, 37''$
at B: horizontal angle $ABE = 82°\, 41'\, 13''$
vertical angle $\qquad\quad = 12°\, 20'\, 23''$

Calculate the height of the ground at E above the ground at A.

Let h_i at $A = h_1 = 1.65$ m; h_i at $B = h_2 = 1.46$ m and $h_t = 1.50$ m.

By equation 3.57, height of ground at E above A is

$$\Delta h_1 = AB \sin \beta_b \operatorname{cosec} (\beta_a + \beta_b) \tan \theta_a + \delta h_1$$
$$\Delta h_2 = AB \sin \beta_a \operatorname{cosec} (\beta_a + \beta_b) \tan \theta_b + \delta h_2$$

i.e.

$$\Delta h_1 = 127.552 \sin 82° \, 41' \, 13'' \operatorname{cosec} 163° \, 01' \, 55'' \tan 12° \, 15' \, 37''$$
$$+ 1.65 - 1.50 \qquad = 94.36 \text{ m}$$
$$\Delta h_2 = 127.552 \sin 80° \, 20' \, 42'' \operatorname{cosec} 163° \, 01' \, 55'' \tan 12° \, 20' \, 23''$$
$$+ 1.46 - 1.50 \qquad = 94.22 \text{ m}$$
$$\Delta h_m = (94.36 + 94.22)/2 \;\; = 94.29 \text{ m}$$

3.8.3   To find the height of an object above the ground when the base and top are visible but not accessible

(a)   Base line in line with object (Fig. 3.27)
Vertical angles measured at A: $\theta_1, \theta_2, \alpha$
at B: δ_1, δ_2

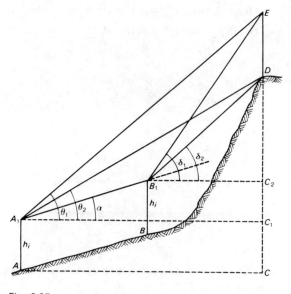

Fig. 3.27

By equation 3.54,

$$EC_1 = AB \sin (\delta_1 - \alpha) \operatorname{cosec} (\delta_1 - \theta_1) \sin \theta_1$$
and $$DC_1 = AB \sin (\delta_2 - \alpha) \operatorname{cosec} (\delta_2 - \theta_2) \sin \theta_2$$
then $$ED = EC_1 - DC_1$$
$$= AB \left[\sin (\delta_1 - \alpha) \operatorname{cosec} (\delta_1 - \theta_1) \sin \theta_1 - \sin (\delta_2 - \alpha) \operatorname{cosec} (\delta_2 - \theta_2) \sin \theta_2 \right] \qquad (3.60)$$

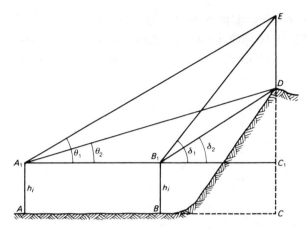

Fig. 3.28

If the ground is level, $\alpha = 0$ (Fig. 3.28) and

$$ED = AB\left[\sin \delta_1 \operatorname{cosec}(\delta_1 - \theta_1)\sin \theta_1 - \sin \delta_2 \operatorname{cosec}(\delta_2 - \theta_2)\sin \theta_2\right]$$

$$= AB\left[\frac{1}{\cot \theta_1 - \cot \delta_1} - \frac{1}{\cot \theta_2 - \cot \delta_2}\right] \qquad (3.61)$$

(b)   Base line not in line with object (Fig. 3.29)

Angles measured at A: horizontal β_a;   vertical $\theta_1, \theta_2, \alpha$

at B: horizontal β_b;   vertical δ_1, δ_2

Horizontal plane $A_1 B_2 C_1$

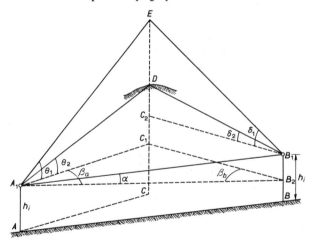

Fig. 3.29

By equation 3.57,

$$EC_1 = AB \cos \alpha \sin \beta_b \operatorname{cosec}(\beta_a + \beta_b)\tan \theta_1$$

$$DC_1 = AB \cos \alpha \sin \beta_b \operatorname{cosec}(\beta_a + \beta_b)\tan \theta_2$$

$$ED = EC_1 - DC_1$$

$$= AB \cos \alpha \sin \beta_b \operatorname{cosec}(\beta_a + \beta_b)(\tan \theta_1 - \tan \theta_2) \qquad (3.62)$$

also $ED = AB \cos \alpha \sin \beta_a \operatorname{cosec}(\beta_a + \beta_b)(\tan \delta_1 - \tan \delta_2) \qquad (3.63)$

If the ground is level, then $\alpha = 0$ (Fig. 3.30) and

$$ED = AB \sin \beta_b \cosec (\beta_a + \beta_b) (\tan \theta_1 - \tan \theta_2) \quad (3.64)$$

also $\quad ED = AB \sin \beta_a \cosec (\beta_a + \beta_b) (\tan \delta_1 - \tan \delta_2) \quad (3.65)$

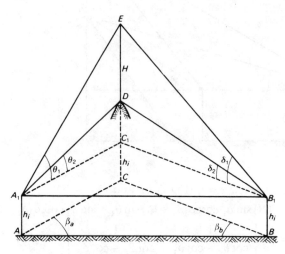

Fig. 3.30

3.8.4 To find the length of an inclined object (e.g. an inclined flagstaff) on the top of a building (Fig. 3.31)

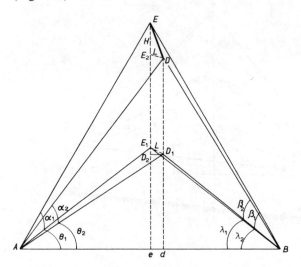

Fig. 3.31

Base line AB is measured and if on sloping ground reduced to horizontal.

Angles measured at A:

horizontal angles θ_1 and θ_2 to top and bottom of pole
vertical angles α_1 and α_2 to top and bottom of pole

Angles measured at B:

horizontal angles λ_1 and λ_2 to top and bottom of pole
vertical angles β_1 and β_2 to top and bottom of pole

In plan the length ED is projected as $E_1D_1 (= E_2D)$.

In elevation the length ED is projected as EE_2, i.e. the difference in height. Then

$$AE_1 = AB \sin \lambda_1 \operatorname{cosec}(\theta_1 + \lambda_1)$$

also $AD_1 = AB \sin \lambda_2 \operatorname{cosec}(\theta_2 + \lambda_2)$

The length $E_1D = L$ is now best calculated using coordinates (see Chapter 4).

Assuming

bearing $AB = 180°\,00'$

$$Ae = AE_1 \sin(90 - \theta_1) = AE_1 \cos \theta_1$$
$$Ad = AD_1 \sin(90 - \theta_2) = AD_1 \cos \theta_2$$

Then $ed = D_2D_1 = Ad - Ae$

$$= AD_1 \cos \theta_2 - AE_1 \cos \theta_1$$

and similarly,

$$E_1D_2 = E_1e - D_2e = AE_1 \sin \theta_1 - AD_1 \sin \theta_2$$

In the triangle $E_1D_1D_2$, the bearing of the direction of inclination (relative to AB)

$$= \tan^{-1}\frac{D_2D_1}{E_1D_2}$$

length $E_1D_1 = \sqrt{[(D_2D_1)^2 + (E_1D_2)^2]}$

To find difference in height EE_2,

height of top above A $= AE_1 \tan \alpha_1$
height of base above A $= AD_1 \tan \alpha_2$
length $EE_2 = AE_1 \tan \alpha_1 - AD_1 \tan \alpha_2$

To find length of pole:

In triangle EDE_2,

$$ED^2 = EE_2^2 + E_2D^2$$

i.e. $ED = \sqrt{(EE_2^2 + E_2D^2)}$

3.8.5 To find the height of an object from three angles of elevation only (Fig. 3.32)

Lengths x and y are measured. Solving triangles ADB and ADC by the cosine rule,

$$\cos\phi = \frac{h^2 \cot^2 \alpha + x^2 - h^2 \cot^2 \beta}{2hx \cot \alpha}$$

$$= \frac{h^2 \cot^2 \alpha + (x + y)^2 - h^2 \cot^2 \theta}{2h(x + y)\cot \alpha} \qquad (3.66)$$

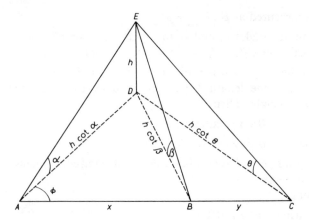

Fig. 3.32

therefore

$$(x+y)\left[h^2\left(\cot^2\alpha-\cot^2\beta\right)+x^2\right]$$
$$= x\left[h^2\left(\cot^2\alpha-\cot^2\theta\right)+(x+y)^2\right]$$

i.e.

$$h^2\left[(x+y)\left(\cot^2\alpha-\cot^2\beta\right)-x\left(\cot^2\alpha-\cot^2\theta\right)\right]$$
$$= x(x+y)^2-x^2(x+y)$$

Then

$$h = \sqrt{\left(\frac{xy(x+y)}{x(\cot^2\theta-\cot^2\beta)+y(\cot^2\alpha-\cot^2\beta)}\right)} \quad (3.67)$$

If $x = y$,

$$h = \frac{\sqrt{2x}}{\sqrt{(\cot^2\theta-2\cot^2\beta+\cot^2\alpha)}} \quad (3.68)$$

Example 3.11 A, B and C are stations on a straight level line of bearing $126°\,03'\,34''$. The distance AB is 523.54 m and BC is 420.97 m. With an instrument of constant height 1.30 m vertical angles were successively measured to an inaccessible up-station D as follows:

At A $7°\,14'\,00''$
      B $10°\,15'\,20''$
      C $13°\,12'\,30''$

Calculate
  (a)   the height of station D above the line ABC
  (b)   the bearing of the line AD
  (c)   the horizontal length AD.   (RICS)

(a)   In Fig. 3.33,

$$x = 523.54 \qquad \alpha = 7°\,14'\,00''$$
$$y = 420.97 \qquad \beta = 10°\,15'\,20''$$
$$\theta = 13°\,12'\,30''$$

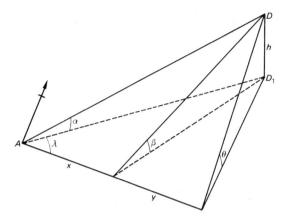

Fig. 3.33

By equation 3.67,

$$xy(x + y) = (523.54 \times 420.97)(523.54 + 420.97) = 2.081\,6493 \times 10^8$$

$$x(\cot^2 \theta - \cot^2 \beta) = 523.54(\cot^2 13.2083 - \cot^2 10.2556) = -6488.57$$

$$y(\cot^2 \alpha - \cot^2 \beta) = 420.97(\cot^2 7.2333 - \cot^2 10.2556) = 13\,273.53$$

Then
$$h = \sqrt{(2.081\,6493 \times 10^8 / 6784.96)} = 175.16\,\text{m}$$

Therefore difference in height of D above ground at A

$$= 175.16 + 1.30 = 176.46\,\text{m}$$

Using equation 3.66,

$$\lambda = \cos^{-1}\left(\frac{h^2(\cot^2 \alpha - \cot^2 \beta) + x^2}{2hx \cot \alpha}\right)$$

$$= \frac{175.16^2(\cot^2 7° 14' 00'' - \cot^2 10° 15' 20'') + 523.54^2}{2 \times 175.16 \times 523.54 \times \cot 7° 14' 00''}$$

$$= 30.7825 = 30° 46' 57''$$

(b) Thus bearing of $AD = 126° 03' 34'' - 30° 46' 57''$

$$= 095° 16' 37''$$

(c) Length of line $AD_1 = h \cot \alpha$

$$= 175.16 \cot 7° 14' = 1380.07\,\text{m}$$

3.8.6 The broken base line problem

Where a base line AD cannot be measured due to some obstacle the following system may be adopted, Fig. 3.34.

Lengths x and z are measured.
Angles α, β and θ are measured at station E.
To calculate $BC = y$:

area of triangle $ABE = \frac{1}{2}xh = \frac{1}{2}AE. EB \sin \alpha$ (1)

area of triangle $BCE = \frac{1}{2}yh = \frac{1}{2}BE. EC \sin \beta$ (2)

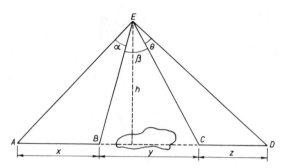

Fig. 3.34

area of triangle $CDE = \frac{1}{2}zh = \frac{1}{2}CE . ED \sin \theta$ (3)

area of triangle $ADE = \frac{1}{2}(x+y+z)h$

$$= \frac{1}{2}AE . ED \sin(\alpha + \beta + \theta) \quad (4)$$

Dividing (1) by (2),

$$\frac{x}{y} = \frac{AE \sin \alpha}{EC \sin \beta} \quad (5)$$

Dividing (3) by (4),

$$\frac{z}{x+y+z} = \frac{CE \sin \theta}{AE \sin(\alpha + \beta + \theta)} \quad (6)$$

Multiplying (5) by (6),

$$\frac{xz}{y(x+y+z)} = \frac{\sin \alpha \sin \theta}{\sin \beta \sin(\alpha + \beta + \theta)}$$

i.e. $y^2 + y(x+z) - xz \dfrac{\sin \beta \sin(\alpha + \beta + \theta)}{\sin \alpha \sin \theta} = 0$

Then $y = -\left(\dfrac{x+z}{2}\right) + \sqrt{\left[\left(\dfrac{x+z}{2}\right)^2 + xz \dfrac{\sin \beta \sin(\alpha + \beta + \theta)}{\sin \alpha \sin \theta}\right]}$ (3.69)

Example 3.12 The measurement of a base line AD is interrupted by an obstacle. To overcome this difficulty two points B and C were established on the line AD and observations made to them from a station E as follows:

$$A\hat{E}B = 20° 18' 20''$$
$$B\hat{E}C = 45° 19' 40''$$
$$C\hat{E}D = 33° 24' 20''$$

Length $AB = 527.43$ m and $CD = 685.29$ m.

Calculate the length of the line AD.   (RICS)

Here

$$\left.\begin{array}{l} \alpha = 20° 18' 20'' \\ \beta = 45° 19' 40'' \\ \theta = 33° 24' 20'' \end{array}\right\} \quad a + \beta + \theta = 99° 02' 20''$$

$$\left.\begin{array}{l} x = 527.43 \\ z = 685.29 \end{array}\right\} \quad \tfrac{1}{2}(x+z) = \tfrac{1}{2}(1212.72) = 606.36$$

By equation 3.69,

$$y = -606.36 + \sqrt{606.36^2 + \frac{527.43 \times 685.29 \sin 45° 19' 40'' \sin 99° 02' 20''}{\sin 20° 18' 20'' \sin 33° 24' 20''}}$$

$$= -606.36 + \sqrt{367\,672 + 1\,328\,614}$$

$$= -606.36 + 1302.415 = 696.055 \text{ m} \quad (696.06 \text{ m})$$

3.8.7 To find the relationship between angles in the horizontal and inclined planes (Fig. 3.35)

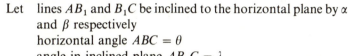

Fig. 3.35

Let   lines AB_1 and B_1C be inclined to the horizontal plane by α
and β respectively
horizontal angle $ABC = \theta$
angle in inclined plane $AB_1C = \lambda$
$B_1B = h$

Then

$$AB = h \cot \alpha \quad AB_1 = h \operatorname{cosec} \alpha$$
$$BC = h \cot \beta \quad B_1C = h \operatorname{cosec} \beta$$

In triangle ABC,

$$AC^2 = AB^2 + BC^2 - 2AB \cdot BC \cos \theta$$
$$= h^2 \cot^2 \alpha + h^2 \cot^2 \beta - 2h^2 \cot \alpha \cot \beta \cos \theta$$

Similarly in triangle AB_1C,

$$AC^2 = h^2 \operatorname{cosec}^2 \alpha + h^2 \operatorname{cosec}^2 \beta - 2h^2 \operatorname{cosec} \alpha \operatorname{cosec} \beta \cos \lambda$$

Then

$$h^2 \cot^2 \alpha + h^2 \cot^2 \beta - 2h^2 \cot \alpha \cot \beta \cos \theta$$
$$= h^2 \operatorname{cosec}^2 \alpha + h^2 \operatorname{cosec}^2 \beta - 2h^2 \operatorname{cosec} \alpha \operatorname{cosec} \beta \cos \lambda$$

i.e. $$\cos \lambda = \frac{(\operatorname{cosec}^2 \alpha - \cot^2 \alpha) + (\operatorname{cosec}^2 \beta - \cot^2 \beta) + 2 \cot \alpha \cot \beta \cos \theta}{2 \operatorname{cosec} \alpha \operatorname{cosec} \beta}$$

as

$$\operatorname{cosec}^2 \alpha - \cot^2 \alpha = \operatorname{cosec}^2 \beta - \cot^2 \beta = 1$$

Then

$$\cos \lambda = \frac{2(1 + \cot \alpha \cot \beta \cos \theta)}{2 \operatorname{cosec} \alpha \operatorname{cosec} \beta}$$

$$= \sin \alpha \sin \beta + \cos \alpha \cos \beta \cos \theta \qquad (3.70)$$

or $$\cos \theta = \frac{\cos \lambda - \sin \alpha \sin \beta}{\cos \alpha \cos \beta} \qquad (3.71)$$

Example 3.13 From a station A observations were made to stations B and C with a sextant and an abney level.
   With sextant—angle $BAC = 84° 30'$
   With abney level—angle of depression (AB) $8° 20'$
                          angle of elevation (AC) $10° 40'$
   Calculate the horizontal angle BA_1C which would have been measured if a theodolite had been used.   (RICS/M)

From equation 3.71,

$$\theta = \cos^{-1}\left(\frac{\cos\lambda - \sin\alpha\sin\beta}{\cos\alpha\cos\beta}\right)$$

$$= \cos^{-1}\left(\frac{\cos 84° 30' - \sin(-8° 20')\sin 10° 40'}{\cos(-8° 20')\cos 10° 40'}\right)$$

$$= 82° 45' 10''$$

Example 3.14 A pipe-line is to be laid along a bend in a mine roadway ABC. If AB falls at a gradient of 1 in 2 in a direction $036° 27'$, whilst BC rises due South at 1 in 3.5, calculate the angle of bend in the pipe. (RICS/M)

Consider Fig. 3.36.

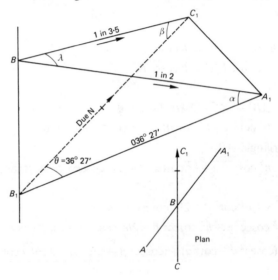

Fig. 3.36

From equation 3.70,

$$\cos\lambda = \sin\alpha\sin\beta + \cos\alpha\cos\beta\cos\theta$$

where $\alpha = \cot^{-1}2 = 26° 33'$

$\beta = \cot^{-1}3.5 = 15° 57'$

$\theta = 036° 27' - 00° = 36° 27'$

therefore

$$\cos\lambda = \sin 26° 33'\sin 15° 57' + \cos 26° 33'\cos 15° 57'\cos 36° 27'$$

$$\lambda = 35° 26' 40'' \text{ i.e. } 35° 27'$$

Exercises 3.3

1 Show that for small angles of slope the difference between horizontal and sloping lengths is $h^2/2l$ (where h is the difference of vertical height of the two ends of a line of sloping length l).

If errors in chaining are not to exceed 1 part in 1000, what is the greatest slope that can be ignored?   (LU/E Ans.   2° 34′)

2   The height of an electricity pylon relative to two stations A and B (at the same level) is to be calculated from the data given below. Find the height from the two stations if at both stations the height of the theodolite axis is 1.52 m.

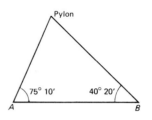

Fig. 3.37

Data: $AB = 60.96$ m

Horizontal angle at $A = 75° 10′$ Vertical angle at $A = 43° 12′$

at $B = 40° 20′$ at $B = 32° 13′$

(RICS Ans.   Mean height 42.62 m)

3 X, Y and Z are three points on a straight survey line such that $XY = 56$ m and $YZ = 80$ m.

From X, a normal offset was measured to a point A and XA was found to be 42 m. From Y and Z respectively, a pair of oblique offsets were measured to a point B, and these distances were as follows: $YB = 96$ m, $ZB = 88$ m. Calculate the distance AB, and check your answer by plotting to some suitable scale, and state the scale used.   (EMEU Ans.   113 m)

4   From the top of a tower 120 m high, the angle of depression of a point A is 15°, and of another point B is 11°. The bearings of A and B from the tower are 205° and 137° respectively. If A and B lie in a horizontal plane through the base of the tower, calculate the distance AB.   (RICS Ans.   612 m)

5 A, B, C, D are four successive km posts on a straight horizontal road.

From a point O due W of A, the direction of B is 84°, and of D is 77°. The post C cannot be seen from O, owing to trees. If the direction in which the road runs from A to D is θ, calculate θ, and the distance of O from the road.

(RICS Ans. $\theta = 60° 06′ 50″$, $OA = 3.874$ km)

6   At a point A, a man observes the elevation of the top of a tower B to be 42° 15′. He walks 200 m up a uniform slope of elevation 12° directly towards the tower, and then finds that the elevation of B has increased by 23° 09′. Calculate the height of B above the level of A.   (RICS Ans.   275 m)

7   At two points, 500 m apart on a horizontal plane, observations at the bearing and elevation of an aeroplane are taken simultaneously. At one point the bearing is 041° and the elevation is 24°, and at another point the bearing is 032° and the elevation is 16°. Calculate the height of the aeroplane above the plane.   (RICS Ans.   1139 m)

8   Three survey stations X, Y and Z lie in one straight line on the same plane. A series of angles of elevation is taken to the top of a colliery chimney, which lies to one side of the line XYZ. The angles measured at X, Y and Z were: at X, $14°\,02'$; at Y, $26°\,34'$; at Z, $18°\,26'$. The lengths XY and YZ are 121.92 m and 73.15 m respectively. Calculate the height of the chimney above station X. (EMEU Ans. 34.19 m)

9   The altitude of a mountain, observed at the end A of a base line AB of 2992.5 m, was $19°\,42'$ and the horizontal angles at A and B were $127°\,54'$ and $33°\,09'$ respectively. Find the height of the mountain. (Ans. 1804 m)

10   It is required to determine the distance between two inaccessible points A and B by observations from two stations C and D, 1000 m apart. The angular measurements give $ACB = 47°$, $BCD = 58°$, $BDA = 49°$; $ADC = 59°$. Calculate the distance AB. (Ans. 2907.4 m)

11   An aeroplane is observed simultaneously from two points A and B at the same level, A being a distance (c) due north of B. From A the aeroplane is S $\theta°$ E and from B N $\phi°$ E. Show that the height of the aeroplane is

$$\frac{c \tan \alpha \sin \phi}{\sin (\theta + \phi)}$$

and find its elevation from B.

$$\left(\text{LU Ans.}   \beta = \tan^{-1} \frac{\sin \phi \tan \alpha}{\sin \theta} \right)$$

12   A straight base line $ABCD$ is sited such that a portion of BC cannot be measured directly. If AB is 575.64 m and CD is 728.56 m and the angles measured from station O to one side of $ABCD$ are

$$DOC = 56°\,40'\,30''   COB = 40°\,32'\,00''   BOA = 35°\,56'\,30''$$

calculate the length BC. (EMEU Ans. 259.32 m)

13   It is proposed to lay a line of pipes of large diameter along a roadway of which the gradient changes from a rise of $30°$ to a fall of $10°$ coincident with a bend in the roadway from a bearing of N $22°$ W to N $25°$ E. Calculate the angle of bend in the pipe. (Ans. $119°\,39'\,30''$)

Bibliography

USILL, G. W., *Practical Surveying*, 16th edition. Technical Press (1973)

4

Coordinates

A point in a plane may be defined by two systems:

(1) polar coordinates;
(2) rectangular or Cartesian coordinates.

4.1 Polar coordinates

This system involves angular and linear values, i.e. bearing and length, the former being plotted by protractor as an angle from the meridian (Fig. 4.1).

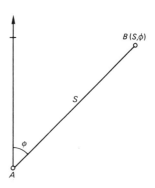

Fig. 4.1 Polar coordinates

A normal 150 mm protractor allows plotting to the nearest $\frac{1}{4}°$; a cardboard protractor with parallel rule to $\frac{1}{8}°$; whilst the special Bocking protractor enables 01′ to be plotted.

The displacement of the point being plotted depends on the physical length of the line on the plan, which in turn depends on the horizontal projection of the ground length and the scale of the plotting (see Fig. 4.2).

If $\delta\phi$ is the angular error, then the displacement

$$BB_1 = S\,\delta\phi$$

If $S = 100.000$ m and $\delta\phi = 0°\,01'\,00''$,

$$BB_1 = \frac{100.000 \times 60}{206\,265} \simeq 0.029\,\text{m}$$

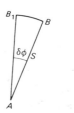

Fig. 4.2 Displacement due to angular error

Thus, 1 minute of arc subtends approximately 30 mm in 100 m,
1 second of arc subtends approximately 1 mm in 200 m
or   1 cm in 2 km

A point plotted on a plan may be assumed to be 0.2 mm, and if this subtends an arc of radius 1 m this represents 40 seconds of arc.

As this represents a possible plotting error on every line, it can be seen how the error may accumulate, particularly as each point is dependent on the preceding point.

4.1.1 Plotting to scale

The length of the plotted line is some definite fraction of the ground length, the 'scale' chosen depending on the purpose of the plan and the size of the area.

Scales are now expressed entirely as representative fractions, i.e. 1 in x; e.g. 1/25 000, which represents 1 m in 25 000 m on this scale.

4.1.2   Scales in common use

Comparison of scales

Scales in common use with the metric system	
Recommended by BSI	Other alternative scales
1 : 1000 000	
1 : 500 000	1 : 625 000
1 : 200 000	1 : 250 000
1 : 100 000	1 : 125 000
1 : 50 000	1 : 62 500
1 : 20 000	1 : 25 000
1 : 10 000	
1 : 5000	
1 : 2000	1 : 2500
1 : 1000	1 : 1250
1 : 500	
1 : 200	
1 : 100	
1 : 50	
1 : 20	
1 : 10	
1 : 5	

4.1.3   Plotting accuracy

Considering 0.2 mm as the size of a plotted point, the following table shows the representative value at the typical scales.

OS scales	Suggested measurement lower precision limit
1/500	$0.2 \times 500 = 0.100$ m
1/1250	$0.2 \times 1250 = 0.250$ m
1/2500	$0.2 \times 2500 = 0.500$ m
1/10 000	$0.2 \times 10 000 = 2.000$ m
1/25 000	$0.2 \times 25 000 = 5.000$ m

Engineering scales

1/50	$0.2 \times 50 = 0.010$ m
1/100	$0.2 \times 100 = 0.020$ m
1/200	$0.2 \times 200 = 0.040$ m

4.2   Bearings

Four meridians may be used; Fig. 4.3:

(1)   True or geographical north.
(2)   Magnetic north.
(3)   Grid north.
(4)   Arbitrary north.

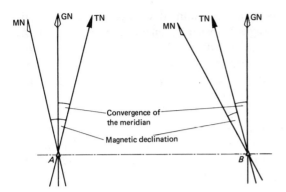

Fig. 4.3 Meridians

4.2.1 True north

The meridian can be obtained precisely only by astronomical observation or by the use of a gyroscopic theodolite. The difference between true bearings at *A* and *B* is the convergence of the meridians to a point, i.e. the north pole. For small surveys the discrepancy is small and can be neglected but where necessary a correction may be computed and applied.

4.2.2 Magnetic north

There is no fixed point and thus the meridian is unstable and subjected to a number of variations (Fig. 4.4), viz.:

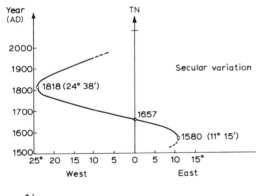

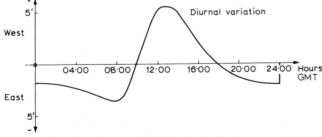

Fig. 4.4 Approximate secular and diurnal variations in magnetic declination in the London area (Abinger Observatory)

(a) *Secular variation*—the annual change in the magnetic declination or angle between magnetic and true north. At present the magnetic meridian in Britain is to the west of true north but moving towards it at the approximate rate of 10 min per annum. (Values of declination and the annual change are shown on certain OS sheets.)

(b) *Diurnal variation*—a daily sinusoidal oscillation effect, with the mean value at approximately 10 a.m. and 6–7 p.m., and maxima and minima at approximately 8 a.m. and 1 p.m.

(c) *Irregular variation*—periodic magnetic fluctuations thought to be related to sun spots.

4.2.3 Grid north (see Section 4.7)

OS sheets are based on a modified Transverse Mercator projection which, within narrow limits, allows:

(a) Constant bearings related to a parallel grid.
(b) A scale factor for conversion of ground distance to grid distance solely dependent on the easterly coordinates of the measurement site (see page 38).

4.2.4 Arbitrary north

It may not be necessary to obtain absolute reference and often the first leg of the traverse is assumed to be 0° 00′.

Example 4.1   True north is 0° 37′ E of Grid North (Fig. 4.5). Magnetic declination in June 1962 was 10° 27′ W.

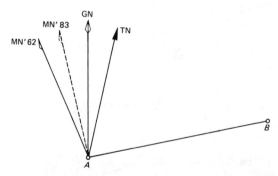

Fig. 4.5

If the annual variation was 10′ per annum towards North and the grid bearings of line *AB* 082° 32′, what was the magnetic bearing of line *AB* in January 1983?

Grid bearing	$082°\,32'$
Correction	$-0°\,37'$
True bearing	$081°\,55'$
Magnetic declination in June 1962	$10°\,27'$
Magnetic bearing in June 1962	$092°\,22'$
Variation for January $1983 - 20\frac{1}{2} \times 10'$	$-3°\,25'$
Magnetic bearing in January 1983	$088°\,57'$

Case (i)

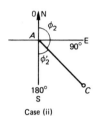

Case (ii)

4.2.5 Types of bearing

There are two types in general use:

(a) Whole circle bearings (WCB), which are measured clockwise from north or $0°$–$360°$.

(b) Quadrant bearings (QB), which are angles measured to the east or west of the N/S meridian.

Case (i) Whole circle bearing in the first quadrant 0–$90°$ (Fig. 4.6):

$$\text{WCB of } AB = \phi_1$$
$$\text{QB of } AB = N\phi_1'E$$

Case (ii) $90°$–$180°$:

$$\text{WCB of } AC = \phi_2$$
$$\text{QB of } AC = S\phi_2'E$$
$$= S(180 - \phi_2)°E$$

Case (iii) $180°$–$270°$:

$$\text{WCB of } AD = \phi_3$$
$$\text{QB of } AD = S\phi_3'W$$
$$= S(\phi_3 - 180)°W$$

Case (iv) $270°$–$360°$:

$$\text{WCB of } AE = \phi_4$$
$$\text{QB of } AE = N\phi_4'W$$
$$= N(360 - \phi_4)°W$$

NB: As the same numbers, 0–$90°$, occur in each of the quadrants, the quadrant bearings are somewhat ambiguous. Now that scientific calculators produce the correct sign for the trigonometrical functions 0–$360°$, whole circle bearings are preferred.

4.2.6 Angles and bearings

The bearing of the line AB (Fig. 4.7) is given as ϕ_{AB}. The reverse or reciprocal bearing BA is given as

$$\phi_{BA} = \phi_{AB} \pm 180$$

NB: As the bearing cannot be negative or greater than 360, 180 is either added or subtracted.

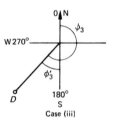

Case (iii)

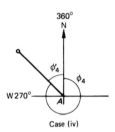

Case (iv)

Fig. 4.6 Comparison of bearings

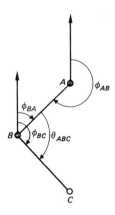

Fig. 4.7 Bearing convention

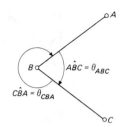

Fig. 4.8 Angle convention

The clockwise angle ABC (Fig. 4.8) is given as

$$\theta_{ABC} = \phi_{BC} - \phi_{BA}$$

or $\phi_{BC} = \phi_{BA} + \theta_{ABC}$

Given the forward bearing AB and the angle ABC (Fig. 4.9) then the forward bearing BC

$$= \phi_{BC} = \phi_{AB} \pm 180 + \theta_{ABC}(-360)$$

NB: 360 is subtracted if the sum exceeds 360.

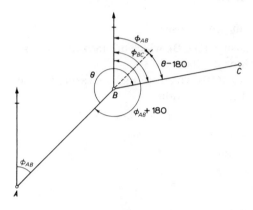

Fig. 4.9 Conversion of horizontal angles into bearings

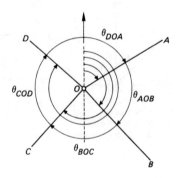

Fig. 4.10

This basic process may always be used but the following rules simplify the process.

(1) To the forward bearing *add* the clockwise angle.
(2) If the sum is less than 180° *add* 180°.
     If the sum is more than 180° *subtract* 180°;
     (In some cases the sum may be more than 540°, then subtract 540°.)

NB: If the angles measured are anticlockwise they must be subtracted.

Example 4.2   See Fig. 4.10.

$$\phi_{OA} = 062° \, 35' = \text{N } 62° \, 35' \text{ E} \qquad \phi_{AO} = 242° \, 35' = \text{S } 62° \, 35' \text{ W}$$
$$\phi_{OB} = 143° \, 45' = \text{S } 36° \, 15' \text{ E} \qquad \phi_{BO} = 323° \, 45' = \text{N } 36° \, 15' \text{ W}$$
$$\phi_{OC} = 212° \, 20' = \text{S } 32° \, 20' \text{ W} \qquad \phi_{CO} = 032° \, 20' = \text{N } 32° \, 20' \text{ E}$$
$$\phi_{OD} = 314° \, 10' = \text{N } 45° \, 50' \text{ W} \qquad \phi_{DO} = 134° \, 10' = \text{S } 45° \, 50' \text{ E}$$

$$\theta_{AOB} = \phi_{OB} - \phi_{OA} = 143° \, 45' - 062° \, 35' = 81° \, 10'$$
$$\theta_{BOC} = \phi_{OC} - \phi_{OB} = 212° \, 20' - 143° \, 45' = 68° \, 35'$$
$$\theta_{COD} = \phi_{OD} - \phi_{OC} = 314° \, 10' - 212° \, 20' = 101° \, 50'$$
$$\theta_{DOA} = \phi_{OA} - \phi_{OD} = 422° \, 35' - 314° \, 10' = \underline{108° \, 25'}$$
$$360° \, 00'$$

Example 4.3   See Fig. 4.11. Let

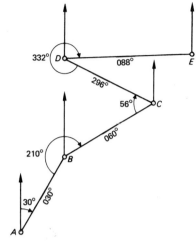

Fig. 4.11

bearing $AB = 030°$ N 30° E

+ angle $ABC = 210°$

$\phantom{+ \text{angle } ABC =}\overline{240°}$

$\phantom{+ \text{angle } ABC =}-180°$

bearing $BC = 060°$ N 60° E

+ angle $BCD = 56°$

$\phantom{+ \text{angle } BCD =}\overline{116°}$

$\phantom{+ \text{angle } BCD =}+180°$

bearing $CD = 296°$ N 64° W

+ angle $CDE = 332°$

$\phantom{+ \text{angle } CDE} = \overline{628°}$

$\phantom{+ \text{angle } CDE} -540°$

bearing $DE = 088°$ N 88° E

Check   bearing $AB = 030°$

 angles $= 210°$

$56°$

$332°$

$\overline{628°}$

$-n \times 180°$, i.e. $-3 \times 180°$, $-540°$

 bearing $DE = \overline{088°}$

The final bearing is checked by adding the bearing of the first line to the sum of the clockwise angles, and then subtracting some multiple of 180°.

Example 4.4   The clockwise angles of a closed polygon are observed to be as follows:

A 223° 46′
B 241° 17′
C 257° 02′
D 250° 21′
E 242° 19′
F 225° 15′

If the true bearings of *BC* and *CD* are 123° 14′ and 200° 16′ respectively, and the magnetic bearing of *EF* is 333° 21′, calculate the magnetic declination.   (TP)

From the size of the angles it may be initially assumed that these are external to the polygon and should sum to $(2n + 4)90°$, i.e.

$$\{(2 \times 6) + 4\}90 = 16 \times 90 = 1440°$$

223° 46′
241° 17′
257° 02′
250° 21′
242° 19′
225° 15′

Check 1440° 00′

To obtain the bearings,

Line BC bearing	123° 14′
+ angle BCD	257° 02′
	380° 16′
	− 180°
bearing CD	200° 16′ (this checks
+ angle CDE	250° 21′ with given value)
	450° 37′
	− 180°
bearing DE	270° 37′
+ angle DEF	242° 19′
	512° 56′
	− 180°
bearing EF	332° 56′
+ angle EFA	225° 15′
	558° 11′
	− 540°
bearing FA	018° 11′
+ angle FAB	223° 46′
	241° 57′
	− 180°
bearing AB	061° 57′
+ angle ABC	241° 17′
	303° 14′
	− 180°
bearing BC	123° 14′ Check
Magnetic bearing EF	333° 21′
True bearing EF	332° 56′
Magnetic declination	0° 25′ W

Exercises 4.1

1 Convert the following whole circle bearings into quadrant bearings:

214° 30′; 027° 15′; 287° 45′; 093° 30′; 157° 30′;
311° 45′; 218° 30′; 078° 45′; 244° 14′; 278° 04′.

      (Ans.   S 34° 30′ W; N 27° 15′ E; N 72° 15′ W; S 86° 30′ E;
              S 22° 30′ E; N 48° 15′ W; S 38° 30′ W; N 78° 45′ E;
              S 64° 14′ W; N 81° 56′ W)

2 Convert the following quadrant bearings into whole circle bearings:

N 25° 30′ E; S 34° 15′ E; S 42° 45′ W; N 79° 30′ W;
S 18° 15′ W; N 82° 45′ W; S 64° 14′ E; S 34° 30′ W.

        (Ans.   025° 30′; 145° 45′; 222° 45′; 280° 30′; 198° 15′;
             277° 15′; 115° 46′; 214° 30′)

3 The following clockwise angles were measured in a closed traverse. What is the angular closing error?

163° 27′ 36″; 324° 18′ 22″; 62° 39′ 27″; 330° 19′ 18″;
181° 09′ 15″; 305° 58′ 16″; 188° 02′ 03″; 292° 53′ 02″;
131° 12′ 50″

                                   (Ans. 09″)

4 Measurement of the interior anticlockwise angles of a closed traverse *ABCDE* have been made with a vernier theodolite reading to 20 seconds of arc. Adjust the measurements and compute the bearings of the sides if the bearing of the line *AB* is N 43° 10′ 20″ E.

Angle   *EAB* 135° 20′ 40″ (RICS Ans. *AB* N 43° 10′ 20″ E
           ABC   60° 21′ 20″   *BC* S 17° 10′ 52″ E
           BCD 142° 36′ 20″   *CD* S 20° 12′ 56″ W
           CDE   89° 51′ 40″   *DE* N 69° 38′ 36″ W
           DEA 111° 50′ 40″   *EA* N 01° 29′ 08″ W)

5 From the theodolite readings given below, determine the angles of a traverse *ABCDE*. Having obtained the angles, correct them to the nearest 10 seconds of arc and then determine the bearing of *BC* if the bearing of *AB* is 45° 20′ 40″.

Back station	Theodolite station	Forward station	Readings	
			Back station	Forward station
E	*A*	*B*	0° 00′ 00″	264° 49′ 40″
A	*B*	*C*	264° 49′ 40″	164° 29′ 10″
B	*C*	*D*	164° 29′ 10″	43° 58′ 30″
C	*D*	*E*	43° 58′ 30″	314° 18′ 20″
D	*E*	*A*	314° 18′ 20″	179° 59′ 10″

                           (RICS Ans. 125° 00′ 20″)

4.3   Rectangular coordinates

A point may be fixed in a plane by linear values measured parallel to the general mathematical axes x and y.

4.3.1   Cartesian coordinates

The Cartesian coordinates of a point P are defined as x_p and y_p. In surveying the x values are referred to as *Eastings* (E) whilst the y values are known as *Northings* (N). These *rectangular coordinates* of the same point P are then defined as E_p and N_p.

It is common practice to try to keep all the coordinates positive, but they may be negative dependent upon the position of the point relative to the origin of the coordinate system.

The following sign convention is used (Fig. 4.12):

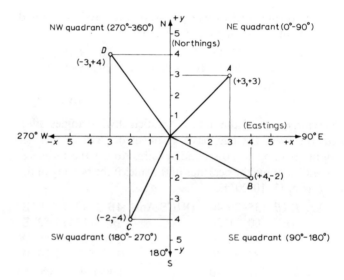

Fig. 4.12 Rectangular coordinates

Direction	Cartesian	Rectangular
East	x	E
West	$-x$	$-$E
North	y	N
South	$-y$	$-$N

i.e.
0 to 90	NE	$+$N	$+$E
90 to 180	SE	$-$N	$+$E
180 to 270	SW	$-$N	$-$E
270 to 360	NW	$+$N	$-$E

This gives a mathematical basis for the determination of a point with no need for graphical representation and is more satisfactory for the following reasons:

(1) Each station can be plotted independently.
(2) In plotting, the point is not dependent on any angular measuring device.
(3) Distances and bearings between points can be computed.

Rectangular coordinates are sub-divided into:

(1) Partial coordinates, which relate to a line.
(2) Total coordinates, which relate to a point.

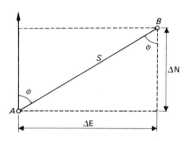

4.3.2   Partial coordinates (ΔE, ΔN)

These relate to a line and are derived from the polar coordinates or from the total coordinates (see Fig. 4.13). They represent the distance travelled East $(+)$/West$(-)$ and North$(+)$/South$(-)$ for a single line or join between any two points.

Fig. 4.13 Partial coordinates

Given a line of length S and bearing ϕ. Then for a line AB,

$$\Delta E_{AB} = E_B - E_A = S_{AB} \sin \phi_{AB} \qquad (4.1)$$

$$\Delta N_{AB} = N_B - N_A = S_{AB} \cos \phi_{AB} \qquad (4.2)$$

4.3.3   Total coordinates

These relate any point to the axes of the coordinate system used (see Fig. 4.14). The following notation is used:

Total Easting of A $= E_A$
Total Northing of A $= N_A$
Total Easting of B $= E_A + \Delta E_{AB}$
Total Northing of B $= N_A + \Delta N_{AB}$
Total Easting of C $= E_B + \Delta E_{BC} = E_A + \Delta E_{AB} + \Delta E_{BC}$
Total Northing of C $= N_B + \Delta N_{BC} = N_A + \Delta N_{AB} + \Delta N_{BC}$

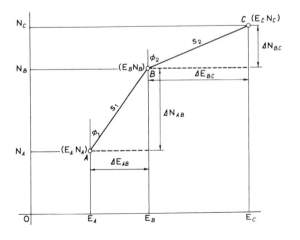

Fig. 4.14 Total coordinates

Thus in general terms

Total Easting of any point $= E_A + \Sigma\Delta E$ (4.3)

$=$ Total easting of the first point + the sum of the partial eastings up to that point.

Total Northing of any point $= N_A + \Sigma\Delta N$ (4.4)

$=$ Total northing of the first point + the sum of the partial northings up to that point.

NB: If a traverse is closed polygonally then

$\Sigma\Delta E = 0$ (4.5)

$\Sigma\Delta N = 0$ (4.6)

i.e. the sum of the partial coordinates should equal zero.

Example 4.5   Given (Fig. 4.15):

AB 045° 00'   100.0 m
BC 120° 00'   150.0 m
CD 210° 00'   100.0 m

Total coordinates of A E 50.0 m N 40.0 m.

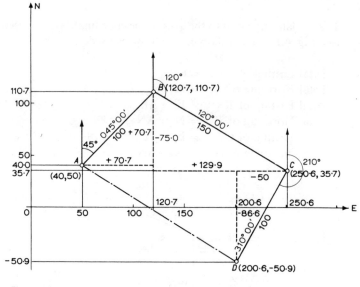

Fig. 4.15

Station	E	N
A	50.0	40.0

Line AB $\phi = 045° 00'$ $S = 100.0$
$\Delta E = 100.0 \sin 45.00$ 70.7
$\Delta N = 100.0 \cos 45.00$ 70.7

B	120.7	110.7

Line BC $\phi = 120° \, 00'$ $S = 150.0$

$\quad \Delta E = 150.0 \sin 120.00$ 129.9

$\quad \Delta N = 150.0 \cos 120.00$ -75.0

                             C   250.6   35.7

Line CD $\phi = 210° \, 00'$ $S = 100.0$

$\quad \Delta E = 100.0 \sin 210.00$ -50.0

$\quad \Delta N = 100.0 \cos 210.00$ -86.6

                             D   200.6 -50.9

Check: $E_D = E_A + \Delta E_{AB} + \Delta E_{BC} + \Delta E_{CD}$

$\qquad\quad = 50.0 + 70.7 + 129.9 - 50.0 = +200.6 \text{ m}$

$\quad\;\; N_D = N_A + \Delta N_{AB} + \Delta N_{BC} + \Delta N_{CD}$

$\qquad\quad = 40.0 + 70.7 - 75.0 - 86.6 \;\; = -50.9 \text{ m}$

Exercises 4.2 (Plotting)

1 The following table shows angles and distances measured in a theodolite traverse from a line *AB* bearing due South and of horizontal length 33.5 m.

Angle	Angle value	Inclination	Inclined distance (m)
ABC	192° 00′	$+15°$	*BC* 45.7
BCD	92° 15′	$0°$	*CD* 61.0
CDE	93° 30′	$-13°$	*DE* 70.1
DEF	170° 30′	$0°$	*EF* 45.7

Compute the whole circle bearing of each line, plot the survey to a scale of 1/1000 and measure the horizontal length and bearing of the closing line. (MQB/M Ans.   79.2 m; 076° 30′)

2 The following notes refer to an underground traverse made from the mouth, *A*, of a surface drift.

Line	Bearing	Distance (m)	
AB	038°	65.4	dipping at 1 in 2.4
BC	111°	41.8	level
CD	006°	73.0	level
DE	308°	47.1	rising at 1 in 3.2

Plot the survey to a scale of 1/1000.

Taking *A* as the origin, measure from your plan, the coordinates of *E*. What is the difference in level between *A* and *E* to the nearest 0.1 m?

                   (MQB/UM Ans.   46.9 m E, 138.4 m N;
                            difference in level *AE* 23.8 m)

4.4 Computation processes

The use of the scientific calculator has altered the basic computational processes and thus the calculations may be more systemised.

4.4.1 Machine logic

(a) *Algebraic* logic, in which the equation is tackled from left to right taking due regard to the application of parentheses; and

(b) *Reverse Polish Notation* (RPN) as adopted by the Hewlett Packard Co. Here no equals sign is used and the variables are entered first followed by the function.

4.4.2 Economy of accuracy

It is essential to use sufficient significant figures at all times during the intermediate stages of any calculation but the final output of a derived value must take due regard of the measured values and the full display used with discretion. For example to quote a reduced length, originally measured with a linen tape, to 4 places of decimals of a metre indicates a lack of understanding.

4.4.3 Machine operations

Dependent upon the type of machine in use the following applies:

(1) *Basic machine* with only trigonometrical functions.

As $\Delta E = S \sin \phi$ and $\Delta N = S \cos \phi$ it is possible to store either S or ϕ (in degree decimals) or both.

(2) *Scientific machine* with polar–rectangular coordinate function key. Here the length (S) and the bearing (ϕ) are entered and with the press of a key the partial coordinates are displayed in turn. Most machines work on the mathematical notation for polar coordinates (Fig. 4.16) where $r \equiv S$ and $\theta \equiv \phi$, the bearing, is measured anticlockwise from the x axis. It is therefore vital to know your own machine and to check:

(i) which of the polar variables is entered first, i.e. r or θ (S or ϕ);

(ii) which of the partial coordinates is displayed first, i.e. x or y (ΔE or ΔN).

NB: The alternative coordinate is usually obtained by pressing an ($x \rightarrow y$) key.

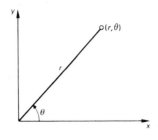

Fig. 4.16

TRENT POLYTECHNIC

TRAVERSE COMPUTATION SHEET No. _1_

Date _16·2·82_

Fixed pts _A 1000·000_ E _1000·000_ N Initial ϕ _AB 039°37·1'_

_A_ " E " N Final ϕ _____ "

Computed by _F.A.S._

Checked by _R.R._

ΔE _____ ΔN _____ $\Delta\phi$ _____

LINE	Sum ($\phi + \alpha$) / ±180/−540 / WCB (ϕ)	Length	ΔE / δE / $\Delta E'$	ΔN / δN / $\Delta N'$	Station Coordinates E	N	STN
A	Angle (α)				1000·000	1000·000	A
t	219° 37·1		72·4623	87·5349			
o	−180		−0·0013	−0·0021			
	039 37·1	113·636	72·4610	87·5328			
B	125 37·0				1072·461	1087·533	B
t	165 14·1		−27·7170	105·1648			
o	+180		−0·0013	−0·0019			
	345 14·1	108·756	−27·7183	105·1629			
C	27·50·2				1044·743	1192·696	C
t	373 04·3		−44·7404	−192·6921			
o	−180		−0·0023	−0·0036			
	193 04·3	197·818	−44·7427	−192·6957			
A	26 32·8				1000·000	1000·000	A
t							
o							
t							
o							
t							
o							
t							
o							
	Σ	420·210	0·0049	0·0076			

$$K_1 = \frac{-0.0049}{420·210} = -1·166 \times 10^{-5}$$

$$K_2 = \frac{-0.0076}{420·210} = -1·809 \times 10^{-5}$$

Fig. 4.17 Traverse tabulation

Examples

(1) Using a Hewlett Packard machine (with RPN) the procedure is as follows:

    Enter (θ) i.e. ϕ : Enter (r) i.e. S : Press [INV] [→R]

    Display : (y) i.e. ΔN : Press $[x \leftrightarrow y]$: Display (x) i.e. ΔE

(2) Using a Commodore machine:

    Enter (r) : Press $[x \to y]$: Enter (θ) : Press $[P \to R]$

    Display : (x) : Press $[x \to y]$: Display (y)

(3) Using a Casio machine:

    Enter (r) : Press $[INV][P \to R]$: Enter (θ) : Press $[=]$

    Display : (x) : Press $[x \to y]$: Display (y)

It can be seen from the above that each manufacturer develops his own system and the operator must become familiar with the set instructions.

NB: At all times it is desirable that some intermediate statements are made in order that an independent check can be made.

The following format is suggested:

Line AB $\phi = 030°\ 26'\ 25''$: $S = 125.361$ m

$$E_A = 1000.000 \quad N_A = 1000.000$$
$$\Delta E = \quad\ \ 63.513 \quad \Delta N = \quad 108.081$$
$$E_B = 1063.513 \quad N_B = 1108.081$$

NB: The value of $30°\ 26'\ 25''$ will have to be converted into $30.4403°$ and the operator should check to see what conversion facilities are available.

Examples

(1) *HP*   Enter DDD.MMSS i.e. 30.26250

        Press [INV] [D.MS →] : Display : 30.4403

                           (no. of figures fixed)

(2) *Casio*   Enter 30 $[°'\,'']$ 26 $[°'\,'']$ Display : 30.433 333 33

                       25.0$[°'\,'']$ Display : 30.440 277 78

(3) If no key is available then in computer language:

    $B = D + M/60 + S/3600$

i.e. $B = 30 + 26/60 + 25.0/3600 = 30.440\ 277°$ etc.

4.4.4 Traverse tabulation (See p. 99.)

4.4.5 To find the bearing and distance between two points given their coordinates

Let the coordinates of A and B be $E_A\,N_A$ and $E_B\,N_B$ respectively (Fig. 4.18).

As $\Delta E_{AB} = S_{AB} \sin \phi_{AB} = E_B - E_A$

and $\Delta N_{AB} = S_{AB} \cos \phi_{AB} = N_B - N_A$

then $\phi_{AB} = \tan^{-1} (\Delta E/\Delta N)_{AB}$ (4.7)

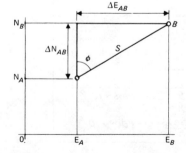

Fig. 4.18 To find the length and bearing between two points

and $\qquad S_{AB} = \sqrt{(\Delta E^2 + \Delta N^2)}$ $\qquad\qquad$ (4.8)

$\qquad\qquad = \Delta E_{AB}/\sin \phi_{AB}$ $\qquad\qquad$ (4.9)

$\qquad\qquad = \Delta N_{AB}/\cos \phi_{AB}$ $\qquad\qquad$ (4.10)

As all scientific calculators have square root keys equation 4.8 is preferred.

NB: As the quotient $Q = \Delta E/\Delta N$ may be

(1) positive, i.e. $+/+$ or $-/-$
(2) negative, i.e. $+/-$ or $-/+$
(3) zero when $\Delta E = 0$
(4) infinity when $\Delta N = 0$

it is necessary to inspect the sign carefully.

Range	Q	ΔE	ΔN	ϕ'	ϕ (in computer terms)
0°–90°	+	+ (E)	+ (N)	+	$\phi = \phi'$
90°–180°	−	+ (E)	− (S)	−	$\phi = \phi' + 180°$
180°–270°	+	− (W)	− (S)	+	$\phi = \phi' + 180°$
270°–360°	−	− (W)	+ (N)	−	$\phi = \phi' + 360°$
		100.00	173.21	30.000°	030.000°
		173.21	− 100.00	− 60.000°	120.000°
		− 100.00	− 100.00	45.000°	225.000°
		− 145.42	301.63	− 25.739°	334.261°

4.4.6 Machine fitted with a rectangular–polar key R → P

(1) *HP*
Enter ΔE (x) : Enter ΔN (y)
Press [→ P] key $\qquad$ Display : $S(r)$
Press [x ↔ y] key $\qquad$ Display : $\phi(\theta)$

(2) *Commodore*
Enter ΔE : Press [x → y] key : Enter ΔN
Press [→ P] key $\qquad$ Display : S
Press [x → y] key $\qquad$ Display : ϕ

(3) *Casio*
Enter ΔE : Press [INV][R → P] keys : Enter ΔN
Press [=] key $\qquad$ Display : S
Press [x → y] key $\qquad$ Display : ϕ

NB: In each case the machines display ϕ as +ve $\qquad$ 0 − 180
$\qquad\qquad\qquad\qquad\qquad\qquad$ and −ve $\qquad$ 0 − 180

When ϕ is −ve then 360 must be added to give the whole circle bearing.

For example:

Given $\Delta E = 173.21$; $\Delta N = -100.00$ then ϕ is displayed as 120.000°
Given $\Delta E = -100.00$; $\Delta N = -100.00$ then ϕ' is displayed as − 135.000°
$\qquad$ and $\phi = \phi' + 360$ i.e. − 135.000 + 360.000 = 225.000°
Given $\Delta E = -145.42$; $\Delta N = 301.63$ then ϕ' is displayed as − 25.739°
$\qquad$ and $\phi = -25.739 + 360.000 = 334.261°$

Example 4.6

	E(m)	N(m)
A	632.165	949.882
B	925.485	421.742
ΔE_{AB} 293.320		ΔN_{AB} − 528.140

Using the arc tan key

$$\phi_{AB} = -29.047\,05° = 150.952\,95°$$

By Pythagoras

$$S_{AB} = \sqrt{(293.320^2 + 528.140^2)} = 604.1262$$

Alternatively

$$S_{AB} = \Delta E/\sin \phi \qquad\qquad = 604.1262$$
$$= \Delta N/\cos \phi \qquad\qquad = 604.1262$$

NB: To obtain this accuracy ϕ must be stored as computed.

Using the R → P keys

$$\phi_{AB} = 150.9530° \quad \text{i.e. } 150° \, 57' \, 10''$$
$$S_{AB} = 604.1262 \quad \text{i.e. } 604.126 \, \text{m}$$

4.4.7 To estimate the required accuracy of the output

Given (1) $\Delta E = S \sin \phi$.

If the length S is subject to a standard error δS and the bearing to an error $\delta\phi$, then the error in the partial coordinate is estimated by partially differentiating the equation successively for each of the variables:

$$\delta(\Delta E)_S = \sin \phi \, \delta S = \Delta E (\delta S/S)$$

and $\quad \delta(\Delta E)_\phi = S \cos \phi \, \delta\phi = \Delta E(\cot \phi \, \delta\phi)$

Similarly, if $\Delta N = S \cos \phi$,

$$\delta(\Delta N)_S = \cos \phi \, \delta S = \Delta N(\delta S/S)$$

and $\quad \delta(\Delta N)_\phi = -S \sin \phi \, \delta\phi = -\Delta N(\tan \phi \, \delta\phi)$

If $S = 300.00 \, \text{m} \pm 0.01 \, \text{m}$ and $\phi = 045° \, 00' \, 00'' \pm 30''$ then $\Delta E = \Delta N = 212.13$.

$$\delta(\Delta E)_S = 212.13 \times 0.01/300.00 \quad = \pm 0.007$$
$$\delta(\Delta E)_\phi = 212.13 \times 30/206\,265 \quad = \pm 0.031$$
$$\delta(\Delta E) = \sqrt{(0.007^2 + 0.031^2)} \quad = \pm 0.032$$

The values for ΔN will be the same and thus $\Delta E = \Delta N = 212.13 \pm 0.03 \, \text{m}$. Thus the output may be quoted to the same accuracy as the measured length but the effect of $\pm 30''$ produces $4\frac{1}{2}$ times the effect from the error in the length. For the measurements to be compatible the bearing should be quoted to $\pm 7''$.

NB: At $300.000 \, \text{m}$, $8''$ subtends $300 \times 7/206\,265 = 0.010 \, \text{m}$, which is then similar to the linear error.

Given (2) $\tan \phi = \Delta E / \Delta N$.

Then

$$\sec^2 \phi \, \delta\phi_E = \delta(\Delta E)/\Delta N$$

$$\delta\phi_E = \cos^2 \phi \, \delta(\Delta E)/\Delta N = \sin \phi \cos \phi \, \delta(\Delta E)/\Delta E$$

$$= \sin 2\phi \, \delta(\Delta E)/2\Delta E$$

and $\quad \delta\phi_N = -\cos^2 \phi \Delta E \, \delta(\Delta N)/\Delta N^2 = -\sin 2\phi \, \delta(\Delta N)/2\Delta N$

$$\delta\phi = \frac{\sin 2\phi}{2} \sqrt{\{[\delta(\Delta E)/\Delta E]^2 + [\delta(\Delta N)/\Delta N]^2\}}$$

Given $\Delta E = \Delta N = 212.13 \pm 0.01$ then $\phi = 45.000$.

$$\delta\phi'' = \frac{206\,265}{2} \sqrt{[2(0.01/212.13)^2]} = \pm 7''$$

Given (3) $S^2 = \Delta E^2 + \Delta N^2$.

$$2S \, \delta S_E = 2\Delta E \, \delta(\Delta E)$$

$$\delta S_E = \Delta E \, \delta(\Delta E)/S$$

and $\quad \delta S_N = \Delta N \, \delta(\Delta N)/S$

$$\delta S = (1/S) \sqrt{\{[\Delta E \, \delta(\Delta E)]^2 + [\Delta N \, \delta(\Delta N)]^2\}}$$

If $S = 300.00$ when $\phi = 45.000$ then

$$\delta S = 0.0033 \sqrt{(2 \times 212.13 \times 0.01)} = \pm 0.01 \text{ m}$$

Thus in general terms,

$$\delta\phi'' \simeq 206\,265 \, \delta S/S \tag{4.11}$$

and $\quad \delta S \simeq \delta E \simeq \Delta N$

For example, if $\Delta E = \Delta N = 100.00 \pm 0.01$ then $S = 141.42\,\text{m}$; $\phi = 45.000$.

By equation 4.11,

$$\delta\phi'' = 206\,265 \times 0.01/141.42 = 14.6'' \text{ say } 15''$$

and $\quad \delta S = 0.01\,\text{m}$

Then $\quad S = 141.42 \pm 0.01$

$$\phi = 45° \, 00' \, 00'' \pm 15''$$

Example 4.7 A disused colliery shaft C, situated in a flooded area, is surrounded by a circular wall and observations are taken from two points A and B of which the coordinates, in metres relative to a local origin, are as follows:

Station	Eastings	Northings
A	1099.75	278.92
B	291.88	551.32

C is approximately NW of A.

Angles measured at A to the tangential points 1 and 2 of the wall are $BAC_1 = 25° \, 55'$ and $BAC_2 = 26° \, 35'$.

Angles measured at B to the tangential points 3 and 4 of the wall are $C_3BA = 40° \, 29'$ and $C_4BA = 39° \, 31'$.

Determine the coordinates of the centre of the shaft and calculate the diameter of the circle formed by the outside of the wall. (TP)

	E	N
A	1099.75	278.92
B	291.88	551.32
ΔE_{AB}	-807.87	ΔN_{AB} 272.40

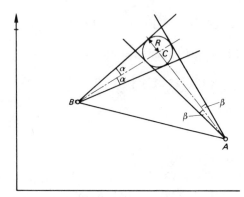

Fig. 4.19

In Fig. 4.19,

$$\phi_{AB} = 288.6332°$$
$$S_{AB} = 852.5583$$

In triangle ABC,

$$\text{angle } A = \tfrac{1}{2}\{25° \, 55' + 26° \, 35'\} \qquad = \quad 26° \, 15'$$
$$\text{angle } B = \tfrac{1}{2}\{40° \, 29' + 39° \, 31'\} \qquad = \quad 40° \, 00'$$
$$\text{angle } C = 180° - (26° \, 15' + 40° \, 00') = 113° \, 45'$$
$$\overline{\phantom{\text{angle } C = 180° - (26° \, 15' + 40° \, 00')} \, 180° \, 00'}$$

By the sine rule,

$$BC = AB \sin A \operatorname{cosec} C$$
$$= 852.56 \sin 26° \, 15' \operatorname{cosec} 113° \, 45' = 411.9665$$
$$AC = BC \sin B \operatorname{cosec} A$$
$$= 411.9665 \sin 40.000/\sin 26.250 = 598.7197$$

$\phi_{AB} = 288.6332$	$\phi_{BA} = 108.6332$
$B\hat{A}C = \underline{26.2500}$	$C\hat{B}A = \underline{-40.0000}$
$\phi_{AC} = 314.8832$	$\phi_{BC} = 068.6332$

To find the coordinates of C

Line BC

$\phi = 068.6332$	$S = 411.9665$
$E_B = 291.88$	$N_B = 551.32$
$\Delta E_{BC} = \underline{383.65}$	$\Delta N_{BC} = \underline{150.10}$
$E_C = 675.53$	$N_C = 701.42$

Line AC

$$\phi = \quad 314.8832 \qquad S = 598.7197$$
$$E_A = \quad 1099.75 \qquad N_A = 278.92$$
$$\Delta E_{AC} = -424.22 \qquad \Delta N_{AC} = 422.50$$
$$E_C = \quad 675.53 \qquad N_C = \overline{701.42}$$

To find the diameter of the wall

$$D = 2R = 2 \cdot BC \tan \alpha = 2 \cdot AC \tan \beta$$
$$= 2 \times 411.966 \tan\tfrac{1}{2}(40° \, 29' - 39° \, 31') = 6.95$$
$$= 2 \times 598.719 \tan\tfrac{1}{2}(26° \, 35' - 25° \, 55') = 6.97$$
$$\text{say} \quad 6.96 \, \text{m}$$

4.5 Computer programming

Computer aids are now so numerous as to be beyond the scope of this book but it may be helpful to consider the principles that might be applied to microcomputers working in simple BASIC.

1) Computers carry out the programs exactly as the 'software' is written, i.e. 'garbage' in – 'garbage' out. Any faults in the output will be either:
(a) operator error   (this is the most likely)
(b) program error
(c) machine error   (this is the least likely)
2) Every program should be tested against a hand solution and must be capable of receiving zeros as input for each variable. (NB: The machine will often not accept zero as a denominator.)
3) Once the program is working efficiently, the input may be punched in by anyone.
4) All input should be capable of being edited.
5) Subroutines should be used where reiteration is involved.
6) Whilst stations may be named it is more flexible if numbers are attributed as arrays which are easier to handle. For example, if stations are given numbers $I = 1$ to N then the coordinates become $E(I)$, $N(I)$; the lengths are then $S(I, J)$ and the bearings $Q(I, J)$ where $J = I + 1$.
7) Flow diagrams help the program writer to define the unambiguous operations (algorithms).

An *algorithm* is a precise set of well-defined rules or procedures for the unambiguous solution of a problem in a finite number of steps.

A *flowchart* is a visual outline of an algorithm in which steps and processes are represented as symbols. It has the following functions:

(a)   to give a diagrammatic solution to the problem;
(b)   to show the logical paths to be taken;
(c)   to show the input and output functions;
(d)   to show all entry points, halts and other actions.

4.5.1   Flowcharting symbols

It is important wherever possible to use standard symbols.

1)   **Start/End**

Initial statements e.g. Degree mode, Title
The termination of the program is important when recording on tape or disk.

2)   **Input/output**
Manual input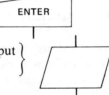

Other types of input ⎱
All output   ⎰

3)   **Labels**

This symbol indicates the segmentation of the program and provides an address to which a loop may return. It is more important in hand held machines.

4)   **Arithmetic or algebraic processes**

5)   **Decisions** (IF statements)

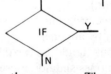

These represent a junction in the program. They represent yes or no answers.

6)   **Flowlines/connectors**

Flowlines are required to join symbols together and where overflow occurs the connector is used.

7)   **Loops**   ----►----

A sequence of instructions obeyed repeatedly is termed a 'loop'. It may be

   unconditional, the program branching back to a label, or conditional, the program branching after an IF statement.

When one loop is contained within another loop it is said to be 'nested'. There is frequently some restriction on the number of loops that can be nested.

4.5.2 Example program
For program see page 108.

4.6   The equation of a straight line

(a)   Given 2 points on the line with Cartesian coordinates (x_1, y_1), (x_2, y_2), then

$$(y - y_1)/(y_2 - y_1) = (x - x_1)/(x_2 - x_1) \qquad (4.12)$$

(b)   Given 1 point (x_1, y_1) and the bearing of the line ϕ, then

$$x - x_1 = (y - y_1)\tan\phi \qquad (4.13)$$

If the line passes through the origin, then

$$x = y\tan\phi \qquad (4.14)$$

Equations 4.12, 4.13 and 4.14 can be reduced to give a general equation (see Fig. 4.20):

$$ax + by + c = 0 \qquad (4.15)$$

or $\qquad\qquad y = mx + k \qquad\qquad (4.16)$

where $\qquad\quad m = -a/b = \cot\phi \qquad (4.17)$

and $\qquad\qquad k = -c/b \qquad\qquad (4.18)$

It can be seen from the above that:

$$\tan\phi = (x - x_1)/(y - y_1) = a/b = \Delta x/\Delta y \qquad (4.19)$$

To find the offset of a point $P(x_p, y_p)$ from a given line having the equation $ax + by + c = 0$, the coordinates of P are inserted into the equation

$$d = K(ax_p + by_p + c) \qquad (4.20)$$

where $K = 1/\sqrt{(a^2 + b^2)} \qquad (4.21)$

(K is of the same sign as b.)

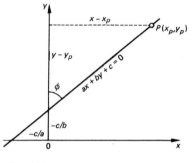

Fig. 4.20

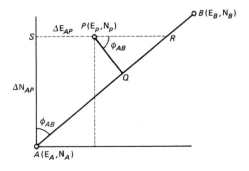

Fig. 4.21

Program

```
100 DIM E(10), N(10), S(10,10), Q(10,10)
110 PRINT "POLY-JOIN"


120 INPUT "NUMBER OF STATIONS "; N


130 FOR I = 1 TO N
140 PRINT "STATION NO. "; I
150 INPUT "COORDINATES: E, N "; E(I), N(I)
160 NEXT I


170 FOR I = 1 TO N-1: FOR J = I+1 TO N
180 D1 = E(J) - E(I): D2 = N(J) - N(I)
190 S(I,J) = SQR(D1↑2 + D2↑2)


200 IF D2 = 0 GOTO 260


210 X = D1/D2: Q = ATAN(X)*180/π


220 IF D2 > 0 GOTO 240


230 Q = Q + 180: GOTO 280


240 IF X > 0 GOTO 280


250 Q = Q + 360: GOTO 280
260 Q = 90
270 IF D1 < 0 THEN Q = 270


280 Q(I,J) = Q


290 PRINT "LINE NO. "; I; " TO "; J
300 PRINT "LENGTH = "; S(I,J), "BEARING = "; Q(I,J)
310 NEXT J: NEXT I
320 END
```

Flowchart

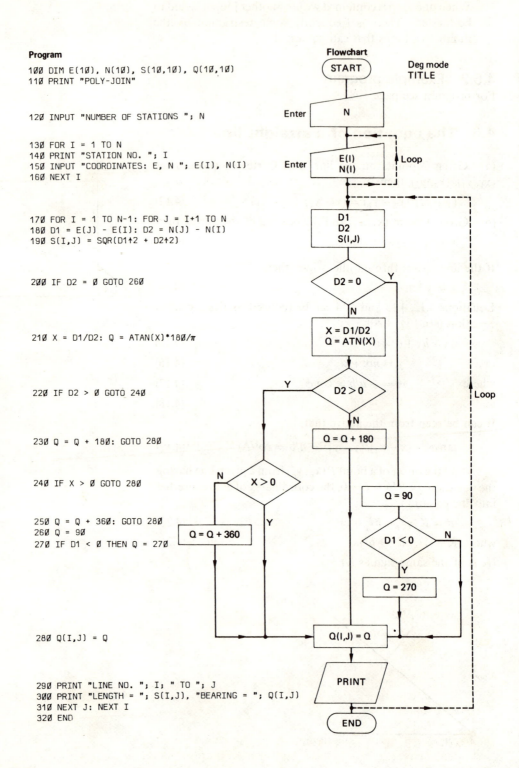

Alternatively, in terms of the rectangular coordinates (Fig. 4.21),

$$SR = \Delta N_{AP} \tan \phi_{AB}$$

$$PR = SR - SP = \Delta N_{AP} \tan \phi_{AB} - \Delta E_{AP}$$

$$PQ = d = PR \cos \phi_{AB}$$

$$= \Delta N_{AP} \sin \phi_{AB} - \Delta E_{AP} \cos \phi_{AB} \qquad (4.22)$$

If the coordinates of A are assumed zero, then

$$d = N_p \sin \phi_{AB} - E_p \cos \phi_{AB} \qquad (4.23)$$

Example 4.8 Given the following coordinates calculate the length of the offset from the point P to the line AB.

	E	N
A	1966.853	1251.183
B	1741.813	1101.889
P	2073.408	809.207

The equation of the line AB is:

$$\frac{(y - 1251.183)}{(1101.889 - 1251.183)} = \frac{(x - 1966.853)}{(1741.813 - 1966.853)}$$

i.e. $149.294\,x - 225.040\,y - 12073.130 = 0$

As b is $-$ve then

$$K = \sqrt{(a^2 + b^2)} = -270.059$$

By equation 4.20,

$$d = -270.059(149.294\,x - 225.040\,y - 12073.130)$$

$$= -0.552\,820\,x + 0.833\,299\,y + 44.706$$

Substituting the values for P gives

$$d = -427.204$$

Alternatively

Let

$$E'_B = N'_B = 0$$

Then $E'_P = 331.595$ and $N'_P = -292.682$

$$\phi_{BA} = 56.4393°$$

By equation 4.23,

$$d = N'_P \sin \phi_{BA} - E'_P \cos \phi_{BA}$$

$$= -427.204$$

Exercises 4.3

1   It is required to re-align a major transport system. Given the direction of the re-alignment between two control stations X and Y as ϕ_{XY} prove that if the coordinates of X are

zero then the offset from a point $P(E_P N_P)$ is given as $d = N_P \sin \phi_{XY} - E_P \cos \phi_{XY}$.

2 The following coordinates refer to the position of theodolite traverse stations along an underground roadway.

	E (m)	N (m)
A	457 467.45	347 040.57
B	457 513.52	347 090.79
C	457 628.00	347 197.50
D	457 775.86	347 331.62
X	457 206.00	346 797.00
Y	457 950.20	347 498.30

Calculate the offsets from A, B, C and D to the line XY.

(TP Ans.   2.04, 2.91, 2.06, 1.74)

Exercises 4.4   (General)

1 If the bearing of the line LM is $246° 15' 20''$ and the observed angle LMN is $48° 26' 30''$ what is the bearing MN?

(Ans. $114° 41' 50''$)

2 If the bearing of the line DE is $076° 41' 30''$ and the bearing of the line FD is $144° 18' 40''$, calculate the angle EDF.

(Ans. $247° 37' 10''$)

3 In a triangle ABC the bearings of the sides are as follows:

$CA = 43° 12' 18''$; $AB = 124° 27' 11''$; $BC = 296° 48' 53''$

Calculate the internal angles of the triangle.

(Ans. $73° 36' 35''$, $98° 45' 07''$, $7° 38' 18''$)

4 Given $\Delta E_{AB} = -46.73$ m and $\Delta N_{AB} = 68.25$ m, calculate the length and bearing of the line AB.

(Ans.   82.71 m; $325° 36' 00''$ to nearest $10''$)

5 Calculate the length and bearing of the line AB from the following coordinates:

		E	N
(a)	A	100.7	29.5
	B	25.8	37.9
(b)	A	100.74	129.53
	B	125.79	37.88
(c)	A	100.743	29.529
	B	25.787	137.884

(Ans.   (a) 75.4; $276° 24'$ (b) 95.01; $164° 42' 50''$
(c) 131.754; $325° 19' 33''$)

6 The coordinates of P are 428.34 m E, 617.48 m N and of M are 314.52 m E, 748.32 m N. Calculate the WCB of PM and MP and the length.

(Ans. $\phi_{MP} = 138° 58' 45''$; $\phi_{PM} = 318° 58' 45''$; 173.42 m)

7 The coordinates of T are 451.31 m E; 321.64 m N and $\Delta E_{TZ} = 47.38$ m and $\Delta N_{TZ} = -62.15$ m. Calculate:

(a) the bearing of the line TZ;
(b) the length of the line TZ;
(c) the coordinates of the point Z.

(Ans.   (a) 142° 40′ 50″; (b) 78.15 m; (c) 498.69 m E 259.49 m N)

8 In a triangle ABC the coordinates are:

	E (m)	N (m)
A	163.147	349.236
B	196.833	219.477
C	104.236	189.761

Calculate the coordinates of the centre of gravity of the triangle.

(Ans.   154.739 m E, 252.825 m N)

9   The following are the coordinates in metres of three stations A, B and C:

	E	N
A	510.63	724.37
B	834.45	977.51
C	1352.27	516.18

Calculate the coordinates of the point X such that BX is the shortest distance from B to AC.   (Ans. 756.772, 663.484)

10   Given

	E (m)	N (m)
A	1263.564	1165.475
B	1186.493	943.421
C	1492.379	1165.475
D	1554.687	822.118

Calculate:

(a) the length of all the lines joining the points;
(b) the bearing of all the lines;
(c) the clockwise angles in the figure $ACDB$.

(Ans.   AC. 228.815; 90.0000, (1) 49° 42′ 23″
AD. 450.163; 139.7063, (8) 59° 26′ 05″
AB. 235.049; 199.1410
BA.   19.1410 (7) 34° 52′ 54″
BC. 377.987; 54.0227 (6) 54° 12′ 43″
BD. 387.661; 108.2347
CD. 348.965; 169.7146 (3) 64° 18′ 29″
CB.   234.0227 (2) 35° 58′ 38″
CA.   270.0000
DB.   288.2347 (5) 31° 28′ 18″
DA.   319.7063 (4) 30° 00′ 30″
DC.   349.7146)

4.7 To find the coordinates of the intersection of two lines

4.7.1 Given their bearings from two known coordinate stations

Given: E_A, N_A; E_B, N_B; ϕ_{AC} and ϕ_{BC}.

(a) *By solving the triangle ABC* (Fig. 4.22)

ϕ_{AB} and S_{AB} are computed from the partial coordinates. Then

$$\text{angle } A = \phi_{AB} - \phi_{AC}$$
$$\text{angle } B = \phi_{BC} - \phi_{BA}$$
$$S_{AC} = b = AB . \sin B / \sin (A + B)$$
$$S_{BC} = a = AB . \sin A / \sin (A + B)$$

then
$$E_C = E_A + b \sin \phi_{AC} = E_B + a \sin \phi_{BC}$$
$$N_C = N_A + b \cos \phi_{AC} = N_B + a \cos \phi_{BC}$$

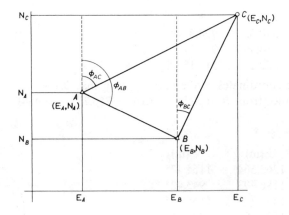

Fig. 4.22

(b) *Direct computation of the coordinates 'Cut by bearings'*

From Fig. 4.22,

$$\tan \phi_{AC} = (E_C - E_A)/(N_C - N_A)$$
$$\tan \phi_{BC} = (E_C - E_B)/(N_C - N_B)$$

Then
$$E_C - E_A = N_C \tan \phi_{AC} - N_A \tan \phi_{AC}$$
$$E_C - E_B = N_C \tan \phi_{BC} - N_B \tan \phi_{BC}$$

therefore
$$\Delta E_{AB} = N_C (\tan \phi_{AC} - \tan \phi_{BC}) - N_A \tan \phi_{AC} + N_B \tan \phi_{BC}$$

Then
$$N_C = \frac{N_A \tan \phi_{AC} - N_B \tan \phi_{BC} + \Delta E_{AB}}{\tan \phi_{AC} - \tan \phi_{BC}} \qquad (4.24)$$

Similarly,
$$N_C - N_A = E_C \cot \phi_{AC} - E_A \cot \phi_{AC}$$
$$N_C - N_B = E_C \cot \phi_{BC} - E_B \cot \phi_{BC}$$

therefore

$$\Delta N_{AB} = E_C (\cot \phi_{AC} - \cot \phi_{BC}) - E_A \cot \phi_{AC} + E_B \cot \phi_{BC}$$

then

$$E_C = \frac{E_A \cot \phi_{AC} - E_B \cot \phi_{BC} + \Delta N_{AB}}{\cot \phi_{AC} - \cot \phi_{BC}} \qquad (4.25)$$

The partial coordinate equations are then derived as:

$$\Delta N_{AC} = N_C - N_A = \frac{N_A \tan \phi_{AC} - N_B \tan \phi_{BC} + \Delta E_{AB} - N_A \tan \phi_{AC} + N_A \tan \phi_{BC}}{\tan \phi_{AC} - \tan \phi_{BC}}$$

$$= \frac{\Delta E_{AB} - \Delta N_{AB} \tan \phi_{BC}}{\tan \phi_{AC} - \tan \phi_{BC}} \qquad (4.26)$$

$$\Delta E_{AC} = \Delta N_{AC} \tan \phi_{AC} \qquad (4.27)$$

Similarly,

$$\Delta N_{BC} = \frac{N_A \tan \phi_{AC} - N_B \tan \phi_{BC} + \Delta E_{AB} - N_B \tan \phi_{AC} + N_B \tan \phi_{BC}}{\tan \phi_{AC} - \tan \phi_{BC}}$$

$$= \frac{\Delta E_{AB} - \Delta N_{AB} \tan \phi_{AC}}{\tan \phi_{AC} - \tan \phi_{BC}} \qquad (4.28)$$

$$\Delta E_{BC} = \Delta N_{BC} \tan \phi_{BC} \qquad (4.29)$$

Example 4.9   Let the coordinates be:

	E	N
A	263 479.263	487 288.142
B	264 827.381	488 199.426

and the bearings ϕ_{AC} 132° 07′ 30″ and ϕ_{BC} 165° 22′ 40″.

To help in the computation, particularly with hand calculators, the coordinates may be reduced by omitting the first two common digits.

	E		N	
A	(26) 3479.263		(48) 7288.142	
B	(26) 4827.381		(48) 8199.426	
ΔE_{AB}	1348.118	ΔN_{AB}	911.284	

	tan	cot
$\phi_{AC} = 132.1250$	−1.105 752	−0.904 362
$\phi_{BC} = 165.3778$	−0.260 895	−3.832 964
	−0.844 857	2.928 602

$$N_A \tan \phi_{AC} = -8058.877; \qquad E_A \cot \phi_{AC} = -3146.513$$
$$N_B \tan \phi_{BC} = -2139.189; \qquad E_B \cot \phi_{BC} = -18503.178$$

By equation 4.25,

$$E_C = \frac{E_A \cot \phi_{AC} - E_B \cot \phi_{BC} + \Delta N_{AB}}{\cot \phi_{AC} - \cot \phi_{BC}}$$

$$= 5554.851, \quad \text{i.e.} \quad 265 554.851$$

By equation 4.24,

$$N_C = \frac{N_A \tan \phi_{AC} - N_B \tan \phi_{BC} + \Delta E_{AB}}{\tan \phi_{AC} - \tan \phi_{BC}}$$

$$= 5411.058, \quad \text{i.e.} \quad 485\,411.058$$

Alternatively, by equation 4.26,

$$\Delta N_{AC} = \frac{\Delta E_{AB} - \Delta N_{AB} \tan \phi_{BC}}{\tan \phi_{AC} - \tan \phi_{BC}}$$

$$= -1877.083$$

$$N_C = 487\,288.142 - 1877.083 = 485\,411.059$$

By equation 4.27,

$$\Delta E_{AC} = \Delta N_{AC} \tan \phi_{AC}$$

$$= 2075.588$$

$$E_C = 263\,479.263 + 2075.588 = 265\,554.851$$

For manual computation the latter is preferred, and this may be laid out in tabulated form.

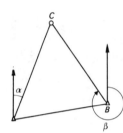

Example 4.10

(1)　$\Delta N_{AC} = \dfrac{\Delta E_{AB} - \Delta N_{AB} \tan \phi_{BC}}{\tan \phi_{AC} - \tan \phi_{BC}}$

(2)　$\Delta E_{AC} = \Delta N_{AC} \tan \phi_{AC}$

(3)　$\Delta N_{BC} = \dfrac{\Delta E_{AB} - \Delta N_{AB} \tan \phi_{AC}}{\tan \phi_{AC} - \tan \phi_{BC}}$

Fig. 4.23　(4)　$\Delta E_{BC} = \Delta N_{BC} \tan \phi_{BC}$

Stations		E	Bearing	N	
E Base	(A)	+13 486.85 m	ϕ_{AC} 278° 13′ 57″	+10 327.36 m	
Igloo	(B)	+12 759.21 m	ϕ_{BC} 182° 27′ 44″	+13 142.72 m	
ΔE_{AB}		−727.64	tan ϕ_{AC} −6.911 745 2 tan ϕ_{BC} +0.043 000 4 tan ϕ_{AC} − tan ϕ_{BC}	+2815.36	ΔN_{AB}
ΔN_{AB} tan ϕ_{BC}		+121.06			
		−848.70	÷ −6.954 7456 ┐		
ΔE_{AC}		−843.44	← × tan ϕ_{AC} ───→	+122.03	ΔN_{AC}
W Base	E_C	12 643.41 m		10 449.39 m	N_C ←
			Check		
ΔE_{AB}		−727.64			
ΔN_{AB} tan ϕ_{AC}		−19 459.05			
		+18 731.41	tan ϕ_{AC} − tan ϕ_{BC} ÷ −6.954 7456 ┐		
ΔE_{BC}		115.81	← × tan ϕ_{BC} ───→	2 693.33	ΔN_{BC}
W Base	E_C	+12 643.40 m		10 449.39 m	N_C ←

4.7.2   Given the angles measured at the two known coordinate stations

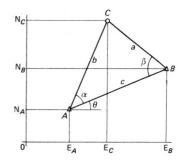

Given: E_A, N_A; E_B, N_B; angle $CAB = \alpha$; angle $ABC = \beta$
(Fig. 4.24).

(a)   *By solving the triangle ABC*

Using partial coordinates, ϕ_{AB} and S_{AB} are computed.

$$\phi_{AC} = \phi_{AB} - \alpha$$

$$\phi_{BC} = \phi_{BA} + \beta$$

$$S_{AC} = b = AB \sin\beta / \sin(\alpha + \beta)$$

$$S_{BC} = a = AB \sin\alpha / \sin(\alpha + \beta)$$

$$E_C = E_A + b\sin\phi_{AC} = E_B + a\sin\phi_{BC}$$

$$N_C = N_A + b\cos\phi_{AC} = N_B + a\cos\phi_{BC}$$

Fig. 4.24

(b)   *By direct computation of the coordinates 'Cut by angles'*

NB: C is assumed to lie to the left of AB.

$$\Delta E_{AC} = E_C - E_A = b\cos(A + \theta)$$

$$= b(\cos A \cos\theta - \sin A \sin\theta)$$

But $b = c\sin B / \sin C$

and $\sin C = \sin[180 - (A + B)] = \sin(A + B)$

therefore

$$\Delta E_{AC} = \frac{c\cos\theta\cos A\sin B - c\sin\theta\sin A\sin B}{\sin A\cos B + \cos A\sin B}$$

$$= \frac{\Delta E_{AB}\cot A - \Delta N_{AB}}{\cot A + \cot B} \tag{4.30}$$

then

$$E_C = \frac{E_B\cot A - E_A\cot A - \Delta N_{AB} + E_A\cot A + E_A\cot B}{\cot A + \cot B}$$

$$= \frac{E_A\cot B + E_B\cot A - \Delta N_{AB}}{\cot A + \cot B} \tag{4.31}$$

Similarly,

$$\Delta N_{AC} = N_C - N_A = b\sin(A + \theta)$$

$$= \frac{c\cos\theta\sin A\sin B + c\sin\theta\cos A\sin B}{\sin A\cos B + \cos A\sin B}$$

$$\Delta N_{AC} = \frac{\Delta N_{AB}\cot A + \Delta E_{AB}}{\cot A + \cot B} \tag{4.32}$$

then

$$N_C = \frac{N_B\cot A - N_A\cot A + N_A\cot A + N_A\cot B + \Delta E_{AB}}{\cot A + \cot B}$$

$$= \frac{N_A\cot B + N_B\cot A + \Delta E_{AB}}{\cot A + \cot B} \tag{4.33}$$

By adding equations 4.31 and 4.33 a useful check may be applied:

$$E_C + N_C = \frac{(E_A\cot B + E_B\cot A - N_B + N_A) + (N_A\cot B + N_A\cot A + E_B - E_A)}{\cot A + \cot B}$$

therefore

$$(E_C + N_C)(\cot A + \cot B) = E_A(\cot B - 1) + E_B(\cot A + 1) + N_A(\cot B + 1) + N_B(\cot A - 1)$$

$$(4.34)$$

Using the values of Example 4.10,

$$\phi_{AB} = 345.5088$$
$$\phi_{AC} = 278.2325 \quad \text{therefore } A = \quad 67.2763$$
$$\phi_{BC} = 182.4622$$
$$\phi_{BA} = 165.5088 \quad \text{therefore } B = \quad 16.9534$$
$$\phi_{CB} = 002.4622$$
$$\phi_{CA} = 098.2325 \quad \text{therefore } C = \quad 95.7703$$
$$\Sigma = 180.0000$$

Check

$$\cot A = 0.418\,795 \qquad E_A\cot B = 44\,242.168$$
$$\cot B = 3.280\,393 \qquad N_A\cot B = 33\,877.799$$
$$\Sigma = 3.699\,188 \qquad E_B\cot A = 5343.493$$
$$\qquad\qquad\qquad\qquad N_B\cot A = 5504.105$$

By equation 4.31,

$$E_C = \frac{E_A\cot B + E_B\cot A - \Delta N_{AB}}{\cot A + \cot B} = 12\,643.40$$

By equation 4.33,

$$N_C = \frac{N_A\cot B + N_B\cot A + \Delta E_{AB}}{\cot A + \cot B} = 10\,449.39$$

or by equation 4.30,

$$\Delta E_{AC} = \frac{\Delta E_{AB}\cot A - \Delta N_{AB}}{\cot A + \cot B} \qquad = -843.45$$

$$E_C = 13\,486.85 - 843.45 \qquad = 12\,643.40$$

By equation 4.32,

$$\Delta N_{AC} = \frac{\Delta N_{AB}\cot A + \Delta E_{AB}}{\cot A + \cot B} \qquad = \quad 122.03$$

$$N_C = 10\,327.36 + 122.03 \qquad = 10\,449.39$$

Check (equation 4.34):

$$E_A(\cot B - 1) = 13\,486.85 \times 2.280\,393 \quad = \quad 30\,755.318$$
$$E_B(\cot A + 1) = 12\,759.21 \times 1.418\,795 \quad = \quad 18\,102.703$$
$$N_A(\cot B + 1) = 10\,327.36 \times 4.280\,393 \quad = \quad 44\,205.159$$
$$N_B(\cot A - 1) = 13\,142.72 \times -0.581\,205 = \quad -7638.615$$
$$= \quad 85\,424.565$$

$$(E_C + N_C)(\cot A + \cot B) = (12\,643.40 + 10\,449.39)(3.699\,188)$$
$$= 85\,424.572$$

Example 4.11   Given the coordinates of four stations (Fig. 4.25),

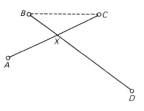

	E	N
A	1250.00	1100.00
B	1320.70	1170.70
C	1520.70	1170.70
D	1652.45	736.88

Fig. 4.25

find the coordinates of the intersection of the lines AC and BD.

Method 1

$$\phi_{BD} = 142.5942$$
$$\phi_{CA} = 255.3628$$

In triangle BCX,

$$BC = 520.70 - 320.70 \quad = 200.00 \text{ (due E)}$$
$$\phi_{BC} = 090.0000 \qquad \phi_{CB} = 270.0000$$
$$\underline{\phi_{BX} = 142.5942 \qquad \phi_{CX} = 255.3628}$$
$$CBX = \quad 52.5942 \qquad XCB = \quad 14.6372$$
$$BX = 200.00 \sin 14.6372/\sin (52.5942 + 14.6372)$$
$$= 54.811$$

To find the coordinates of X:

Line BX

$$\phi = \quad 142.5942 \qquad S = \quad 54.811$$
$$E_B = 1320.70 \qquad N_B = 1170.70$$
$$\underline{\Delta E_{BX} = \quad 33.30 \qquad \Delta N_{BX} = -43.54}$$
$$E_X = 1354.00 \qquad N_X = 1127.16$$

Method 2 (Cut by bearings)

$$\tan \phi_{BD} = \frac{331.75}{-433.82} = -0.764\,718$$

$$\tan \phi_{CA} = \frac{-270.70}{-70.70} = \underline{3.828\,854}$$

$$\tan \phi_{CA} - \tan \phi_{BD} = \quad 4.593\,572$$

Using equation 4.26,

$$\Delta N_{CX} = \frac{\Delta E_{CB} - \Delta N_{CB} \tan \phi_{BD}}{\tan \phi_{CA} - \tan \phi_{BD}}$$

and $\Delta E_{CX} = \Delta N_{CX} \tan \phi_{CA}$

	E	N
B	1320.70	1170.70
C	1520.70	1170.70
	$\Delta E_{CB} - 200.00$	$\Delta N_{CB} \quad 0.00$

$$\Delta N_{CX} = \frac{-200.00}{4.593\,57} - 0.00 = -43.54 \quad \text{as before}$$

$$\Delta E_{CX} = -43.539 \times 3.828\,854 = -166.70$$

$$E_C = \underline{1520.70}$$
$$E_X = \underline{1354.00}$$

Method 3 (Cut by angles)

As before, $CBX = \beta = 52.5942$

$$XCB = \theta = 14.6372$$

NB: X lies to the left of CB.

By equation 4.31,

$$E_X = \frac{E_C \cot \beta + E_B \cot \theta - \Delta N_{CB}}{\cot \theta + \cot \beta}$$

	E	N
C	1520.70	1170.70
B	1320.70	1170.70
$\Delta E_{BC} =$	200.00 $\Delta N_{BC} = 0.00$	

$$\cot \beta = 0.764\,718$$
$$\cot \theta = \underline{3.828\,866}$$
$$\cot \theta + \cot \beta = 4.593\,584$$

$$E_C \cot \beta = 1162.907; \qquad N_C \cot \beta = 895.255$$
$$E_B \cot \theta = 5056.783; \qquad N_B \cot \theta = 4482.453$$

By equation 4.25,

$$E_X = \frac{E_C \cot \beta + E_B \cot \theta - \Delta N_{CB}}{\cot \theta + \cot \beta} = 1354.00$$

By equation 4.24,

$$N_X = \frac{N_C \cot \beta + N_B \cot \theta + \Delta E_{CB}}{\cot \theta + \cot \beta} = 1127.16$$

Check:

$$E_C(\cot \beta - 1) = -357.793$$
$$E_B(\cot \theta + 1) = 6377.483$$
$$N_C(\cot \beta + 1) = 2065.955$$
$$N_B(\cot \theta - 1) = \underline{3311.753}$$
$$11\,397.398$$

$$(E_X + N_X)(\cot \theta + \cot \beta) = 11\,397.417$$

Method 4

By normal coordinate geometry, using equation 4.12, the equation of line AC is

$$\frac{y - y_1}{x - x_1} = \frac{y_2 - y_1}{x_2 - x_1}$$

i.e. $\dfrac{y-1100.00}{x-1250.00} = \dfrac{1170.70-1100.00}{1520.70-1250.00} = \dfrac{70.70}{270.70} = 0.2612$

therefore $y - 1100.00 = 0.2612(x - 1250.00)$ (1)

Similarly, the equation of line BD is

$$\dfrac{y-1170.70}{x-1320.70} = \dfrac{736.88-1170.70}{1652.45-1320.70}$$

$$= \dfrac{-433.82}{331.75} = -1.3076$$

i.e. $y - 1170.70 = -1.3076(x - 1320.70)$ (2)

Subtracting (1) from (2),

$$70.7 = 1.5688x - (1250.00 \times 0.2612) - (1320.70 \times 1.3076)$$

$$= 1.5688x - 326.50 - 1726.95$$

$$x = \dfrac{2124.15}{1.5688} = 1354.00$$

Substituting in equation (1),

$$y = 0.2612(x - 1250.00) + 1100.00$$

$$= 0.2612(1354.00 - 1250.00) + 1100.00$$

$$= (0.2612 \times 104.00) + 1100.00$$

$$= 1127.16$$

Ans. $X =$ E 1354.00, N 1127.16

NB: All four methods are mathematically sound but the first has two advantages in that no formulae are required beyond the solution of triangles, and that additional information is derived which might be required in setting-out processes.

4.8   Transposition of grid

It is sometimes necessary to transpose a set of rectangular coordinates from one grid system to another. This may involve:

(a) swing (or slew), i.e. a rotation of the axes through an angle α;
(b) shift, a change in the position of the origin of the grid axes;
(c) scale, a change in units, e.g. ft to m.

Swing   Given: a line AB of length S and of bearing ϕ_{AB} and the coordinates of A, i.e. (E_A, N_A), and of B, i.e. (E_B, N_B), on the initial system, the same line AB having a bearing ϕ'_{AB} as part of a final system with coordinates of A (E'_A, N'_A) and B (E'_B, N'_B); see Fig. 4.26.

Let α, the angle of swing, be given by

$$\phi_{AB} - \phi'_{AB} \quad \text{(old bearing} - \text{new bearing)} \quad (4.35)$$

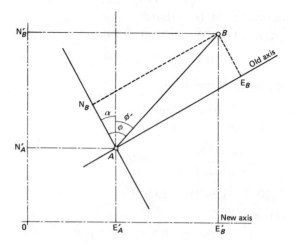

Fig. 4.26 Transposition of grid

Assuming the swing occurs at A, then

$$E'_B = E'_A + S \sin \phi'$$
$$= E'_A + S \sin (\phi - \alpha)$$
$$= E'_A + S \sin \phi \cos \alpha - S \cos \phi \sin \alpha$$
$$= E'_A + \Delta E_{AB} \cos \alpha - \Delta N_{AB} \sin \alpha \qquad (4.36)$$

Similarly,

$$N'_B = N'_A + S \cos \phi'$$
$$= N'_A + \Delta N_{AB} \cos \alpha + \Delta E_{AB} \sin \alpha \qquad (4.37)$$

Shift If the final coordinates of the origin of the temporary system are $O\,(0,0)$ and this is the point of swing on the final system with coordinates (E'_0, N'_0) then the general equations for a point $P(E_P, N_P)$ are given as:

$$E'_P = E'_0 + \Delta E_{OP} \cos \alpha - \Delta N_{OP} \sin \alpha$$
$$= E'_0 + E_P \cos \alpha - N_P \sin \alpha \qquad (4.38)$$

Similarly,

$$N'_P = N'_0 + \Delta N_{OP} \cos \alpha + \Delta E_{OP} \sin \alpha$$
$$= N'_0 + N_P \cos \alpha + E_P \sin \alpha \qquad (4.39)$$

If the lengths on each system are S and S' respectively then

$$S' = kS$$

where $k = S'/S$ is the scale factor. The above equations then become

$$E'_P = E'_0 + k(E_P \cos \alpha - N_P \sin \alpha)$$
$$= E'_0 + mE_P - nN_P \qquad (4.40)$$

where $m = k \cos \alpha$ and $n = k \sin \alpha$.

Similarly,

$$N'_P = N'_0 + mN_P + nE_P \qquad (4.41)$$

NB: It is important that the accuracy of m and n is compatible with that of the initial coordinates; for example, taking the largest value of the initial coordinates as $50\,000.01$, this represents an accuracy of $1/5\,000\,000$ and the number of digits in the coefficients should be at least 7.

To find the coordinates of the point of swing:

$$E'_0 = E'_A - (mE_A - nN_A) = E'_B - (mE_B - nN_B) \qquad (4.42)$$
$$N'_0 = N'_A - (mN_A + nE_A) = N'_B - (mN_B + nE_B) \qquad (4.43)$$

If the coordinates of two points A and B are known on both systems, then the values of m, n, E'_0 and N'_0 can be found and thus the transposition of other points on the temporary system made. Given

$$A \qquad (E_A, N_A) \quad \text{and} \quad (E'_A, N'_A)$$
$$B \qquad (E_B, N_B) \quad \text{and} \quad (E'_B, N'_B)$$

then $E'_A = E'_0 + mE_A - nN_A \qquad N'_A = N'_0 + mN_A + nE_A$

and $E'_B = E'_0 + mE_B - nN_B \qquad N'_B = N'_0 + mN_B + nE_B$

Subtracting,

$$\Delta E'_{AB} = m\Delta E_{AB} - n\Delta N_{AB} \qquad (4.44)$$
and $$\Delta N'_{AB} = m\Delta N_{AB} + n\Delta E_{AB} \qquad (4.45)$$

Solving these equations simultaneously for m and n, e.g. by determinants,

$$m = \frac{\Delta N \Delta N' + \Delta E \Delta E'}{s^2} = k \cos \alpha \qquad (4.46)$$

$$n = \frac{\Delta E \Delta N' - \Delta N \Delta E'}{s^2} = k \sin \alpha \qquad (4.47)$$

where $s^2 = \Delta E^2 + \Delta N^2$

Then

$$\alpha = \tan^{-1}(n/m) \qquad (4.48)$$
$$k = \sqrt{(m^2 + n^2)} \qquad (4.49)$$
$$E'_0 = E'_A - mE_A + nN_A \qquad (4.50)$$
$$N'_0 = N'_A - mN_A - nE_A \qquad (4.51)$$

Alternatively, the values of swing and scale factor may be obtained as follows:

ϕ_{AB}, S_{AB} are computed from the initial partial coordinates.
ϕ'_{AB}, S'_{AB} are computed from the final partial coordinates.

Then

$$\alpha = \phi - \phi'$$
$$k = S'/S$$

Example 4.12 Given the coordinate values of two points A and B on each of the systems:

	E	N	E'	N'
A	11 000.00 ft	10 000.00 ft	437 528.46 m	387 294.16 m
B	14 537.39 ft	4 860.75 ft	438 142.78 m	385 495.24 m
	ΔE 3537.39	ΔN $-5\,139.25$	ΔE' 614.32	ΔN' $-1\,798.92$

Method 1

$$m = \frac{\Delta N \Delta N' + \Delta E \Delta E'}{\Delta E^2 + \Delta N^2} = 0.293\,338\,04 \qquad \text{(Equation 4.46)}$$

$$n = \frac{\Delta E \Delta N' - \Delta N \Delta E'}{\Delta E^2 + \Delta N^2} = -0.082\,372\,15 \quad \text{(Equation 4.47)}$$

$$\alpha = \tan^{-1}(n/m) \qquad = -15.685\,25° \qquad \text{(Equation 4.48)}$$
$$k = \sqrt{(m^2 + n^2)} \qquad = 0.304\,684\,06 \qquad \text{(Equation 4.49)}$$

This scale factor is made up of:

(a) a conversion factor (ft to m);
(b) a local scale factor (ground to grid);
(c) a survey error to produce compatibility.

i.e.

conversion factor = 0.3048

local scale factor $= 0.999\,6013(1 + 38^2 \times 1.228 \times 10^{-8})$

$$\text{(Equation 2.46)}$$
$$= 0.999\,6190$$

$$\text{compatibility factor} = \frac{0.304\,684\,06}{0.3048 \times 0.999\,6190} = 1.000\,0006$$

To find the point of swing:

$$E'_o = E'_A - (mE_A - nN_A) = 433\,478.02 \quad \text{(Equation 4.42)}$$
$$N'_o = N'_A - (mN_A + nE_A) = 385\,266.87 \quad \text{(Equation 4.43)}$$

Method 2

Converting rectangular coordinates to polar coordinates:

$$\phi_{AB} = 145.460\,03 \qquad S_{AB} = 6238.9918$$
$$\phi'_{AB} = 161.145\,28 \qquad S'_{AB} = 1900.9214$$
$$\alpha = -15.685\,25 \qquad k = 0.304\,684\,06$$
$$m = k \cos\alpha = 0\,293\,338\,04$$
$$n = k \sin\alpha = -0.082\,372\,12$$

Tabulation of the computation of the other points:

Pt	E	N	mE	$-n$N	mN	nE	E'	N'
O	0.00	0.00					433 478.02	385 266.87
C	1000.00	1000.00	293.34	82.37	293.34	-82.37	433 853.73	385 477.84
D	5124.26	4182.47	1503.14	344.52	1226.88	-422.10	435 325.68	386 071.65

4.9 The National Grid reference system

Based on the Davidson Committee's recommendations, all British Ordnance Survey Maps are based on the National Grid Reference System with the metre as the unit.

The origin of the 'Modified Transverse Mercator Projection' for the British Isles is

Latitude, 49° N
Longitude, 2° W

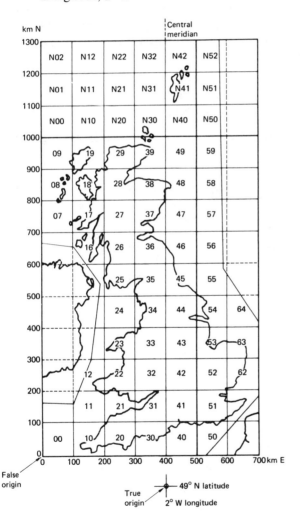

Fig. 4.27 Old OS grid reference system

To provide positive coordinates for the reference system a 'False Origin' was produced by moving the origin **100 km North** and **400 km West**.

The basic grid is founded upon a 100 km square, commencing from the false origin which lies to the SW of the British Isles, and all squares are referenced by relation to this corner of the square.

'*Eastings are always quoted first.*'

Originally the 100 km squares were given a reference based on the number of 100 kilometres East and North from the origin (see Fig. 4.27).

Subsequently, 500 km squares were given prefix letters of S, N and H, and then each square was given a letter of the alphabet (neglecting I). To the right of the large squares the next letter in the alphabet gives the appropriate prefixes, T, O and J (see Fig. 4.28).

Square 32 becomes SO
43 becomes SK
17 becomes NM

The basic reference map is to the scale 1/25 000, Fig. 4.29.

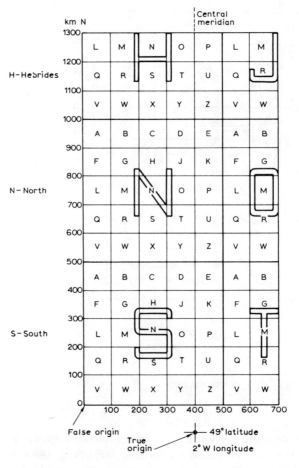

Fig. 4.28 New OS grid reference system

Each map is prefixed by the reference letters followed by two digits representing the reference numbers of the SW corner of the sheet; see the example in Fig. 4.29, i.e. SK 54. This shows the relationship between the various scaled maps and the manner in which each sheet is referenced.

A point P in Trent Polytechnic, Nottingham, has the grid coordinates E 457 076.32 m, N 340 224.19 m. Its full 'Grid Reference' to the nearest metre is written as SK/5740/076 224 and the sheets on which it will appear are:

Reference	Scale	Sheet size	Grid size
SK 54	1/25 000	10 km	1 km
SK 54 SE	1/10 560	5 km	1 km
SK 57 40	1/2500	1 km	100 m
SK 57 40 SW	1/1250	500 m	100 m

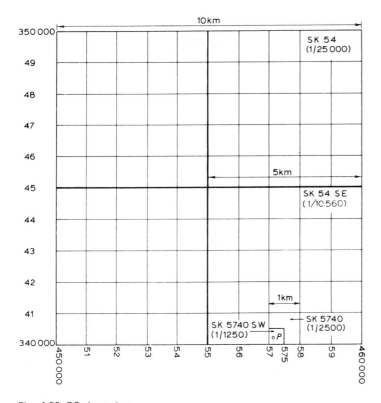

Fig. 4.29 OS sheet sizes

4.10 Cutting points on grid lines

A 'cutting point' is the point where a line of known bearing ϕ cuts a selected coordinate grid line.

Any given line cuts both the 'Easting' and 'Northing' lines of a coordinate grid, and thus there are two cases:

(1)   to find the 'easting' which corresponds to a specified northing line;

(2)   to find the 'northing' which corresponds to a specified easting line.

Consider a line AP of bearing ϕ from a station $A(E_A, N_A)$ which cuts the grid lines passing through a point $G(E_G, N_G)$, Fig. 4.30.

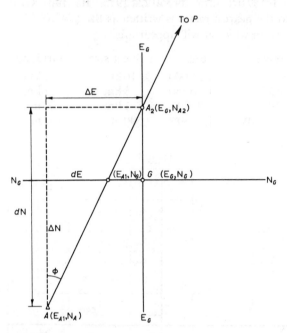

Fig. 4.30

Case 1   Northing specified, i.e. given N_G to find E_{A_1}.

$$\Delta N = N_G - N_A \quad \text{(known values)}$$

$$dE = \Delta N \tan \phi$$

$$E_{A_1} = E_A + dE = E_A + \Delta N \tan \phi \qquad (4.52)$$

Case 2   Easting specified, i.e. given E_G to find N_{A2}.

$$\Delta E = E_G - E_A \quad \text{(known values)}$$

$$dN = \Delta E \cot \phi$$

$$N_{A_2} = N_A + dN = N_A + \Delta E \cot \phi \qquad (4.53)$$

This process is useful for:

(a)   accurate plotting of a line of known bearing, from a point where the line cuts the OS sheet edges;

(b)   accurate plotting, to a large scale, of rays close to a given point, as in the survey process known as 'semi-graphic intersection/resection'.

(a)   *To calculate the Grid 'Cut' coordinates*

Given OS sheet edges with coordinate grid lines E_W, E_E, N_N, N_S.

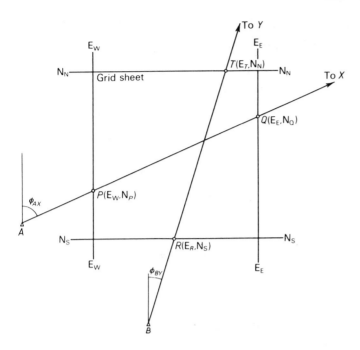

Fig. 4.31

Line AX bearing ϕ_{AX} from $A(E_A, N_A)$ cuts the grid lines at $P(E_W, N_P)$ and $Q(E_E, N_Q)$, Fig. 4.31.

Line BY bearing ϕ_{BY} from $B(E_B, N_B)$ cuts the grid lines at $R(E_R, N_S)$ and $T(E_T, N_N)$.

Then

$$N_P = N_A + (E_W - E_A)\cot\phi_{AX}$$
$$N_Q = N_A + (E_E - E_A)\cot\phi_{AX}$$
$$E_R = E_A + (N_S - N_A)\tan\phi_{BY}$$
$$E_T = E_A + (N_N - N_A)\tan\phi_{BY}$$

(b)   *To calculate the cutting points close to a given point P*
Here a portion of the ray adjacent to P is required to be drawn to a large scale. The axes are drawn through $P(E_P, N_P)$ and the cutting points plotted.

NB: As an alternative, only one cutting point may be plotted and the bearing drawn with the aid of a protractor, Fig. 4.32. This will normally give the required accuracy.

By convention, for greater accuracy with limited trigonometrical functions, the larger partial value is chosen and the trigonometrical function will be less than 1.

The line will not pass through P unless the coordinates are compatible with the bearing.

Let ΔN be greater than ΔE and the cut can be said to be 'Northing specified'. Then

$$E_{P_1} = E_A + (N_P - N_A)\tan\phi_{AP}$$

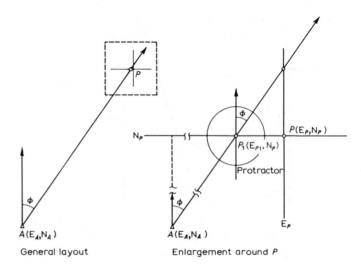

General layout   Enlargement around P

Fig. 4.32

If ΔN were less than ΔE then the cut would be 'Easting specified'.

$$N_{P_2} = N_A + (E_P - E_A) \cot \phi_{AB}$$

Example 4.13 From a station A (457 040.55 m E, 341 195.15 m N) a station P was observed having a grid bearing of $178° 45' 17''$, and from another station B (454 137.32 m E, 338 013.72 m N) the same point was observed having a grid bearing of $052° 38' 17''$. These two lines are known to cut at approximately 457 050 m E and 340 250 m N. Calculate the coordinates of the intersection of these lines with the sheet edges of an OS plan to a scale of 1/1250. State the reference for this particular sheet.   (TP)

The OS sheet 1/1250 reference is obtained by considering the approximate coordinates as: 43/5740/050 250.

The sheet will thus be SK 5740 SW (see Section 4.9) with the sheet corners as shown in Fig. 4.33. By plotting the approximate coordinates of P, the lines can be seen to cut the sheet edges as shown.

For the line AP
Coordinates:

	E	N	ϕ
A	457 040.55	341 195.15	178° 45' 17''
A_1		340 500.00	
A_2		340 000.00	
$\Delta N_{A/A_1}$		− 695.15	
$\Delta N_{A/A_2}$		− 1195.15	

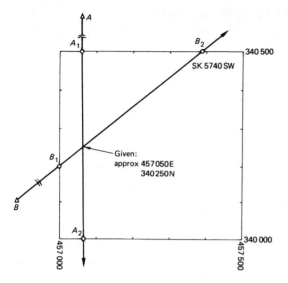

Fig. 4.33

$$\Delta E_{A/A_1} = \Delta N_{A/A_1} \tan \phi_{A/A_1}$$
$$= -695.15 \tan 178° 45' 17'' = 15.11$$
$$E_{A_1} = 457\,040.55 + 15.11 = 457\,055.66$$

Similarly,
$$E_{A_2} = 457\,040.55 + 25.98 = 457\,066.53$$

For the line BP

	E	N	ϕ
B	454 137.32	338 013.72	052° 38′ 17″
B_1	457 000.00		
B_2		340 500.00	
ΔE	2862.68	ΔN 2486.28	

$$N_{B_1} = 338\,013.72 + 2862.68 \cot 52° 38' 17'' = 340\,199.39$$
$$E_{B_2} = 454\,137.32 + 2486.28 \tan 52° 38' 17'' = 457\,393.72$$

Example 4.14 Given the same data as in the previous example, find the 'cut' coordinates on the axis of a grid passing through the provisional coordinates of *P*.

Tabulating the values and following the previous processes:

Station	*A*		*B*	
Bearing to P	178° 45′ 17″		052° 38′ 17″	
Station coordinates	457 040.55	341 195.15	454 137.32	338 013.72
Provisional coordinates	457 050.00	340 250.00	457 050.00	340 250.00
Partial coordinates (larger)		− 945.15	2912.68	
Tan or cot of bearing		−0.021 738	0.763 506	
Computed partial coordinates	20.55			2223.85
Cut coordinates	457 061.10			340 237.57

The table may be explained as follows:

Considering ϕ_{AP}, $\Delta N > \Delta E$ therefore $\tan \phi < 1$ is used. Then

$$dE = \Delta N \tan \phi$$
$$= -945.15 \tan 178° 45' 17'' = 20.55$$
$$E_{PA} = E_A + dE \qquad\qquad = 457\,061.10$$

Considering ϕ_{BP}, ΔE is greater than ΔN therefore $\cot \phi$ is used. Then

$$dN = \Delta E \cot \phi$$
$$= 2912.68 \cot 52° 38' 17'' = 2223.85$$
$$N_{PB} = N_B + dN \qquad\qquad = 340\,237.57$$

Plotting these 'cut' coordinates on the provisional coordinate axes and setting off the bearing with a protractor gives an intersection at P' of 457 061.2 m E and 340 246.1 m N.

By computation, the 'true' values are 457 061.18 m E and 340 246.10 m N.

Example 4.15   Horizontal angles are measured at C and D to two control points A and B. The following data was recorded:

$$ADB = 57° 56' 32'' \qquad DCA = 32° 12' 55''$$
$$BDC = 46° 51' 19'' \qquad ACB = 57° 56' 32''$$

	Fixed coordinates		Assumed coordinates	
	E	N		E	N
A	8568.24	10 360.73	C	2000.00	1000.00
B	12 423.80	12 606.60	D	1000.00	1000.00

Calculate the coordinates of C and D.

$$\Delta E'_{AB} = 3855.56 \qquad \Delta N'_{AB} = 2245.87$$
$$\Delta E_{DC} = 1000.00 \qquad \Delta N_{DC} = 0$$
angle $ADC = 57.9422 + 46.8553 = 104.7975$
angle $DCA \qquad\qquad\qquad = 32.2153$
angle $BDC \qquad\qquad\qquad = 46.8553$
angle $DCB = 32.2153 + 57.9422 = 90.1575$

In the triangle ADC, using the cut by angles,

$$E_A = 800.309 \qquad N_A = 1755.932$$

In the triangle BCD,

$$E_B = 2002.942 \qquad N_B = 2070.091$$

Then

$$\Delta E_{AB} = 1202.633 \qquad \text{and} \quad \Delta N_{AB} = 314.159$$
$$\phi_{AB} = 75.3600 \qquad \text{and} \quad S_{AB} = 1242.989$$
$$\phi'_{AB} = 59.7791 \qquad \text{and} \quad S'_{AB} = 4461.981$$

Thus

$$\alpha = 15.5809 \quad \text{and} \quad k = 3.589\,7188$$

$$m = k\cos\alpha = 3.457\,804$$

$$n = k\sin\alpha = 0.964\,194$$

Using the transposition of coordinates for swing and scale factor:

$$E'_D = E'_A + m\Delta E_{AD} - n\Delta N_{AD} = \quad 9987.59$$

$$N'_D = N'_A + m\Delta N_{AD} + n\Delta E_{AD} = \quad 7939.40$$

$$E'_C = E'_A + m\Delta E_{AC} - n\Delta N_{AC} = \quad 13\,445.41$$

$$N'_C = N'_A + m\Delta N_{AC} + n\Delta E_{AC} = \quad 8903.59$$

Exercises 4.5

1   Survey station X (65.14 m E, 129.36 m N) is 79.49 m AOD. Station Y (2823.48 m E, 526.88 m N) has a depth of 232.62 m below Ordnance datum. Find the length, bearing and inclination of a line joining X and Y.

(Ans.   2786.84 m; 081° 47′ 57″; 1 in 8.92)

2   A shaft is sunk to a certain seam in which the workings to the dip have reached a level DE. It is proposed to deepen the shaft and connect the point E in the dip workings to a point X by a cross-measures drift, dipping at 1 in 200 towards X. The point X is to be 40.8 m from the centre of the shaft A and due East from it, AX being level.

The following are the notes of a traverse made in the seam from the centre of the shaft A to the point E:

Line	Azimuth	Distance (m)	Vertical angle
AB	270° 00′	38.7	level
BC	184° 30′	167.7	dipping 21°
CD	159° 15′	222.5	dipping $18\frac{1}{2}°$
DE	90° 00′	25.3	level

Calculate (a) the azimuth and horizontal length of the drift EX and (b) the amount by which it is necessary to deepen the shaft.

(Ans.   (a) 358° 40′ 353.4 m; (b) 132.4 m)

3   The notes of a traverse between two points A and E in a certain seam are as follows:

Line	Azimuth	Distance (m)	Angle of inclination
AB	89°	182.9	+6°
BC	170°	137.2	−30°
CD	181°	167.6	level
DE	280°	108.2	level

It is proposed to drive a cross-measures drift from a point E to another point F exactly midway between A and B. Calculate the azimuth and length EF.

(Ans.   359° 33′; 264.3 m, 270.8 inclined)

4 In order to set out the curve connecting two straights of a road to be constructed, the coordinates on the National Grid of *I*, the point of intersection of the centre lines of the straights produced, are required.

A is a point on the centre line of one straight, the bearing *AI* being 72° 00′ 00″, and *B* is a point on the centre line of the other straight, the bearing *IB* being 49° 26′ 00″.

Using the following data, calculate with full checks the coordinates of *I*.

	E (m)	N (m)
A	13 337.535	15 911.322
B	13 766.587	16 116.148

The length *AB* is 475.436 m and the bearing 64° 28′ 50″.

(NU Ans.　13 643.429 m E, 16 010.713 m N)

5 From an underground traverse between 2 shaft wires *A* and *D* the following partial coordinates in feet were obtained:

AB	E	150.632 ft,	S　327.958 ft
BC	E	528.314 ft,	N　82.115 ft
CD	E	26.075 ft,	N　428.862 ft

Transform the above partials to give the total Grid coordinates of *station B* given that the Grid coordinates of *A* and *D* were:

A	E 520 163.462 metres,	N 432 182.684 metres
D	E 520 378.827 metres,	N 432 238.359 metres

(TP Ans.　B 520 209.363 E, 432 082.481 N)

6 Two fixed points *A* and *B* have the following coordinates:

	E (m)	N (m)
A	10 000.00	10 000.00
B	13 462.56	11 373.62

Measurements at *A* and *B* to *C* were recorded as follows:

AC = 3051.63 m	*BC* = 1980.27 m
CAB = 32° 03′ 56″	*ABC* = 54° 53′ 27″

Compute the coordinates of *C* by
(a) using the lengths measured;
(b) using the angles measured;
(c) converting the angles to bearings and using the 'cut by bearings'.　(Ans.　E_C = 11 806.52 m, N_C = 12 459.46 m)

7 Given national grid coordinates as follows:

	E	N
A	416 732.14	318 492.38
B	417 215.33	319 558.52
C	419 387.49	316 476.87

Compute
(a) the grid length and bearing of all the connecting lines;
(b) the ground lengths;
(c) the angles of the triangle;
(d) the area formed by the triangle.

>(Ans. (a) *AB* 1170.52 m, 024.3807°,
>   *BC* 3770.26 m, 144.8212°,
>   *CA* 3333.64 m, 307.1999°.
>   (b) 1170.99, 3771.75, 3334.96 m.
>   (c) *BAC* 102° 49′ 09″,
>   *CBA* 59° 33′ 34″,
>   *ACB* 17° 37′ 17″.
>   (d) 1903 928 m².)

8 Given fixed points *A* and *B* as:

	E (m)	N (m)
A	141 796.12	616 832.04
B	154 042.55	620 919.02

The following angles were measured:

ADB	46° 49′ 53″	*BDC*	59° 57′ 16″
DCA	19° 12′ 48″	*ACB*	54° 42′ 42″

By assuming coordinates for *D* and *C* calculate the coordinates of *A* and *B* relative to *D* and then transpose the coordinates to obtain the grid coordinates of *D* and *C*.

(Ans.   *C* 151 654.80, 606 972.19; *D* 141 046.03, 612 098.40)

Bibliography

CLARK, D., *Plane and Geodetic Surveying for Engineers*, 6th edition, **1**. Constable (1972)

CLENDINNING, J., and OLLIVER, J. G., *Principles of Surveying*, 4th edition. Van Nostrand Reinhold (1978)

HOLLAND, J. L., WARDELL, K., and WEBSTER, A. C., Surveying, *Coal Mining Series*. Virtue

ORDNANCE SURVEY Booklet No. 1/45, *A Brief Description of the National Grid and Reference System*. HMSO

SCHOFIELD, W., *Engineering Surveying*, 3rd edition. Butterworth (1978)

5

Instrumental optics

5.1 Reflection at plane surfaces

5.1.1 Laws of reflection

(1) The incident ray, the reflected ray, and the normal to the mirror at the point of incidence all lie in the same plane.
(2) The angle of incidence (i) = the angle of reflection (r).

The ray AO, Fig. 5.1, is inclined at α (glancing angle MOA) to the mirror MN. Since $i = r$, angle $BON = MOA = \alpha$. If AO is produced to C,

$$\text{angle } MOA = NOC = BON = \alpha$$

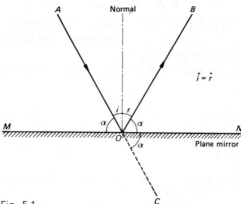

Fig. 5.1

Thus the deviation of the ray AO is 2α. Therefore *the deviation angle is twice the glancing angle*, i.e.

$$D = 2\alpha \qquad (5.1)$$

5.1.2 Deviation by successive reflections on two inclined mirrors

Ray AB is incident on mirror $M_1 N_1$ at a glancing angle α (Fig. 5.2). It is thus deflected by reflection $+2\alpha$.

The reflected ray BC incident upon mirror $M_2 N_2$ at a glancing angle β is deflected by reflection -2β (here clockwise is assumed $+$ve). The total deflection D is thus $(2\alpha - 2\beta) = 2(\alpha - \beta)$.

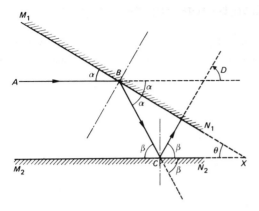

Fig. 5.2

In triangle BCX,

$$\beta = \alpha + \theta$$

therefore $\theta = \beta - \alpha$

i.e. $D = 2(\alpha - \beta) = 2\theta$ (5.2)

As θ is constant, *the deflection after two successive reflections is constant and equal to twice the angle between the mirrors.*

5.1.3   The optical square

This instrument (Fig. 5.3), used for setting out right angles, employs the above principle.

By equation 5.2, the deviation of any ray from O_2 incident on mirror M_2 at an angle α to the normal $= 2\theta$, i.e. $2 \times 45° = 90°$.

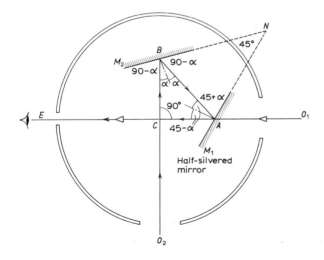

Fig. 5.3 Optical square

5.1.4   Deviation by rotating the mirror

Let the incident ray AO be constant, with a glancing angle α
(Fig. 5.4). The mirror M_1N_1 is then rotated by an anticlockwise
angle β to M_2N_2.

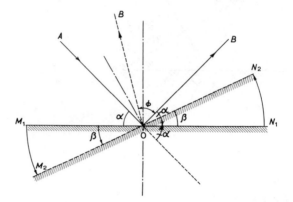

Fig. 5.4

When the glancing angle is α the deviation angle is 2α.

After rotation the glancing angle is $(\alpha + \beta)$ and the deviation
angle is therefore $2(\alpha + \beta)$.

Thus the reflected ray is rotated by

$$\phi = 2(\alpha + \beta) - 2\alpha = 2\beta \tag{5.3}$$

*If the incident ray remains constant the reflected ray deviates
by twice the angular rotation of the mirror.*

5.1.5   The sextant

The principles of the sextant are shown in Figs 5.5 and 5.6.
Mirror M_1, silvered, is connected to a pointer P. As M_1 is
rotated the pointer moves along the graduated arc.

Mirror M_2 is only half-silvered and is fixed.

When the reading at P is zero, Fig. 5.5, the image K reflected
from both mirrors should be seen simultaneously with K

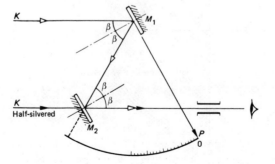

Fig. 5.5 Zero setting on the sextant

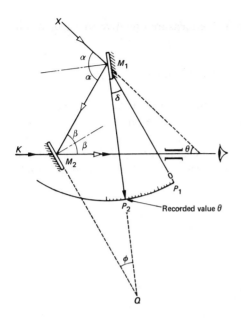

Fig. 5.6 Principles of the sextant

through the plain glass part of M_2. With a suitable object K as the horizon, the mirrors should be parallel.

When observing an elevated object X, Fig. 5.6, above the horizon K, the mirror M_1 is rotated through angle δ until S is simultaneously observed with K. The angle being measured is therefore θ.

From triangle $M_1 E M_2$,

$$2\alpha = 2\beta + \theta$$

therefore

$$\theta = 2(\alpha - \beta)$$

From triangle $M_1 Q M_2$,

$$90 + \alpha = 90 + \beta + \delta$$

therefore

$$\alpha = \beta + \delta$$

therefore

$$\delta = \alpha - \beta = \tfrac{1}{2}\theta \qquad (5.4)$$

i.e. *the rotation of the mirror is half the angle of elevation.*

When the mirrors are parallel, $\delta = 0$ with the index pointer MP at zero on the graduated arm. When the angle θ is being observed, the mirror is turned through angle δ, but the recorded value $MP_2 = \theta$, i.e. 2δ.

Notes

(1)   Horizontal angles are only measured if the objects are at the same height relative to the observer, which means that in most cases the angle measured is in an inclined plane; see p. 81.

(2) Vertical angles must be measured relative to a true or an artificial horizon.

5.1.6 Use of the true horizon

(a) As the angle of deviation, after two successive reflections, is independent of the angle of incidence on the first mirror, the object will continue to be seen on the horizon no matter how much the observer moves. Once the mirror M_1 has been set, the angle between the mirrors is set, and the observed angle recorded (see Fig. 5.7).

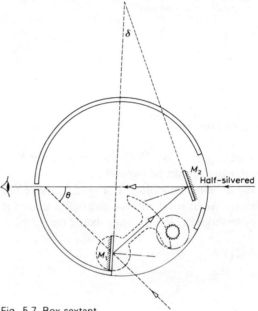

Fig. 5.7 Box sextant

This is the main advantage of the sextant as a hand instrument, particularly in marine and aerial navigation where the observer's position is unstable.

(b) If the observer is well above the horizon, a correction $\delta\theta$ is required for the dip of the horizon, Fig. 5.8.

The Nautical Almanac contains tables for the correction factor $\delta\theta$ due to the dip of the horizon based on the equation:

$$\delta\theta = -0.97 \sqrt{h} \text{ minutes} \tag{5.5}$$

where h = height in feet above sea level; or

$$\delta\theta = -1.756 \sqrt{H} \text{ minutes}$$

where H = height in metres above sea level.

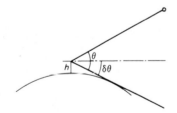

Fig. 5.8 Dip of the horizon

5.1.7 Artificial horizon

On land, no true horizon is possible, so an 'artificial horizon' is employed (Fig. 5.9). This consists essentially of a trough of

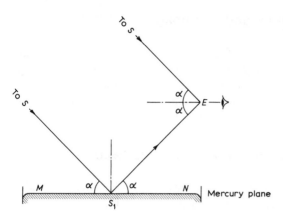

Fig. 5.9 Artificial horizon

mercury, the surface of which assumes a horizontal plane forming a mirror.

The vertical angle observed between the object S and the reflection of the image S_1 in the mercury is twice the angle of altitude (α) required.

$$\text{Observed angle} = S_1 ES = 2\alpha$$

$$\text{True altitude} = MS_1 S = \alpha$$

Rays SE and SS_1 are assumed parallel due to the distance of S from the instrument.

5.1.8 Images in plane mirrors

Object O in front of the mirror is seen at E as though it were situated at I, Fig. 5.10. From the glancing angles α and β it can be seen that

(a) triangles OFC and ICF are congruent;
(b) triangles OFD and IDF are congruent.

Thus the point I (image) is the same perpendicular distance from the mirror as O (object), i.e. $OF = FI$.

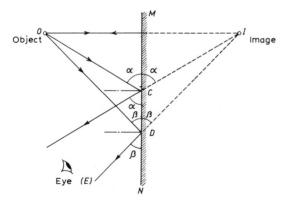

Fig. 5.10 Images in plane mirrors

5.1.9 Virtual and real images

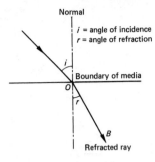

Normal

i = angle of incidence
r = angle of refraction

Boundary of media

B
Refracted ray

Fig. 5.11

As above, the rays reflected from the mirror appear to pass through *I*, the image thus being *unreal* or *virtual*. For the image to be real, the object would have to be virtual.

The real test is whether the image can be received on a screen: if it can be, it is real – if not, it is virtual.

5.2 Refraction at plane surfaces

The incident ray *AO*, meeting the boundary between two media, e.g. air and glass, is refracted to *B*, Fig. 5.11.

5.2.1 Laws of refraction

(1) The incident ray, the refracted ray, and the normal to the boundary plane between the two media at the point of incidence all lie in the same plane.

(2) For any two given media the ratio $\dfrac{\sin i}{\sin r}$ is a constant known as the refractive index (the light assumed to be monochromatic). Thus

$$\text{refractive index } (\mu) = \frac{\sin i}{\sin r} \tag{5.6}$$

5.2.2 Total internal reflection

If a ray *AO* is incident on a glass/air boundary (Fig. 5.12) the ray may be refracted or reflected according to the angle of incidence.

When the angle of refraction is 90°, the critical angle of incidence is reached, i.e.

$$_g\mu_a = \frac{\sin c}{\sin 90°} = \sin c \tag{5.7}$$

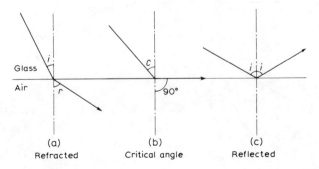

Glass
Air

i

r

(a)
Refracted

c

90°

(b)
Critical angle

i *i*

(c)
Reflected

Fig. 5.12

For crown glass the refractive index $_a\mu_g \simeq 1.5$. Therefore

$$c = \sin^{-1} 1/1.5 \simeq 41° 30'$$

If the angle of incidence (glass/air) $i > 41° 30'$, the ray will be internally reflected, and this principle is employed in optical prisms within such surveying instruments as optical squares, reflecting prisms in binoculars, telescopes and optical scale-reading theodolites.

NB: Total internal reflection can only occur when light travels from one medium to an optically less dense medium, e.g. glass/air.

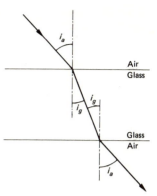

5.2.3 Relationships between refractive indices
(Fig. 5.13)

Fig. 5.13

(a) If the refractive index from air to glass is $_a\mu_g$, then the refractive index from glass to air is $_g\mu_a$. Therefore

$$_g\mu_a = \frac{1}{_a\mu_g} \tag{5.8}$$

For example, if $_a\mu_g = 1.5$ (taking air as 1), then

$$_g\mu_a = \frac{1}{1.5} = 0.66$$

(b) Given parallel boundaries of air, glass, air, then

$$\sin i = \text{constant} \tag{5.9}$$

$$_a\mu_g = \frac{\sin i_a}{\sin i_g}$$

(c) The emergent ray is parallel to the incident ray when returning to the same medium although there is relative displacement.

This factor is used in the parallel plate micrometer.

5.2.4 Refraction through triangular prisms

When the two refractive surfaces are not parallel the ray may be bent twice in the same direction, thus deviated from its former direction by an angle D.

It can be seen from Fig. 5.14 that

$$A = \beta_1 + \beta_2 \tag{5.10}$$

and $$D = (\alpha_1 - \beta_1) + (\alpha_2 - \beta_2) \tag{5.11}$$

$$= (\alpha_1 + \alpha_2) - (\beta_1 + \beta_2) \tag{5.12}$$

i.e. $$D = (\alpha_1 + \alpha_2) - A \tag{5.13}$$

Thus the minimum deviation occurs when

$$\alpha_1 + \alpha_2 = A \tag{5.14}$$

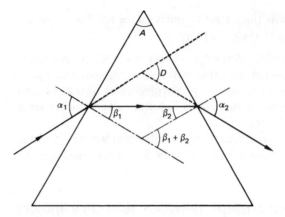

Fig. 5.14 Refraction through a triangular prism

If A is small, then

$$\alpha = \mu\beta$$

and $\quad D = \mu\beta_1 + \mu\beta_2 - A$

$$= \mu(\beta_1 + \beta_2) - A$$

$$= A(\mu - 1) \tag{5.15}$$

5.2.5 Instruments using refraction through prisms

The line ranger (Fig. 5.15)

$$\alpha + \beta = 90°$$

$$2(\alpha + \beta) = 180°$$

Thus $O_1 C O_2$ is a straight line.

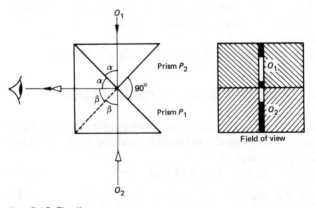

Fig. 5.15 The line ranger

The prism square (Fig. 5.16)

This is precisely the same mathematically as the optical square (Fig. 5.3), but light is internally reflected, the incident ray being greater than the critical angle of the glass.

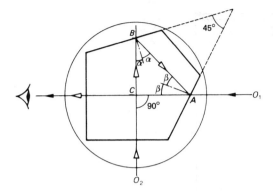

Fig. 5.16 The prism square

The double prismatic square (Fig. 5.17) combines the advantages of both the above hand instruments.

Images O_2 and O_3 are reflected through the prisms. O_1 is seen above and below the prisms.

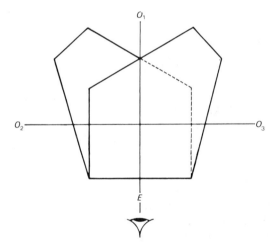

Fig. 5.17 The double prismatic square

The parallel plate micrometer (Fig. 5.18)
A parallel-sided disc of glass of refractive index μ and thickness t is rotated through an angle θ. Light is refracted to produce displacement of the line of sight by an amount x.

$$x = DB = AB \sin(\theta - \delta)$$

$$= \frac{t(\sin\theta\cos\delta - \cos\theta\sin\delta)}{\cos\delta}$$

$$= t(\sin\theta - \cos\theta\tan\delta)$$

but refractive index

$$\mu = \frac{\sin\theta}{\sin\delta}$$

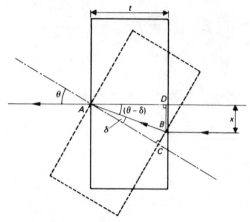

Fig. 5.18 The parallel plate micrometer

therefore

$$\sin \delta = \frac{\sin \theta}{\mu} \quad \text{and} \quad \cos \delta = \sqrt{(1 - \sin^2 \delta)}$$

$$\tan \delta = \sin \theta / \sqrt{(\mu^2 - \sin^2 \theta)}$$

Therefore

$$x = t \sin \theta \left[1 - \frac{\cos \theta}{\sqrt{(\mu^2 - \sin^2 \theta)}} \right]$$

$$= t \sin \theta \left[1 - \sqrt{\left(\frac{1 - \sin^2 \theta}{\mu^2 - \sin^2 \theta} \right)} \right] \tag{5.16}$$

If θ is small, then $\sin \theta \simeq \theta_{\text{rad}}$ and $\sin^2 \theta$ may be neglected. Therefore

$$x \simeq t \theta \left(1 - \frac{1}{\mu} \right) \tag{5.17}$$

When the ratio x/t is so large that the degree of approximation for the required linearity is not acceptable, the drum divisions have to be so spaced to rectify this defect or some form of gearing introduced. As an alternative the drum may be divided on both sides of the normal, i.e. $+$ or $-$ values.

If θ varies by $\pm 3°$ the error ratio is less than $1/10\,000$.

Example 5.1   A parallel plate micrometer attached to a level is to show a displacement of 2.5 mm when rotated through $15°$ on either side of the vertical. Calculate the thickness of glass required if its refractive index is 1.6.

Transposing equation 5.16,

$$t = \frac{x}{\sin \theta \left[1 - \sqrt{\left(\frac{1 - \sin^2 \theta}{\mu^2 - \sin^2 \theta} \right)} \right]}$$

$$= \frac{x}{0.258\,82 \times 0.388\,24} = 0.0249 \, \text{m}$$

The micrometer is geared to the parallel plate and must be correlated. Precise levelling staves are usually graduated in either 5 or 10 mm divisions and the micrometer subdivided such that 1 division represents 1 mm and thus estimation may be made to 0.1 mm.

To avoid confusion, the micrometer should be set to zero before each sight is taken and the micrometer reading is then added to the staff reading as the parallel plate refracts the line of sight to the next lower reading.

Exercise 5.1

1 Describe the method of operation of a parallel plate micrometer in precise levelling. If the index of refraction from air to glass is 1.6 and the parallel plate prism is 16 mm thick, calculate the angular rotation of the prism to give a vertical displacement of the image of 1 mm. (LU Ans. $9°\,33'$)

5.3 Spherical mirrors

5.3.1 Concave or converging mirrors (Fig. 5.19)

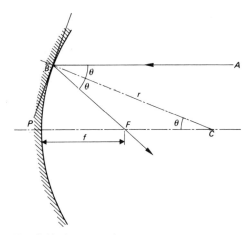

Fig. 5.19 Concave mirror

A narrow beam of light produces a real principal focus F. P is called the *pole* of the mirror and C is the *centre of curvature*. PF is the *focal length* of the mirror.

A ray AB, parallel to the axis, will be reflected to F. BC will be normal to the curve at B, so that

$$\text{angle } ABC = \text{angle } CBF = PCB = \theta$$

As the beam is assumed narrow

$$PF \simeq BF \simeq FC$$

therefore

$$PC \simeq 2PF = 2f \simeq r \tag{5.18}$$

5.3.2   Convex or diverging mirrors (Fig. 5.20)

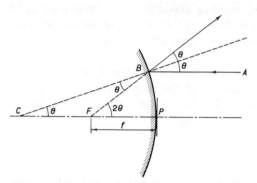

Fig. 5.20 Convex mirror

A narrow beam of light produces a virtual principal point F, being reflected away from the axis.

The angular principles are the same as for a concave mirror and

$$r \simeq 2f$$

5.3.3 The relationship between object and image in curved mirrors

Assuming a *narrow* beam, the following rays are considered in all cases (Fig. 5.21).

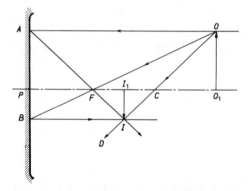

Fig. 5.21

(a)   Ray OA, parallel to the principal axis, is reflected to pass through the focus F.

(b)   Ray OB, passing through the focus F, is then reflected parallel to the axis.

(c)   Ray OD, passing through the centre of curvature C, and thus a line normal to the curve.

NB: In graphical solutions, it is advantageous to exaggerate the vertical scale, the position of the image remaining in the true position. As the amount of curvature is distorted, it should be represented as a straight line perpendicular to the axis.

Any two of the above rays produce, at their intersection, the position of the image I.

The relationships between object and image for *concave mirrors* are:

(a) When the object is at infinity, the image is small, real, and inverted.
(b) When the object is at the centre of curvature C, the image is also at C, real, of the same size and inverted.
(c) When the object is between C and F, the image is real, enlarged and inverted.
(d) When the object is at F, the image is at infinity.
(e) When the object is between F and P, the image is virtual, enlarged and erect.

For *convex mirrors*, in all cases the image is virtual, diminished and erect; Fig. 5.22.

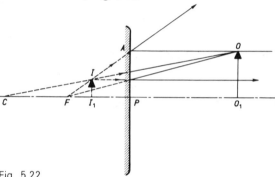

Fig. 5.22

5.3.4 Sign convention

There are several sign conventions but here the convention *real-is-positive* is adopted. This has many advantages provided the work is not too advanced.

All real distances are treated as positive values whilst virtual distances are treated as negative values—in all cases distances are measured from the pole.

NB: In the diagrams real distances are shown as solid lines whilst virtual distances are dotted.

5.3.5 Standard formulae

Consider the concave mirror shown in Fig. 5.23. The ray OA is reflected at A to AI making an angle α on either side of the normal AC.

$$\alpha = \theta - \beta \quad \text{and} \quad \delta = 2\alpha + \beta = \theta + \alpha$$

then $2\theta = \delta + \beta$

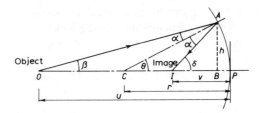

Fig. 5.23 Concave mirror (image real)

As the angles α, β, θ and δ are all small, and B is adjacent to P,

$$\delta = \sin \delta = h/IP \quad (I \text{ is real so } IP \text{ is } +\text{ve})$$
$$\beta = \sin \beta = h/OP \quad (O \text{ is real so } OP \text{ is } +\text{ve})$$
$$\theta = \sin \theta = h/CP$$

Then

$$\frac{h}{IP} + \frac{h}{OP} = \frac{2h}{CP}$$

i.e. $\qquad \dfrac{1}{v} + \dfrac{1}{u} = \dfrac{2}{r} = \dfrac{1}{f} \quad \text{as } f = r/2 \qquad (5.19)$

It can be shown that this equation is standard for all curved mirrors.

In the concave mirror with the image virtual (Fig. 5.24) and the convex mirror (Fig. 5.25) the value of v is $-$ve.

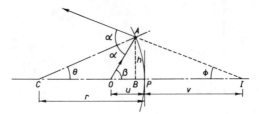

Fig. 5.24 Concave mirror (image virtual)

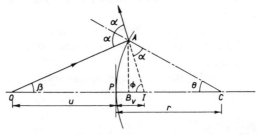

Fig. 5.25 Convex mirror

5.3.6 Magnification in spherical mirrors

Consider Fig. 5.26. IB is the image of OA.

In the right angled triangles OPO_1 and IPI_1 the angle α is common, being the angles of incidence and of reflection, and

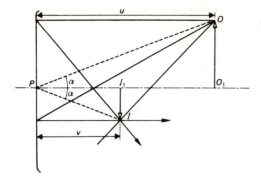

Fig. 5.26 Magnification in spherical mirrors

therefore the triangles are similar. Thus

$$\text{magnification} = \frac{II_1 \text{ (image size)}}{OO_1 \text{ (object size)}} = \frac{I_1 P(v)}{O_1 P(u)}$$

therefore $\qquad M = \dfrac{v}{u} \quad$ neglecting signs $\qquad\qquad$ (5.20)

Example 5.2   An object 24 mm high is placed on the principal axis 500 mm from a concave mirror which has a radius of curvature of 400 mm. Find the position, size and nature of the image.

As the mirror is concave,

$$f = r/2 = 400/2; \quad u = 500$$

As

$$\frac{1}{v} = \frac{1}{f} - \frac{1}{u}$$

$$= \frac{2}{400} - \frac{1}{500} = 0.0030$$

therefore

$$v = 333.3 \text{ mm}$$

Magnification

$$M = \frac{v}{u} = 0.67$$

therefore

$$\text{size of image} = 24 \times 0.67 = 16 \text{ mm}$$

5.4   Refraction through thin lenses

5.4.1   Definitions

(a)   *Types of lens*
Convex (converging), Fig. 5.27.
Concave (diverging), Fig. 5.28.

Double convex   Plano-convex   Convex meniscus

Fig. 5.27 Types of convex lens

Double concave   Plano-concave   Concave meniscus

Fig. 5.28 Types of concave lens

(b) *Focal points*

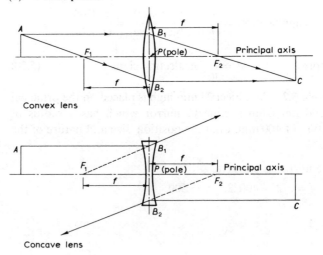

Convex lens

Concave lens

Fig. 5.29 Conjugate foci

5.4.2 Formation of images (Fig. 5.30)

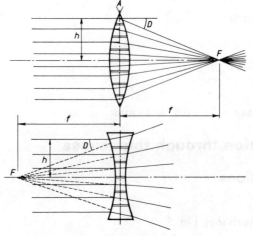

Fig. 5.30 Formation of images

If a thin lens is assumed to be split into a series of small prisms, any ray incident on the face will be refracted and will deviate by an angle

$$D = A(\mu - 1) \qquad \text{(Equation 5.15)}$$

NB: The deviation angle D is also related to the height h and the focal length f, i.e.

$$D = h/f \qquad\qquad\qquad (5.21)$$

5.4.3 The relationship between object and image in a thin lens

The position of the image can be drawn using three rays; Fig. 5.31. Note that two principal foci, F_1 and F_2, exist.

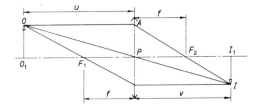

Fig. 5.31

(a)   Ray OA parallel to the principal axis is refracted to pass through principal focus F_2.
(b)   Ray OB passes through the principal focus F_1 and is then refracted parallel to the principal axis.
(c)   Ray OPI passes from object to image through the pole P without refraction.

Convex lens
(a)   When the object is at infinity, the image is at the principal focus F_2, real and inverted.
(b)   When the object is between infinity and F_1, the image is real and inverted.
(c)   When the object is between F_1 and P, the image is virtual, magnified, and erect, i.e. a simple magnifying glass.

Concave lens
The image is always virtual, erect and diminished.

5.4.4   Derivation of formulae

The real-is-positive sign convention is again adopted, but for convex lenses the real distances and focal lengths are considered positive, whilst for concave lenses the virtual distances and focal lengths are considered negative.

   As with mirrors, thin lens formulae depend on small angle approximations.

Convex lens

(a)   Image real, Fig. 5.32:

$$D = \alpha + \beta$$

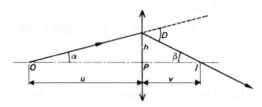

Fig. 5.32

By equation 5.21,

$$D = h/f$$

therefore

$$\frac{h}{f} = \frac{h}{u} + \frac{h}{v}$$

i.e. $$\frac{1}{f} = \frac{1}{u} + \frac{1}{v}$$

(b)   Image virtual, i.e. object between *F* and *P*, Fig. 5.33:

$$D = \alpha - \beta$$

i.e. $$\frac{h}{f} = \frac{h}{u} - \frac{h}{v}$$

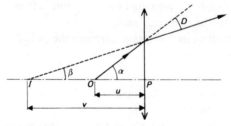

Fig. 5.33

but *v* is virtual, i.e. negative, therefore

$$\frac{1}{f} = \frac{1}{u} - \frac{1}{v}$$

Concave lens (Fig. 5.34):

$$D = \beta - \alpha$$

i.e. $$\frac{h}{f} = \frac{h}{v} - \frac{h}{u}$$

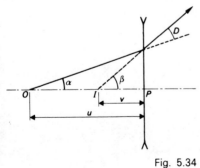

Fig. 5.34

but *v* and *f* are negative, being virtual distances, therefore

$$\frac{1}{f} = \frac{1}{u} + \frac{1}{v}$$

5.4.5 Magnification in thin lenses (Fig. 5.35)

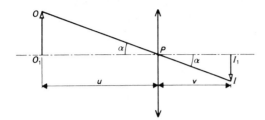

Fig. 5.35

As with special spherical mirrors, OPO_1 and IPI_1 are similar right angled triangles with angle α common. Therefore

$$\text{magnification } M = \frac{II_1}{OO_1} \quad \begin{array}{l}\text{(image size)}\\ \text{(object size)}\end{array}$$

$$M = \frac{I_1 P}{O_1 P} \quad \begin{array}{l}\text{(image distance } v)\\ \text{(object distance } u)\end{array}$$

$$= \frac{v}{u} \quad \text{as before}$$

NB: This should not be confused with angular magnification or magnifying power (M), which is defined as

$$\frac{\text{the angle subtended at the eye by the image}}{\text{the angle subtended at the eye by the object}}$$

For the astronomical telescope, with the image at infinity,

$$M = \frac{\text{focal length of objective}}{\text{focal length of eyepiece}} = \frac{f_0}{f_e} \tag{5.22}$$

5.5 Telescopes

5.5.1 Kepler's astronomical telescope (Fig. 5.36)

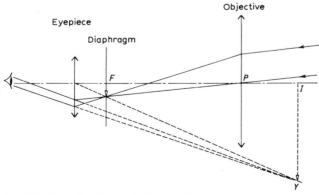

Fig. 5.36 Kepler's astronomical telescope

The telescope is designed to increase the angle subtending distant objects and thus apparently to bring them nearer.

The objective lens, converging and of long focal length, produces an image *FX*, inverted but real, of the object at infinity.

The eyepiece lens, converging but of short focal length, is placed close to *F* so as to produce from the real object *FX* a virtual image *I Y*, magnified but similarly inverted.

5.5.2   Eyepieces

Ideally, eyepieces should reduce *chromatic* and *spherical aberration*.

Lenses of the same material are *achromatic* if their distance apart is equal to the average of their focal lengths, i.e.

$$d = \tfrac{1}{2}(f_1 + f_2) \qquad (5.23)$$

If their distance apart is equal to the differences between their focal lengths, spherical aberration is reduced, i.e.

$$d = f_1 - f_2 \qquad (5.24)$$

For surveying purposes the diaphragm must be between the eyepiece and the objective. The most suitable is *Ramsden's eyepiece*, Fig. 5.37.

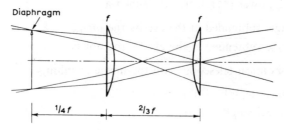

Fig. 5.37 Ramsden's eyepiece

The focal length of each lens is the same, namely *f*. Neither of the conditions (4.23) or (4.24) is satisfied.

chromatic $\tfrac{1}{2}(f_1 + f_2) = f$ compared with $2/3f$

spherical $f_1 - f_2 = 0$ compared with $2/3f$

Huyghen's eyepiece, Fig. 5.38, satisfies the conditions but the

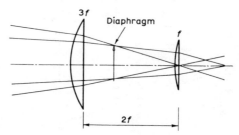

Fig. 5.38 Huyghen's eyepiece

focal plane lies between the lenses. It is used in the Galileo telescope.

chromatic condition $\quad \frac{1}{2}(3f+f) = 2f = d$

spherical condition $\quad\quad 3f-f = 2f = d$

Example 5.3 An astronomical telescope consists of two thin lenses 611 mm apart. If the magnifying power is × 12, what are the focal lengths of the two lenses?

With reference to Fig. 5.39,

$$\text{magnifying power} = f_0/f_e = 12$$

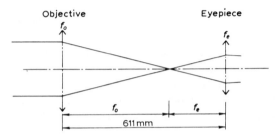

Fig. 5.39

But

$$12 f_e + f_e = 611$$
$$f_e = 611/13 = 47 \text{ mm eyepiece lens}$$
$$f_0 = 12 \times 47 = 564 \text{ mm objective lens}$$

Check: $564 + 47 = 611$

5.5.3 The internal focussing telescope (Fig. 5.40)

The eyepiece and objective are fixed and an internal concave lens is used for focussing.

For the convex lens, by equation 5.19,

$$\frac{1}{f_1} = \frac{1}{u_1} + \frac{1}{v_1}$$

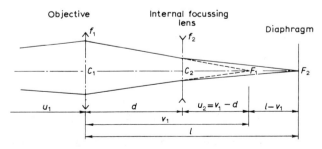

Fig. 5.40 Internal focussing telescope

i.e. $\dfrac{1}{v_1} = \dfrac{1}{f_1} - \dfrac{1}{u_1}$ or $\dfrac{1}{u_1} = \dfrac{1}{f_1} - \dfrac{1}{v_1}$

For the concave lens,

$$u_2 = -(v_1 - d)$$

therefore

$$\frac{1}{f_2} = \frac{1}{u_2} + \frac{1}{v_2}$$

$$-\frac{1}{f_2} = -\frac{1}{v_1 - d} + \frac{1}{l - d}$$

therefore

$$\frac{1}{f_2} = \frac{1}{v_1 - d} - \frac{1}{l - d} \tag{5.25}$$

An internal focussing telescope has a length l from the objective to the diaphragm. The respective focal lengths of the objective and the internal focussing lens are f_1 and f_2.

To find the distance d of the focussing lens from the objective when the object focussed is u_1 from the objective, Fig. 5.40.

For the objective,

$$\frac{1}{v_1} = \frac{1}{f_1} - \frac{1}{u_1}$$

therefore

$$v_1 = \frac{u_1 f_1}{u_1 - f_1} \tag{5.26}$$

For the focussing lens,

$$-\frac{1}{f_2} = -\frac{1}{u_2} + \frac{1}{v_2}$$

i.e. $\dfrac{1}{f_2} = \dfrac{1}{u_2} - \dfrac{1}{v_2}$

$$= \frac{1}{v_1 - d} - \frac{1}{l - d}$$

therefore

$$(v_1 - d)(l - d) = f_2(l - d) - f_2(v_1 - d)$$

i.e. $d^2 - d(l + v_1) + \{ lv_1 - f_2(l - v_1) \} = 0$

$$d^2 - d(l + v_1) + \{ v_1(l + f_2) - f_2 l \} = 0 \tag{5.27}$$

Writing the solution to this gives:

$$d = \tfrac{1}{2}[(l + v_1) \pm \sqrt{\{(l + v_1)^2 - 4(v_1 l + v_1 f_2 - l f_2)\}}]$$
$$= \tfrac{1}{2}[(l + v_1) \pm \sqrt{\{(l - v_1)^2 + 4 f_2(l - v_1)\}}]$$
$$= \tfrac{1}{2}[(l + v_1) \pm \sqrt{\{(l - v_1)(1 + 4 f_2 - v_1)\}}] \tag{5.28}$$

This is a quadratic equation in d and its value will vary according to the distance u_1 of the object from the instrument.

Example 5.4   Describe, with the aid of a sketch, the function of an internal focussing lens in a surveyors' telescope and state the advantages and disadvantages of internal focussing as compared with external focussing.

In a telescope, the object glass of focal length 180 mm is located 230 mm away from the diaphragm. The focussing lens is midway between these when the staff 18 m away is focussed. Determine the focal length of the focussing lens.   (LU)

For the convex objective lens, Fig. 5.41,

$$f_1 = 0.180 \text{ m}, \quad u_1 = 18.000 \text{ m}$$

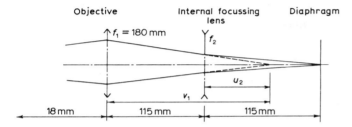

Fig. 5.41

Then, by equation 5.19,

$$\frac{1}{v_1} = \frac{1}{f_1} - \frac{1}{u_1} = 5.500$$

$$v_1 = 0.182$$

For the focussing lens,

$$u_2 = -(v_1 - 0.115) = -0.067$$
$$v_2 = 0.115$$

therefore

$$\frac{1}{f_2} = \frac{1}{u_2} + \frac{1}{v_2}$$

$$\frac{1}{f_2} = -1/0.067 + 1/0.115 = -6.2297$$

$$f_2 = -0.1605, \text{ i.e. } 0.161 \text{ m—the lens is concave.}$$

Example 5.5   In an internally focussing telescope, Fig. 5.40, the objective of focal length 130 mm is 200 mm from the diaphragm. If the internal focussing lens is of focal length 260 mm, find its distance from the diaphragm when focussed at infinity.

For the objective, $f_1 = 130$ mm and thus the position of F_1 will be 130 mm from C_1. Therefore

$$C_2 F_1 = 130 - d$$

For the internal focussing lens,

$$f_2 = -260$$
$$u_2 = -(130-d)$$
$$v_2 = 200-d$$

therefore

$$\frac{1}{f_2} = \frac{1}{u_2} + \frac{1}{v_2}$$

i.e.

$$-\frac{1}{260} = -\frac{1}{130-d} + \frac{1}{200-d}$$

$$-(130-d)(200-d) = -260(200-d) + 260(130-d)$$

i.e.

$$-26\,000 + 330d - d^2 = -52\,000 + 260d + 33\,800 - 260d$$
$$= -18\,200$$

$$d^2 - 330d + 7800 = 0$$
$$d = 25.6 \quad (\text{or } 304.4)$$

therefore

$$v_2 = 200.0 - 25.6 = 174.4$$

i.e. the internal focussing lens will be 174 mm away from the diaphragm when focussed to infinity.

5.5.4 The tacheometric telescope (external focussing) (Fig. 5.42)

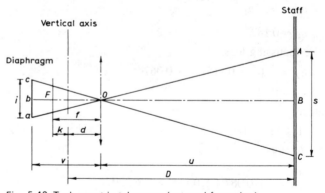

Fig. 5.42 Tacheometric telescope (external focussing)

Let a, b and c represent the three horizontal cross hairs of the diaphragm, a and c being a distance i apart and b midway between a and c.

With the telescope in focus, these lines will coincide with the image of the staff observed at A, B and C respectively; the distance $AC = s$ is known as the staff intercept. The line bOB represents the line of collimation of the telescope, with bO and OB conjugate focal lengths of the lens, v and u, respectively. The principal focal length of the lens is $FO(f)$, whilst the vertical axis is a distance k from the principal focus F.

Because the triangles acO and ACO are similar,

$$\frac{AC}{ac} = \frac{OB}{ob} \quad \text{or} \quad \frac{s}{i} = \frac{u}{v} \tag{5.29}$$

Using the lens formula, equation 5.19,

$$\frac{1}{f} = \frac{1}{u} + \frac{1}{v}$$

and multiplying both sides by uf gives

$$u = f + \frac{uf}{v}$$

Substituting the value of u/v from equation 5.29,

$$u = s\frac{f}{i} + f$$

Thus the distance from the vertical axis to the staff is given as

$$D = s\frac{f}{i} + (f + d) \tag{5.30}$$

This is the formula which is applied for normal stadia observations with the telescope horizontal and the staff vertical.

The ratio $f/i = m$ is given a convenient value of, say, 100 (occasionally 50), whilst the additive constant $(f+d) = K$ will vary depending upon the instrument.

The formula may thus be simplified as

$$D = ms + K \tag{5.31}$$

Example 5.6 The constants m and K for a certain instrument were 100 and 0.1 m respectively. Readings taken on to a vertical staff were 0.960, 1.298, 1.636 respectively, the telescope being horizontal. Calculate

(a)   the horizontal distance from the instrument to the staff,
(b)   the reduced level of the staff station if the reduced level of the instrument station was 31.583 m AOD and the height of the trunnion axis 1.472 m.

(a)   The stadia intercept $s = 1.636 - 0.960 = 0.676$
By equation 5.31,

$$D = ms + K$$
$$= 100 \times 0.676 + 0.1 = 67.7 \text{ m}$$

(b)   If the instrument was set at 31.583 m then the level of the staff station

$$= 31.583 + 1.472 - 1.298 = 31.757 \text{ m}$$

5.5.5   The anallatic lens

To overcome the need for the additive constant K in some forms of external focussing telescopes, a convex lens (known as an anallatic lens) was introduced between the objective and the diaphragm.

As most telescopes are now internal focussing, the concept of an anallatic telescope is only of academic interest and thus the tacheometric formula

$$D = f(s/i) + (f+d)$$

is no longer applicable as the internal focussing instrument cannot be considered anallatic even though the variation in the focal length of the objective system is generally considered negligible for most practical purposes.

The manufacturers aim at a low value for K and in many cases the telescopes are so designed that when focussed at infinity, the focussing lens is midway between the objective and the diaphragm.

5.5.6   The tacheometric telescope (internal focussing) (Fig. 5.43)

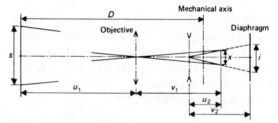

Fig. 5.43 Tacheometric telescope (internal focussing)

To find the spacing of the stadia lines to give a multiplying factor m for a given sight distance: let i = stadia interval and s = stadia intercept.

For the convex lens (objective),

$$\frac{x}{s} = \frac{v_2}{u_2} = M_1, \quad \text{i.e. } x = \frac{sv_1}{u_1} = M_1 s$$

where M_1 is the magnifying power.

For the concave lens (focussing),

$$\frac{i}{x} = \frac{v_2}{u_2} = M_2, \quad \text{i.e. } i = x\frac{v_2}{u_2} = M_2 x$$

therefore $\qquad\qquad\qquad i = M_1 M_2 s \qquad\qquad\qquad (5.32)$

But

$\qquad\qquad$ distance $D = ms$

therefore $\qquad\qquad s = D/m$

therefore $\qquad\qquad i = D M_1 M_2/m \qquad\qquad\qquad (5.33)$

Example 5.7   An internally focussing telescope has an objective 150 mm from the diaphragm. The respective focal lengths of the objective and the internal focussing lens are 125 mm and 250 mm. Find the distance apart the stadia lines should be to have a multiplying factor of 100 for an observed distance of 150 m.

As $l = 150$ mm the axis lies at 75 mm from the objective.

At 150.000 m the object will give

$$u_1 = 150\,000 - 75 = 149\,925 \text{ mm}; \quad f_1 = 125; \quad f_2 = 250 \text{ mm}$$

By equation 5.26,

$$v_1 = u_1 f_1/(u_1 - f_1) = 125.1 \text{ mm}$$

By equation 5.28,

$$d = \tfrac{1}{2}\{(l+v_1) \pm \sqrt{[(l-v_1)(l+4f_2-v_1)]}\}$$
$$= \tfrac{1}{2}\{275.1 \pm \sqrt{(24.9 \times 1024.9)}\}$$
$$= 57.7 \quad \text{or} \quad 217.4$$

(But 217.4 does not fit the physical condition.) Then

$$v_2 = l - d = 150.0 - 57.7 = 92.3 \text{ mm}$$
$$u_2 = v_1 - d = 125.1 - 57.7 = 67.4 \text{ mm}$$
$$i = Dv_1v_2/mu_1u_2$$
$$= \frac{150 \times 10^3 \times 125.1 \times 92.3}{100 \times 149\,925 \times 67.4}$$
$$= 1.714, \text{ say } 1.71 \text{ mm}$$

Example 5.8   What errors will be introduced if the previous instrument is used for distances varying from 30 to 150 m?

At 30 m
$$u_1 = 30\,000 - 75 = 29\,925$$

$$v_1 = \frac{29\,925 \times 125}{29\,925 - 125} = 125.524$$

$$d = \tfrac{1}{2}[(l+v_1) \pm \sqrt{(l-v_1)(l+4f_2-v_1)}]$$
$$= \tfrac{1}{2}[275.524 \pm \sqrt{(24.476 \times 1024.476)}]$$
$$= 58.59$$
$$v_2 = 150.00 - 58.59 = 91.41$$
$$u_2 = 125.52 - 58.59 = 66.93$$
$$s = i \times u_1 u_2/v_1 v_2$$

$$= \frac{1.714 \times 29\,925 \times 66.93}{125.524 \times 91.41}$$

$$= 299.188 \text{ mm}$$

The value should $= 300.000$ mm

Error $= \overline{\quad 0.812 \text{ mm}}$

$= \quad 0.081$ m per 30 m

Therefore $\quad D = \quad 29.919$ m and should be 30.000 m

Using the same procedures gives:

at   30 m, error = 0.081 m
at   50 m, error = 0.062 m
at 100 m, error = 0.038 m
at 150 m, error = 0.000 m

Example 5.9   An internal focussing telescope has an object glass of 200 mm focal length. The distance between the object glass and the diaphragm is 250 mm. When the telescope is at infinity focus, the internal focussing lens is exactly midway between the objective and the diaphragm. Determine the focal length of the focussing lens.

At infinity focus the optical centre of the focussing lens lies on the line joining the optical centre of the objective and the cross-hairs, but deviates laterally 0.025 mm from it when the telescope is focussed at 7.5 m. Calculate the angular error in seconds due to this cause.   (LU)

We refer to Fig. 5.44.

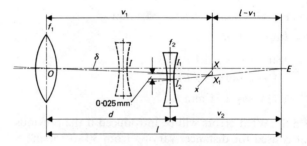

Fig. 5.44

With the telescope focussed at infinity, $v_1 = f_1$.
For the focussing lens,

$$\frac{1}{f_2} = \frac{1}{u_2} - \frac{1}{v_2} = \frac{1}{v_1 - d} - \frac{1}{l - d}$$

$$= \frac{1}{f_1 - d} - \frac{1}{l - d}$$

$$= \frac{1}{200 - 125} - \frac{1}{250 - 125}$$

$f_2 = 187.5$ mm (focal length of focussing lens)

With focus at 7.5 m (assuming 7.5 m from object lens):

$$u_1 = 7.5 \times 10^3$$

therefore

$$v_1 = \frac{u_1 f_1}{u_1 - f_1} = \frac{7.5 \times 10^3 \times 200}{7.5 \times 10^3 - 200} = 205.48 \text{ mm}$$

$$l + v_1 = 250.00 + 205.48 = 455.48$$
$$l - v_1 = 250.00 - 205.48 = 44.52$$
$$d = \tfrac{1}{2}\{455.48 \pm \sqrt{[44.52(750.00 + 44.52)]}\}$$

(Equation 5.28)

$$= \tfrac{1}{2}(455.48 \pm 188.07)$$
$$= 321.78 \quad \text{or} \quad 133.71 \,\text{mm}$$

With the focus at 7.5 m the image would appear at x, neglecting the internal focussing lens, i.e. $OX = v_1$. With the focussing lens moving off line, the line of sight is now $EX_1 I_2$ and all images produced by the objective appear as on this line.

The line of sight through the objective is thus displaced XX_1 in the length v_1.

To calculate XX_1,

$$\frac{XX_1}{I_1 I_2} = \frac{XE}{I_1 E}$$

i.e.
$$x = \frac{0.025 \times (l - v_1)}{l - d}$$

$$= \frac{0.025 \times (250.00 - 205.48)}{250.00 - 133.71} = 0.009\,57 \,\text{mm}$$

To calculate the angular error (δ),

$$\tan \delta = \frac{XX_1}{OX} = \frac{x}{v_1}$$

$$\delta'' = \frac{206\,625 \times 0.009\,57}{205.48} = 9.6''$$

5.6 Instrumental errors in the theodolite

5.6.1 Eccentricity of the horizontal circle

In Fig. 5.45, let O_1 = vertical axis

$\qquad\qquad O_2$ = graduated circle axis

$\qquad O_1 O_2 = e$ = eccentricity

$\qquad O_2 A_1 = O_1 A_2 = r$

If the graduated circle (centre O_2) is not concentric with the vertical axis (centre O_1) containing the readers A and B, the recorded value θ will be in error by the angle α.

As the instrument is rotated, the readers will successively occupy positions $A_1 B_1$, $A_2 B_2$, $A_3 B_3$.

$$\alpha_1 = \theta_1 - \phi$$

$$= \tan^{-1} \frac{O_2 E}{A_2 E}$$

$$= \tan^{-1} \frac{e \sin \phi}{r - e \cos \phi} \tag{5.34}$$

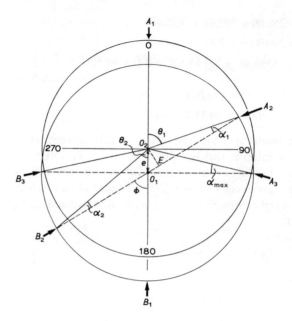

Fig. 5.45

therefore

$$\alpha_1 \simeq \frac{e \sin \phi}{r} \qquad (5.35)$$

Since e is small compared with r and as α is small, $\alpha_{rad} \simeq \tan \alpha$.
Similarly

$$\tan \alpha_2 = \frac{e \sin \phi}{r + e \cos \phi} \qquad (5.36)$$

$$\alpha_2 \simeq \frac{e \sin \phi}{r} \qquad (5.37)$$

If the readers are $180°$ apart, $A_1 O_1 B_1$ is a straight line and the mean of the recorded values θ gives the true value of the angle ϕ.

i.e. $\qquad \phi = \theta_1 - \alpha_1 = \theta_2 + \alpha_2$

therefore

$$2\phi = \theta_1 + \theta_2 \quad \text{as } \alpha_1 \simeq \alpha_2$$
$$\phi = \tfrac{1}{2}(\theta_1 + \theta_2) \qquad (5.38)$$

Notes
(1)   On the line $O_1 O_2$, $\alpha = 0$.
(2)   At $90°$ to this line, $\alpha = $ maximum.
(3)   If the instrument has only one reader, the angle should be repeated by transitting the telescope and rotating anticlockwise, thus giving recorded values $180°$ from original values. This is of particular importance with glass arc theodolites in which the graduated circle is of small radius.

To determine the amount of eccentricity and index error on the horizontal circle:

(1) Set index A to $0°$ and read displacement of index B from $180°$, i.e. δ_1.
(2) Set index B to $0°$ and read displacement of index A from $180°$, i.e. δ_2.
(3) Repeat these operations at a constant interval around the plate, i.e. zeros at multiples of $10°$.

If the readers A and B are diametrically opposed, let $\delta_1 =$ displacement of reader B_1 from $180°$, Fig. 5.46.

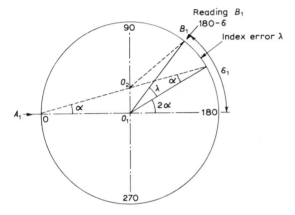

Fig. 5.46

Index A_1 at $0°$.
Index B_1 at $180° - \delta_1$, i.e. $180° - (2\alpha + \lambda)$.
Let $\delta_2 =$ displacement of reader A_2 from $180°$, Fig. 5.47.
Index B_2 at $0°$.
Index A_2 at $180° - \delta_2$, i.e. $180° - (2\alpha + \lambda)$.
If there is no eccentricity and A and B are $180°$ apart, then $\delta_1 = \delta_2 = 0$.

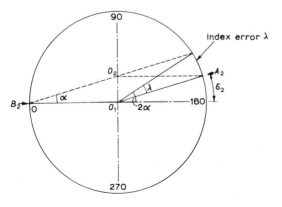

Fig. 5.47

If there is eccentricity and A and B are 180° apart, then $\delta_2 = \delta_2 = $ a constant.

If there is no eccentricity and A and B are not 180° apart, then $+\delta_1 = -\delta_2$, i.e. equal, but opposite in sign.

If there is eccentricity and A and B are not 180° apart, then δ_1 and δ_2 will vary in magnitude as the zero setting is consecutively changed around the circle of centre O_2, but their difference will remain constant.

A plotting of the values using a different zero for each pair of index settings will give the results shown in Fig. 5.48.

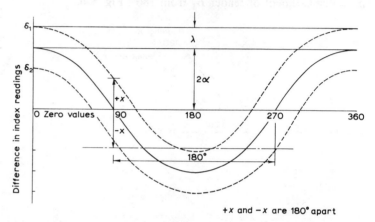

+x and −x are 180° apart

Fig. 5.48

5.6.2 The line of collimation not perpendicular to the trunnion axis (Fig. 5.49)

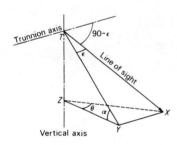

Fig. 5.49 Line of collimation not perpendicular to the trunnion axis

Let the line of sight make an angle of $90° \pm \varepsilon$ with the trunnion axis inclined at an angle α.

The angular error θ in the horizontal plane due to the error ε may be found by reference to Fig. 5.49.

$$\tan\theta = \frac{XY}{YZ}$$

$$\tan\varepsilon = \frac{XY}{TY}$$

i.e. $XY = TY\tan\varepsilon$

But $TY = YZ\sec\alpha$

i.e. $YZ = TY\cos\alpha$

therefore

$$\tan\theta = \frac{TY\tan\varepsilon}{TY\cos\alpha} = \tan\varepsilon\sec\alpha \tag{5.39}$$

If θ and ε are small, then

$$\theta = \varepsilon\sec\alpha \tag{5.40}$$

If the observations are made on the same face to two stations of elevations α_1 and α_2, then the error in the horizontal angle will be

$$\pm(\theta_1 - \theta_2) = \pm\tan^{-1}(\tan\varepsilon\sec\alpha_1) - \tan^{-1}(\tan\varepsilon\sec\alpha_2) \quad (5.41)$$

$$\pm(\theta_1 - \theta_2) \simeq \pm\varepsilon(\sec\alpha_1 - \sec\alpha_2) \quad (5.42)$$

On changing face, the error will be of equal value but opposite in sign. Thus the mean of face left and face right eliminates the error due to collimation in azimuth. The sign of the angle, i.e. elevation or depression, is ignored in the equation.

The extension of a straight line, Fig. 5.50. If this instrument is used to extend a straight line by transitting the telescope, the following conditions prevail.

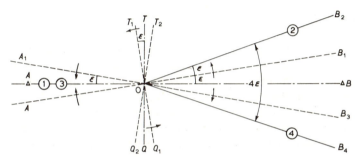
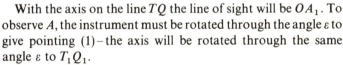

Fig. 5.50

With the axis on the line TQ the line of sight will be OA_1. To observe A, the instrument must be rotated through the angle ε to give pointing (1) – the axis will be rotated through the same angle ε to T_1Q_1.

On transitting the telescope the line of sight will be $(180° - 2\varepsilon)$ $A_1OB_1 = AOB_2$. B_2 is thus fixed – pointing (2).

On changing face the process is repeated – pointing (3) – and then pointing (4) will give position B_4.

The angle $B_2OB_4 = 4\varepsilon$, but the mean position B will be the correct extension of the line AO.

The method of adjustment follows the above process, B_2B_4 being measured on a horizontal scale.

The collimation error may be corrected by moving the telescope graticule to read on B_3, i.e. $\frac{1}{4}B_2B_4$.

5.6.3   The trunnion axis not perpendicular to the vertical axis (Fig. 5.51)

The trunnion (horizontal or transit) axis should be at right angles to the vertical axis; if the plate bubbles are centralised, the trunnion axis will not be horizontal if a trunnion axis error occurs. Thus the line of sight, on transitting, will sweep out a plane inclined to the vertical by an angle equal to the tilt of the trunnion axis.

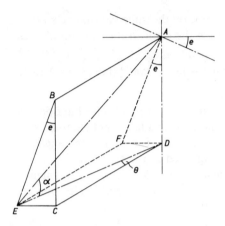

Fig. 5.51 Trunnion axis not perpendicular to the vertical axis

If the instrument is in correct adjustment, the line of sight sweeps out the vertical plane $ABCD$, Fig. 5.51.

If the trunnion axis is tilted by an angle e, the line of sight sweeps out the inclined plane $ABEF$.

The line of sight is assumed to be AE. To correct for the tilt of the plane it is necessary to rotate the horizontal bearing of the line of sight by an angle θ, to bring it back to its correct position. Thus

$$\sin \theta = \frac{EC}{ED} = \frac{BC \tan e}{ED}$$

$$= \frac{AD}{ED} \tan e = \tan \alpha \tan e$$

i.e. $\sin \theta = \tan \alpha \tan e$ $\hspace{2em}$ (5.43)

and if θ and e are small, then

$$\theta = e \tan \alpha \hspace{2em} (5.44)$$

where θ = correction to the horizontal bearing

e = trunnion axis error

α = angle of inclination of sight

On transiting the telescope, the inclination of the trunnion axis will be in the opposite direction but of equal magnitude. Thus the mean of face left and face right eliminates the error.

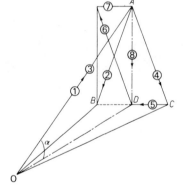

Fig. 5.52 Field test of trunnion axis error

Method of adjustment (Fig. 5.52)

(1)   Observe a highly elevated target A, e.g. a church spire.

(2)   With horizontal plates clamped, depress the telescope to observe a horizontal scale B.

(3)   Change face and re-observe A.

(4)   As before, depress the telescope to observe the scale at C.

(5)   Rotate horizontally to D midway between B and C.

(6)   Elevate the telescope to the altitude of A.

(7)   Adjust the trunnion axis until A is observed.

(8)   On depressing the telescope, D should now be observed.

5.6.4 Vertical axis not truly vertical (Fig. 5.53)

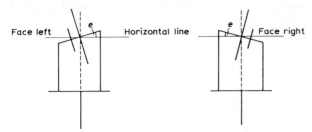

Trunnion axis not perpendicular to the vertical axis

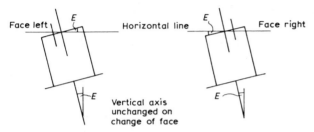

Vertical axis not truly vertical

Fig. 5.53

If the instrument is in correct adjustment but the vertical axis is not truly vertical by an angle E, then the horizontal axis will not be truly horizontal by a maximum angle E.

Thus the error in bearing due to this will be

$$E' \tan \alpha \qquad (5.45)$$

This is a variable error, E', dependent on the direction of pointing relative to the direction of tilt of the vertical axis, and its effect is *not* eliminated on change of face, as the vertical axis does not change in position or inclination.

In Fig. 5.54, the true horizontal angle (θ) $A_1 OB_1$ = angle (ϕ) $A_2 OB_2 - (c_1) E_1 \tan \alpha_1 + (c_2) E_2 \tan \alpha_2$.

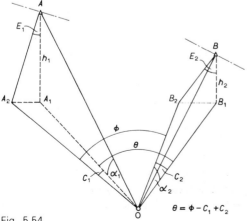

$$\theta = \phi - C_1 + C_2$$

Fig. 5.54

Thus the error in pointing (θ) is dependent on (1) the tilt of the axis E, which itself is dependent on the direction of pointing, varying from maximum (E) when on the line of tilt of the vertical axis to zero when at 90° to this line, and (2) the angle of inclination of the line of sight.

To measure the values of e and E' a *striding level* is used, Fig. 5.55.

Let the vertical axis be inclined at an angle E.

Let the trunnion axis be inclined at an angle e.

Let the bubble be out of adjustment by an angle β.

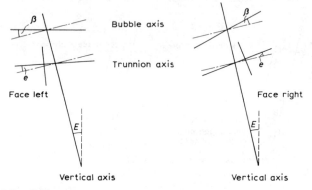

Fig. 5.55

Then *face left*

tilt of the trunnion axis $= E' - e$

tilt of the bubble axis $= E' - \beta$

face right

tilt of the trunnion axis $= E' + e$

tilt of the bubble axis $= E' + \beta$

mean tilt of the trunnion axis $= \frac{1}{2}[(E' - e) + (E' + e)] = E'$

$$(5.46)$$

mean tilt of the bubble axis $= \frac{1}{2}[(E' - \beta) + (E' + \beta)] = E'$

$$(5.47)$$

Therefore the mean correction taking all factors into account is

$$E' \tan \alpha \qquad (5.48)$$

NB: The value of E' is related to the direction of observation and its effective value will vary from maximum to nil. Tilting level readings should be taken for each pointing.

If E is the maximum tilt of the axis in a given direction, then

$$E' = E \cos \theta \qquad (5.49)$$

where θ is the angle between pointing and direction of maximum tilt.

Then the bubble recording the tilt does not strictly need to be in adjustment, nor is it necessary to change it end for end as some authors suggest, the mean of face left and face right giving the true value.

If the striding level is graduated from the centre outwards for n pointings and $2n$ readings of the bubble, then the correction to the mean observed direction is given by

$$c = \frac{d}{2n}(\Sigma L - \Sigma R)\tan\alpha \qquad (5.50)$$

where c = the correction in seconds
  d = the value of one division of the bubble in seconds
  ΣL = the sum of readings of the left-hand end of the bubble
  ΣR = the sum of readings of the right-hand end of the bubble
  α = the angle of inclination of sight
  n = the number of pointings

The sign of the correction is positive as stated. Any changes depend upon the sign of $\Sigma L - \Sigma R$ and that of α.

NB: The greater the change in the value of α the greater the effect on the horizontal angle.

In many modern theodolites, e.g. Wild T2, an automatic index system is used and the manufacturer suggests that the tilt of the trunnion axis (E') for any given pointing may be obtained as follows:

The object X is sighted and the horizontal reading taken as usual (Fig. 5.56). With the instrument on face left and the vertical circle clamped the instrument is turned through an anticlockwise angle of 90° and the zenith angle Z_L read. Similarly turning clockwise through 90° from the original pointing and Z_R is read.

The value of E' (at right angles to the original pointing) is given as:

$$E' = (Z_R - Z_L)/2 \qquad (5.51)$$

or $E' = (\alpha_L - \alpha_R)/2$ for vertical angles (5.52)

If E' is +ve then the trunnion axis will be high on the left-hand side.

As with the striding level the correction to the horizontal circle reading will be

$$\delta\theta = E'\tan\alpha$$

where α = the vertical angle of the pointing.

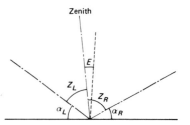

Fig. 5.56

5.6.5   Vertical circle index error (Fig. 5.57)

When the telescope is horizontal, the altitude bubble should be central and the circle index reading zero (90° or 270° on whole circle reading instruments).

If
  the true angle of altitude   = α
  the recorded angles of altitude = α_1 and α_2
  the vertical collimation error   = ϕ
  the circle index error   = θ,

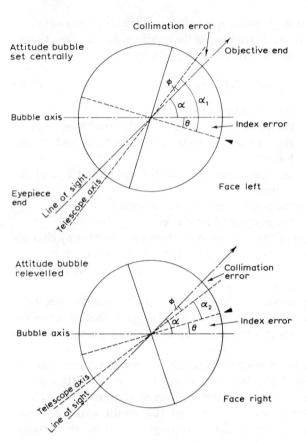

Fig. 5.57

then

$$\text{recorded value (FL)} = \alpha = \alpha_1 - \phi - \theta$$

$$\text{(FR)} = \alpha = \alpha_2 + \phi + \theta$$

therefore

$$\alpha = \tfrac{1}{2}(\alpha_1 + \alpha_2) \qquad (5.53)$$

Thus, provided the altitude bubble is centralised for each reading, the mean of face left and face right will give the true angle of altitude.

If the bubble is not centralised then bubble error will occur, and, depending on the recorded displacement of the bubble at the objective and eyepiece ends, the sensitivity will indicate the angular error. As the bubble is rotated the index is also rotated.

Thus θ will be subjected to an error of $\pm\tfrac{1}{2}(0-E)\delta''$, where $\delta'' = $ the angular sensitivity of the bubble.

If the objective end of the bubble is higher than the eyepiece end and on face left, i.e. $O_L > E_L$, then θ will be decreased by $\tfrac{1}{2}(O_L - E_L)\delta''$, i.e.

$$\text{FL} \quad \alpha = \alpha_1 - \phi - \{\theta - \tfrac{1}{2}(O_L - E_L)\delta''\}$$

and   FR $\alpha = \alpha_2 + \phi + \{\theta + \frac{1}{2}(O_R - E_R)\delta''\}$

$$\alpha = \frac{1}{2}(\alpha_1 + \alpha_2) + \frac{\delta''}{2}\{(O_L + O_R) - (E_L - E_R)\}$$

$$= \frac{1}{2}(\alpha_1 + \alpha_2) + \frac{\delta''}{4}(\Sigma O - \Sigma E) \qquad (5.54)$$

To test and adjust the index error
(1)   Centralise the altitude bubble and set the telescope to read zero (face left).
(2)   Observe a card on a vertical wall—record the line of sight at A.
(3)   Transit the telescope and repeat the operation. Record the line of sight at B.
(4)   Using the slow motion screw (vertical circle) observe the midpoint of AB. (The line of sight will now be horizontal.)
(5)   Bring the reading index to zero and then adjust the bubble to its midpoint.

Example 5.10   *Trunnion axis error.* The following are the readings of the bubble ends A and B of a striding level which was placed on the trunnion axis of a theodolite and then reversed ('Left' indicates the left-hand side of the trunnion axis when looking along the telescope from the eyepiece end with the theodolite face right.)

A on left 11.0, B on right 8.4
B on left 10.8, A on right 8.6

One division of the striding level corresponds to 15″. All adjustments other than the horizontal trunnion axis adjustment of the theodolite being presumed correct, determine the true horizontal angle between P and Q in the following observations (taken with the theodolite face left).

Object	Horizontal circle	Vertical circle	
P	158° 20′ 30″	42° 24′	
Q	218° 35′ 42″	15° 42′	(LU)

By equation 5.50,

$$\text{correction to bearing} = \frac{d}{2n}(\Sigma L - \Sigma R)\tan\alpha$$

to P $c = \dfrac{15}{4}[(11.0 + 10.8) - (8.4 + 8.6)]\tan 42° 24'$

$\qquad = \dfrac{15 \times 4.8}{4}\tan 42° 24'$

$\qquad = -18''\tan 42° 24' = -16''$

to Q $c = -18''\tan 15° 42' = -5''$

NB: The correction would normally be positive when using

the general notation, but the face is changed by the definition given in the problem.

true bearing to $P = 158° 20' 30'' - 16'' = 158° 20' 14''$
true bearing to $Q = 218° 35' 42'' - 5'' = \underline{218° 35' 37''}$
true horizontal angle $\qquad = \underline{60° 15' 23''}$

Example 5.11 The mean horizontal readings to measure an angle ABC have been recorded with a theodolite fitted with an automatic indexing system as:

to A $84° 38' 33''$ to C $120° 39' 35''$

At the time additional observations were taken to compensate for vertical axis error.

Vertical circle readings
To A		FL	$059° 31' 02''$	FR $300° 29' 04''$
$-90°$ to A			$059° 31' 00''$	
$+90°$ to A			$059° 31' 35''$	
To C		FL	$091° 03' 30''$	FR $268° 56' 47''$
$-90°$ to C			$091° 03' 03''$	
$+90°$ to C			$091° 04' 05''$	

Calculate
(a) the most probable value of the angle ABC;
(b) the maximum tilt of the trunnion axis and the direction of the pointing of the telescope at this time.

(a) Reduction of the vertical circle readings:

			Vertical angles (α)	Mean
To A	FL	$059° 31' 02''$	$30° 28' 58''$	
	FR	$300° 29' 04''$	$30° 29' 04''$	$30° 29' 01''$
To C	FL	$091° 03' 30''$	$-01° 03' 30''$	
	FR	$268° 56' 47''$	$-01° 03' 13''$	$-01° 03' 22''$

Horizontal angle $ABC = 120° 39' 35'' - 84° 38' 33''$
$\qquad\qquad\qquad\qquad = 36° 01' 02''$
To determine the correction to the angle ABC

$$E' = (Z_R - Z_L)/2$$

$-90°$ to A	$059° 31' 00''$	
$+90°$ to A	$059° 31' 35''$	$E'_A = +17.5''$
$-90°$ to C	$091° 03' 03''$	
$+90°$ to C	$091° 04' 05''$	$E'_C = +31.0''$

By equation 5.48,

$$\delta\theta = E' \tan \alpha$$

$$\delta\theta_A = +17.5 \tan 30° 29' 01'' \qquad = +10.3''$$
$$\delta\theta_C = +31.0 \tan (-01° 03' 22'') \; = - \; 0.6''$$
$$\delta\theta_C - \delta\theta_A = -10.9''$$

Then $\theta_{ABC} = 36° 01' 02'' - 11'' = 36° 00' 51''$

(b) To find the maximum tilt of the vertical axis, and thus the maximum tilt of the horizontal plate.

Station observed	Horizontal circle	Tilt (E')	Direction of dip
A	084° 39'	17.5"	354° 39'
B	120° 40'	31.0"	030° 40'

By equation 7.5,

$$\tan \lambda = \tan 17.5'' \cot 31.0'' \operatorname{cosec} 36° 01' - \cot 36° 01'$$

$\lambda = -26° 36'$ (angle between full dip and apparent dip)

Bearing of full dip of the horizontal plate $= 030° 40' + 26° 36'$
$$= 056° 16'$$

By equation 7.7,

$$\cot \delta = \cot 31'' \cos (-26° 36')$$

$$\delta = 37'' \text{(maximum tilt)}   = E$$

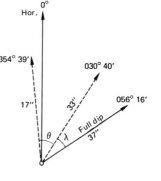

Fig. 5.58

Thus the vertical axis has a tilt of 37" and this is equivalent to the maximum tilt of the trunnion axis when the telescope is pointing with a circle reading of 146° 16' (Fig. 5.58).

Example 5.12 In an underground traverse the following mean values were recorded from station B on to stations A and C:

Station observed	Horizontal angle	Vertical angle	Striding level readings (L)	(R)
	136° 21' 32"			
A		−13° 25' 20"	17.4	5.8
C		+47° 36' 45"	14.5	2.7

Striding level: 1 division = 10 seconds
bubble graduated 0 to 20
height of instrument at B 1.411 m
height of target at A 1.042 m
at C 1.570 m
ground length AB 78.126 m
BC 138.026 m
Calculate the gradient of the line AC. (RICS)

Striding level corrections

to A $(L+R)/2 = (17.4 - 5.8)/2 = 11.6$

i.e. the centre of bubble is 1.6 to left of centre of graduations.

to C $(L+R)/2 = (14.5 - 2.7)/2 = 8.6$

i.e. centre of bubble is 1.4 to right of centre of graduations.

The same results may be obtained by using the basic equation 5.50 and transposing the readings as though the graduation were from the centre of the bubble. That is,

to A 17.4 (L) becomes 7.4 (L)
5.8 (R) becomes 4.2 (R)

$$\frac{\Sigma L - \Sigma R}{2} = \frac{7.4 - 4.2}{2} = +1.6$$

to *C*   14.5 (*L*) becomes 4.5 (*L*)

2.7 (*R*) becomes 7.3 (*R*)

$$\frac{\Sigma L - \Sigma R}{2} = \frac{4.5 - 7.3}{2} = -1.4$$

Applying these values to equation 5.48,

correction to $A = +1.6 \times 10 \times \tan(-13° 25')$

$= -3.8''$

correction to $C = -1.4 \times 10 \times \tan(+47° 37')$

$= -15.3''$

total angle correction $= -11.5''$

corrected horizontal angle $= 136° 21' 32'' - 11.5''$

$= 136° 21' 20''$

To find true inclination of the ground and true distances (Fig. 5.59)

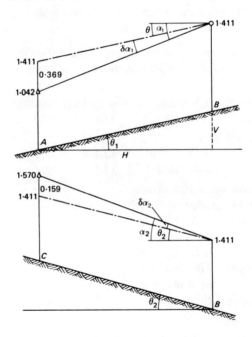

Fig. 5.59

Line *AB*

$$\delta\alpha_1 = \sin^{-1} \frac{0.369 \cos 13° 25' 20''}{78.126} = -0° 15' 47''$$

$\alpha_1 = 13° 25' 20''$

$\theta = 13° 09' 33''$

horizontal length (*H*) $AB = 78.126 \cos 13° 09' 33'' = 76.075$ m

vertical difference (*V*) $AB = 78.126 \sin 13° 09' 33'' = 17.786$ m

Line BC

$$\delta\alpha_2 = \sin^{-1} \frac{-0.159 \cos 47° 36' 45''}{138.026} = -0° 02' 40''$$

$$\alpha_2 = 47° 36' 45''$$
$$\theta = 47° 34' 05''$$

horizontal length $BC = 138.026 \cos 47° 34' 05'' = 93.128 \text{ m}$

vertical difference $BC = 138.026 \sin 47° 34' 05'' = 101.874 \text{ m}$

difference in height $AC = 17.786 + 101.874 \qquad = 119.660 \text{ m}$

To find the horizontal length AC:
In triangle ABC,

$$\tan \frac{A - C}{2} = \frac{93.128 - 76.075}{93.128 + 76.075} \tan \frac{(180° - 136° 21' 20'')}{2}$$

$$\frac{A - C}{2} = 2° 18' 40''$$

$$\frac{A + C}{2} = 21° 49' 20''$$

therefore

$$A = 24° 08' 00''$$

Then

$$AC = 93.128 \sin 136° 21' 20'' \operatorname{cosec} 24° 08' 00'' = 157.207 \text{ m}$$

Gradient

$$AC = 119.660 \text{ in } 157.207$$
$$= \underline{1 \text{ in } 1.314}$$

Example 5.13 (a)   Show that when a pointing is made to an object which has a vertical angle h with a theodolite having its trunnion axis inclined at a small angle i to the horizontal, the error introduced into the horizontal circle reading as a result of the trunnion axis tilt is $i \tan h$.

(b)   The observations set out below have been taken at a station P with a theodolite, both circles of which have two index marks. On face left, the vertical circle nominally records $90°$ minus the angle of elevation. The plate bubble is mounted parallel to the trunnion axis and is graduated with the zero of the scale at the centre of the tube. One division represents $20''$.

The intersection of the telescope cross-hairs was set on signals A and B on both faces of the theodolite. The means of the readings of the circle and the plate bubble readings were:

Signal	Face	Horizontal circle	Vertical circle	Midpoint of bubble
A	left	166° 39' 15"	90° 00' 15"	1.0 division towards circle
	right	346° 39' 29"	270° 00' 17'	1.0 division towards circle

Signal	Face	Horizontal circle	Vertical circle	Midpoint of bubble
B	left	301° 18′ 36″	80° 03′ 52″	central
	right	121° 18′ 30″	279° 56′ 38″	2.0 division towards circle

The vertical axis was then rotated so that the horizontal circle reading with the telescope in the face left position was 256° 40′; the reading of the midpoint of the bubble was then 0.4 division away from the circle.

If the effect of collimation error c on a horizontal circle reading is $c \sec h$, calculate the collimation error, the tilt of the trunnion axis and the index error of the theodolite, the attitude of the vertical axis when the above observations were taken, and the value of the horizontal angle APB. (NU)

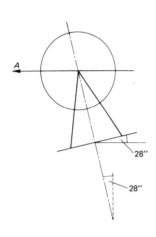

Fig. 5.60

At A (Fig. 5.60): As the bubble reading is equal and opposite, on change of face the horizontal plate is horizontal at 90° to the line of sight. The bubble is out of adjustment by 1 division = 20″ FL. At 90° to A the corrected bubble reading gives

$$\text{FL} \quad 0.4 + 1.0 = +1.4 \text{ div.} = +28''$$

The horizontal plate is thus inclined at 28″ as is the vertical axis, in the direction A.

At B (Fig. 5.61): The corrected bubble readings give

$$\left.\begin{array}{l} \text{FL} \quad 0.0 + 1.0 = +1.0 \\ \text{FR} \quad 2.0 - 1.0 = +1.0 \end{array}\right\} \text{ i.e. } +20''$$

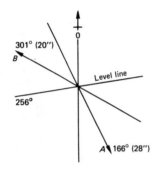

Fig. 5.61

NB: This value may be checked in equation 7.7 as follows:

$$\begin{array}{l} \text{apparent dip} = \text{full dip cosine angle between (Fig. 5.61)} \\ \qquad\qquad = 28'' \cos(301 - 256) \\ \qquad\qquad = 28'' \cos 45° = 20'' \end{array}$$

The effect of instrumental errors in pointings (Fig. 5.62).

Horizontal

Collimation $\theta_c = \pm c \sec h$ (if $h = 0, \theta_c = c$); the mean of faces left and right gives the correct value.

Trunnion axis $\theta_i = \pm i \tan h$ (if $h = 0, \theta_i = 0$); the mean of faces left and right gives the correct value.

Vertical axis $\theta_v = v \tan h$ (if $h = 0, \theta_v = 0$); the sign is dependent on the inclination of the axis. (FL inclination towards circle, with $+h$, θ_v is $-ve$.)

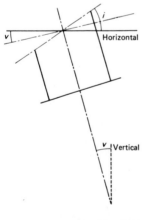

Fig. 5.62

Vertical

Index error $\delta h = \frac{1}{2}\{h_l - h_r\}$; the mean of faces left and right gives the correct value.

Application to given values:

At A FL $166° 39′ 15″ + c \sec h - v \tan h + i \tan h$

i.e. $166° 39′ 15″ + c$

     FR $346° 39′ 29″ - c$

FL must equal FR, therefore

$$166° \ 39' \ 15'' + c = 346° \ 39' \ 29'' - 180° - c$$
$$2c = +14''$$
$$c = +7'' \quad \text{(collimation error)}$$

At *B* FL $301° \ 18' \ 36'' + 7 \sec 9° \ 57' - 20 \tan 9° \ 57' + i \tan 9° \ 57''$

i.e. $301° \ 18' \ 36'' + 7.1'' - 3.5'' + 0.175i$

FR $121° \ 18' \ 30'' - 7.1'' - 3.5'' - 0.176$

FL must equal FR, therefore

$301° \ 18' \ 36'' + 7.1 - 3.5 + 0.175i = 121° \ 18' \ 30'' + 180 - 7.1 - 3.5 - 0.175i$

i.e. $0.35i = -20.2''$

$$i = -58'' \quad (\theta_i = -10.1'') \quad \text{(trunnion axis error)}$$

Corrected readings

FL *A* $166° \ 39' \ 15'' + 7''$ $= 166° \ 39' \ 22''$

   B $301° \ 18' \ 36'' + 7.1'' - 3.5'' - 10.1 = 301° \ 18' \ 29.5''$

                         Angle *APB* = $134° \ 39' \ 07.5''$

FR *A* $346° \ 39' \ 29'' - 7''$ $= 346° \ 39' \ 22''$

   B $121° \ 18' \ 30'' - 7.1 - 3.5 + 10.1$ $= 121° \ 18' \ 29.5''$

                         Angle *APB* = $134° \ 39' \ 07.5''$

Vertical angles

At *A* $\delta h = \frac{1}{2}\{(90 - 90° \ 00' \ 15'') - (270° \ 00' \ 17'' - 270)\} = -16''$

At *B* $\delta h = \frac{1}{2}\{(90 - 80° \ 03' \ 52'') - (279° \ 56' \ 38'' - 270)\} = -15''$

NB: The discrepancy is assumed to be an observational error.

5.7 The auxiliary telescope

This is used where steep sights are involved and in two possible forms:

(1) side telescope;
(2) top telescope.

5.7.1 Side telescope

There are two methods of using this form of telescope: (a) in adjustment and (b) out of adjustment with the main telescope.

Adjustment
(a) *Alignment* (Fig. 5.63). Observe a point *A* with the main telescope. Turn in azimuth to observe with the side telescope without altering the vertical circle. Raise or lower the side telescope until the horizontal cross-hair coincides with the target *A*. The horizontal hairs are now in the same plane.
(b) *Parallel lines of sight* (Fig. 5.64). If *x* is the eccentricity of the telescope at the instrument this should be constant between the lines of sight.

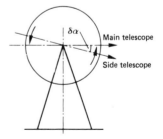

Fig. 5.63 Alignment of telescopes

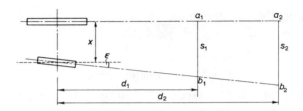

Fig. 5.64

At a distance d_1 from the instrument, a scale set horizontally may be read as a_1b_1 giving an intercept s_1, and then at d_2 readings a_2b_2 give intercept s_2.

If the lines of sight are parallel,

$$s_1 = s_2 = x$$

If not, the angle of convergence/divergence ε is given as

$$\varepsilon = \tan^{-1}\frac{s_2 - s_1}{d_2 - d_1} \tag{5.55}$$

If $s_2 > s_1$, the angle is $+ve$, i.e. diverging.
If $s_2 < s_1$, the angle is $-ve$, i.e. converging.
The amount of eccentricity x can be obtained from the same readings.

$$\frac{d_1}{d_2 - d_1} = \frac{s_1 - x}{s_2 - s_1}$$

i.e.
$$x = \frac{s_1(d_2 - d_1) - d_1(s_2 - s_1)}{(d_2 - d_1)}$$

$$= \frac{s_1 d_2 - s_2 d_1}{d_2 - d_1} \tag{5.56}$$

By making the intercept $s_2 = x$, the collimation of the auxiliary telescope can be adjusted to give parallelism of the lines of sight.

Observations with the side telescope
(a) *Vertical angles.* If the alignment is adjusted, then the true vertical angle will be observed.

If an angular error of $\delta\alpha$ exists between the main and the side telescope, then the mean of face left and face right observations is required, i.e.

$$\mathrm{FL} = \alpha_1 + \delta\alpha = \alpha$$

therefore $\mathrm{FR} = \alpha_2 - \delta\alpha = \alpha$

$$\alpha = \tfrac{1}{2}(\alpha_1 + \alpha_2) \tag{5.57}$$

(b) *Horizontal angles* (Fig. 5.65)
If θ = the true horizontal angle
                ϕ = the recorded horizontal angle
        δ_1 and δ_2 = errors due to eccentricity

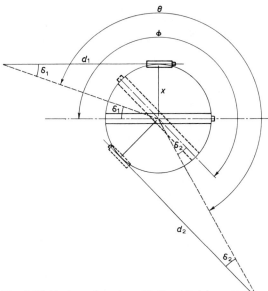

Fig. 5.65 Horizontal angles with the side telescope

then $\qquad \theta = \phi_1 - \delta_1 + \delta_2 \quad$ say $\quad$ FL

$\qquad\qquad \theta = \phi_2 + \delta_1 - \delta_2 \qquad$ FR

i.e. $\qquad \theta = \frac{1}{2}(\phi_1 + \phi_2)$ $\qquad\qquad\qquad$ (5.58)

Example 5.14 In testing the eccentricity of a side telescope, readings were taken on to levelling staves placed horizontally at X and Y, 30.48 m and 60.96 m respectively from the instrument.

$\qquad$ Readings at X $\quad$ 1.652 m $\quad$ 1.527 m

$\qquad\qquad\qquad$ at Y $\quad$ 1.003 m $\quad$ 0.850 m

Calculate (a) the collimation error (ε), (b) the eccentricity (x).

$\qquad s_1 = 1.652 - 1.527 = 0.125 \text{ m}$

$\qquad s_2 = 1.003 - 0.850 = 0.153 \text{ m}$

Then $\quad \varepsilon'' = \dfrac{206\,265 \times (0.153 - 0.125)}{60.96 - 30.48}$

$\qquad\qquad = 189.5'' = 03'\,10''$

and $\quad x = \dfrac{0.125 \times 60.96 - 0.153 \times 30.48}{30.48}$

$\qquad\qquad = 0.097 \text{ m}$

The effect of eccentricity x and collimation error ε. In Fig. 5.66, assuming small angles,

$\qquad$ angle $AOB = \delta''$

$\qquad\qquad = \dfrac{206\,265x}{d}$ $\qquad\qquad\qquad$ (5.59)

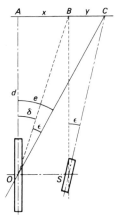

Fig. 5.66

angle $BSC = \varepsilon''$

$$= \frac{206\,265y}{d} \tag{5.60}$$

angle $AOC = e$

$$= \frac{206\,265(x+y)}{d} \tag{5.61}$$

$$e = \delta + \varepsilon \tag{5.62}$$

therefore

angle $BOC \simeq$ angle $BSC = \varepsilon$

As eccentricity x is constant, the angle δ is dependent upon the length of sight d.

As the collimation angle ε is constant, it has the same effect as the collimation error in the main telescope. It affects the horizontal angle by $\varepsilon \sec \alpha$, where α is the vertical angle.

Assuming the targets are at different altitudes, the true horizontal angle θ, Fig. 5.65, is given as

$$\theta = \phi_1 - (\delta_1 + \varepsilon \sec \alpha_1) + (\delta_2 + \varepsilon \sec \alpha_2) \text{ say FL} \tag{5.63}$$

Also $\quad \theta = \phi_2 + (\delta_2 + \varepsilon \sec \alpha_1) - (\delta_2 + \varepsilon \sec \alpha_2) \qquad \text{FR} \tag{5.64}$

therefore

$$\theta = \tfrac{1}{2}(\phi_1 + \phi_2) \tag{5.65}$$

Thus the mean of the face left and right values eliminates errors from all the above sources.

Example 5.15   Using the instrument of Example 5.13, the following data were recorded ($\varepsilon = 03'\,06''$ $x = 0.097$ m):

Station set at	Station observed	Horizontal circle	Vertical circle	Remarks
B	A	$0°\,05'\,20''$	$+30°\,26'$	Horizontal lengths
	C	$124°\,10'\,40''$	$-10°\,14'$	$AB = 30$ m $BC = 90$ m side telescope on right

Calculate the true horizontal angle ABC.   (RICS/M)

By equation 5.62,

true horizontal angle $(\theta) = \phi - (\delta_1 + \varepsilon \sec \alpha_1) + (\delta_2 + \varepsilon \sec \alpha_2)$

$$\delta_1'' = \frac{206\,265 \times 0.097}{30} = 667''$$

$$\delta_2'' = \frac{\delta_1}{3} = 222''$$

$\varepsilon \sec \alpha_1 = 186'' \sec 30°\,26' = 215.7''$, say $216''$

$\varepsilon \sec \alpha_2 = 186'' \sec 10°\,14' = 189.0''$

$\phi = 124°\,10'\,40'' - 0°\,05'\,20'' = 124°\,05'\,20''$

therefore

$$\theta = 124°\,05'\,20'' - (667'' + 216'') + (222'' + 189'')$$
$$= 124°\,05'\,20'' - 0°\,07'\,52'' = 123°\,57'\,28''$$

5.7.2 Top telescope

In this position the instrument can be used to measure horizontal angles only if it is in correct adjustment, as it is not possible to change face.

Adjustment
(a) *Alignment*. The adjustment is similar to that of the side telescope but observations are required by both telescopes on to a plumb line to ensure that the cross-hairs are in the same place.
(b) *Parallel lines of sight* (Fig. 5.67). Here readings are taken on vertical staves with the vertical circle reading zero.

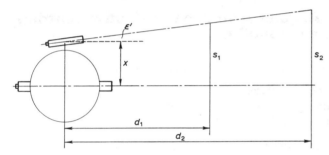

Fig. 5.67

The calculations are the same as for the side telescope:

$$\varepsilon' = \tan^{-1}\frac{s_2 - s_1}{d_2 - d_1} \qquad (5.66)$$

and

$$x' = \frac{s_1 d_2 - s_2 d_1}{d_2 - d_1} \qquad (5.67)$$

Measurement of vertical angles (Fig. 5.68). The angular error (δ) due to eccentricity is given as

$$\delta' = \tan^{-1}\frac{x}{d \sec \alpha} \qquad (5.68)$$

where $x =$ eccentricity
$\quad\quad\ d =$ horizontal distance
$\quad\quad\ \alpha =$ recorded vertical angle

If ε' is the error due to collimation then the true vertical angle (θ) is given as

$$\pm\theta = \pm(\alpha + \delta' + \varepsilon') \qquad (5.69)$$

assuming δ' and ε' small.

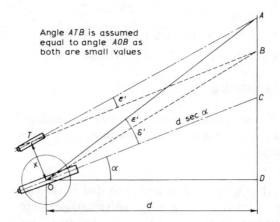

Angle *ATB* is assumed
equal to angle *AOB* as
both are small values

Fig. 5.68 Measurement of vertical angles with the top telescope

5.8 Angular error due to defective centring of the theodolite

The angular error depends on the following:

(a) linear displacement x;
(b) direction of the instrument B_1 relative to the station B;
(c) length of lines a and c.

The instrument may be set on the circumference of the circle of radius x.

No error will occur if the instrument is set up at B_1 or B_2 (Fig. 5.69), where A, B_1, B, B_2 and C lie on the arc of a circle.

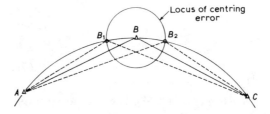

Fig. 5.69 Minimum error due to defective centring of the theodolite

In Fig. 5.70, let the instrument be set at B_1 instead of B. Therefore angle θ_1 is measured instead of θ, i.e.

$$\theta = \theta_1 - (\alpha + \beta)$$

Assume the misplumbing x to be in a direction ϕ relative to the line AB.

In triangle ABB_1,

$$\sin \alpha = \frac{x \sin \phi}{AB_1} \tag{5.70}$$

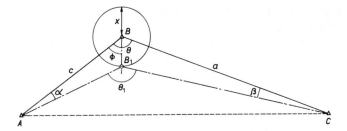

Fig. 5.70 The effects of centring errors

As the angle α is small,

$$\alpha'' = \frac{206\,265x \sin \phi}{c} \tag{5.71}$$

Similarly,

$$\beta'' = \frac{206\,265x \sin (\theta - \phi)}{a} \tag{5.72}$$

Therefore

total error $E = \alpha + \beta = 206\,265x \left(\dfrac{\sin \phi}{c} + \dfrac{\sin (\theta - \phi)}{a} \right)$ (5.73)

For maximum and minimum values,

$$\frac{dE}{d\phi} = 206\,265x \left(\frac{\cos \phi}{c} - \frac{\cos (\theta - \phi)}{a} \right) = 0$$

i.e. $\dfrac{\cos \phi}{c} = \dfrac{\cos (\theta - \phi)}{a}$

$$\cos \phi = \frac{c}{a} (\cos \theta \cos \phi + \sin \theta \sin \phi)$$

Divide by $\sin \phi$:

$$\cot \phi = \frac{c}{a} (\cos \theta \cot \phi + \sin \theta)$$

$$\cot \phi \left(1 - \frac{c \cos \theta}{a} \right) = \frac{c \sin \theta}{a}$$

therefore $\cot \phi = \dfrac{c \sin \theta}{a - c \cos \theta}$ (5.74)

Notes
(1)   If $\phi = 90°$, $\theta = 0$ or $180°$.
(2)   If $a \gg c$, then $\phi \to 90°$, i.e. the maximum error exists when ϕ tends towards $90°$ relative to the shorter line.
(3)   If $a = c$, $\phi = \theta/2$.
   Professor Briggs proves that the probable error in the measured angle is

$$e_{rad} = \pm \frac{2x}{\pi} \sqrt{\left(\frac{1}{a^2} + \frac{1}{c^2} - \frac{2 \cos \theta}{ac} \right)} \tag{5.75}$$

Example 5.16   The centring error in setting up the theodolite at station B in an underground traverse survey is 5 mm. Compute the maximum and minimum errors in the measurement of the clockwise angle ABC induced by the centring error if the magnitude of the angle is approximately $120°$ and the lengths of the lines AB and BC are approximately 24.36 m and 24.24 m respectively.   (RICS)

(1)   The minimum error as before will be nil.
(2)   The maximum error on the bisection of the angle ABC is $AB \simeq BC$, i.e.

$$\phi = \frac{\theta}{2} = 60° \quad a \simeq c \simeq 24.3\,\text{m}$$

$$x = 0.005\,\text{m}$$

Therefore

$$E = 206\,265 \times 0.005 \left(\frac{\sin 60}{24.3} + \frac{\sin (120 - 60)}{24.3} \right)$$

$$= 74''$$

By Professor Briggs' equation 5.75, the probable error

$$e'' = \pm \frac{206\,265 \times 2 \times 0.005}{3.1416} \sqrt{\left(\frac{1}{24.3^2} + \frac{1}{24.3^2} - \frac{2\cos 120}{24.3^2} \right)}$$

$$= 49'' \quad \text{i.e. } 68\% \text{ of the maximum}$$

5.9   The vernier

This device (Fig. 5.71) for determining the decimal parts of a graduated scale may be of two types:

(1)   direct reading
(2)   retrograde

both of which may be single or double.

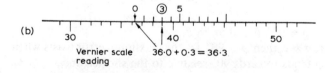

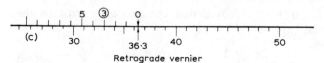

Fig. 5.71 Verniers

5.9.1 Direct reading vernier

Let d = the smallest value on the main scale
v = the smallest value on the vernier scale
n = number of spaces on the vernier
n vernier spaces occupy $(n-1)$ main scale spaces, i.e.

$$nv = (n-1)d$$

$$v = \frac{(n-1)d}{n} \qquad (5.76)$$

Therefore the least count of the reading system is given by:

$$d - v = d - \frac{d(n-1)}{n}$$

$$= d\left(1 - \frac{n-1}{n}\right)$$

$$d - v = d\left(\frac{n-n+1}{n}\right) = \frac{d}{n} \qquad (5.77)$$

Thus the vernier enables the main scale to be read to $\frac{1}{n}$th of 1 division.

Example 5.17 If the main scale value $d = 1\,\text{mm}$ and the number of spaces on the vernier $(n) = 10$, the vernier will read to $1\,\text{mm} \times 1/10 = 0.1\,\text{mm}$.

5.9.2 Retrograde vernier (Fig. 5.71 (c))

In this type, n vernier divisions occupy $(n+1)$ main scale divisions, i.e.

$$nv = (n+1)d$$

$$v = d\left(\frac{n+1}{n}\right) \qquad (5.78)$$

the least count $= v - d = d\left(\frac{n+1}{n}\right) - d$

$$= d\left(\frac{n+1}{n} - 1\right)$$

$$v - d = d/n \quad \text{as before} \qquad (5.79)$$

5.9.3 Special forms used in vernier theodolites

In order to provide a better break down of the graduations, the vernier may be extended in such a way that n vernier spaces occupy $(mn-1)$ spaces on the main scale. (m is frequently 2.)

$$nv = (mn-1)d$$

$$v = d\left(\frac{mn-1}{n}\right) \qquad (5.80)$$

the least count $= md - v = md - d\left(\frac{mn-1}{n}\right)$

$$= d\left(m - \frac{mn-1}{n}\right)$$

$$md - v = d/n \quad \text{as before} \qquad (5.81)$$

5.9.4 Geometrical construction of the vernier scale

In Fig. 5.72 (a) the main scale and vernier zeros are coincident.

For the direct reading vernier 10 divisions on the vernier must occupy 9 divisions on the main scale. Therefore
(1) Set off a random line OR of 10 units.
(2) Join R to V i.e. the end of the random line R to the end of the vernier V.
(3) Parallel through each of the graduated lines or the random line to cut the main scale so that 1 division of the vernier $= 0.9$ divisions of the main scale.

To construct a vernier to a given reading. In Fig. 5.72 (b) the vernier is required to read 36.3. It is thus required to coincide at the 3rd division, i.e. $3 \times 0.9 = 2.7$ main scale division beyond the vernier index.

Therefore coincidence will occur at $(36.3 + 2.7) = 39.0$ on the main scale and 3 on the vernier scale.

The vernier is constructed as above in the vicinity of the point of coincidence. The appropriate vernier coincidence line (i.e.

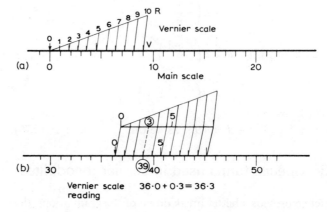

Fig. 5.72 Construction of a direct reading vernier

3rd) is joined to the main scale coincidence line (i.e. 39.0) and lines drawn parallel as before will produce the appropriate position of the vernier on the main scale.

In the case of the retrograde vernier, Fig. 5.73, 10 divisions on the vernier equal 11 divisions on the main scale, and therefore the point of coincidence of 3 on the vernier with the main scale value is

$$36.3 - (3 \times 1.1) = 36.3 - 3.3 = 33.0$$

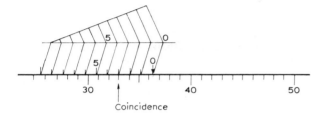

Fig. 5.73 Construction of a retrograde vernier

Example 5.18 Show how to construct the following verniers:
(1) To read to 10″ on a limb divided to 10 minutes.
(2) To read to 20″ on a limb divided to 15 minutes.
(3) The arc of a sextant is divided to 10 minutes. If 119 of these divisions are taken as the length of the vernier, into how many divisions must be vernier be divided in order to read to (a) 5 seconds (b) 10 seconds?   (ICE)

(1) The least count of the vernier is given by equation 5.75 as d/n. Therefore

$$10'' = \frac{10 \times 60}{n}$$

therefore

$$n = \frac{600}{10} = 60$$

Therefore the number of spaces on the vernier is 60 and the number of spaces on the main scale is 59.

(2) Similarly,

$$20'' = \frac{15 \times 60}{n}$$

therefore

$$n = \frac{15 \times 60}{20} = 45$$

i.e. the number of spaces on the vernier is 45 and the number of spaces on the main scale is 44.

(3) The number of divisions 119 is not required, and the calculation is exactly as above.

(a) $n = \dfrac{10 \times 60}{5} = 120$

(b) $n = \dfrac{15 \times 60}{10} = 90$

Exercises 5.2

1   The eccentricity of the line of collimation of a theodolite telescope in relation to the azimuth axis is 0.625 mm. What will be the difference, attributable to this defect, between face right and face left measurement of an angle if the lengths of the drafts adjacent to the instrument are 6.096 m and 36.576 m respectively?   (MQB/S Ans.   17.6″)

2   A horizontal angle is to be measured having one sight elevated to 32° 15′ whilst the other is horizontal. If the vertical axis is inclined at 40″ to the vertical, what will be the error in the recorded value?   (Ans.   25″)

3   In the measurement of a horizontal angle the mean angle of elevation of the backsight is 22° 12′ whilst the foresight is a depression of 37° 10′. If the lack of verticality of the vertical axis causes the horizontal axis to be inclined at 50″ and 40″ respectively in the same direction, what will be the error in the recorded value of the horizontal angle as the mean of face left and right observations?   (Ans.   51″)

4   In a theodolite telescope the line of sight is not perpendicular to the horizontal axis but in error by 5 minutes. In measuring a horizontal angle on one face, the backsight is elevated at 33° 34′ whilst the foresight is horizontal. What error is recorded in the measured angle?   (Ans.   60″)

5   The instrument above is used for producing a level line AB by transitting the telescope, setting out C_1, and then by changing face the whole operation is repeated to give C_2. If the foresight distance is 100 m, what will be the distance between the face left and face right positions, i.e. $C_1 C_2$?   (Ans.   0.582 m)

6   Describe with the aid of a sketch, the function of an internal focussing lens in a surveyor's telescope and state the advantages and disadvantages of internal focussing as compared to external focussing.

In a telescope, the object glass of focal length 150 mm is located 200 mm from the diaphragm. The focussing lens is midway between these when a staff 25 m away is focussed. Determine the focal length of the focussing lens.

(LU Ans.   103.7 mm)

7 In testing the trunnion axis of a vernier theodolite, the instrument was set up at '0', 30 m from the base of the vertical wall of a tall building where a well-defined point *A* was observed on face left at a vertical angle of 36° 52′. On lowering the telescope horizontally with the horizontal plate clamped, a mark was placed at *B* on the wall. On changing face, the whole operation was repeated and a second position *C* was fixed.

If the distance *BC* measured 0.043 m, calculate the inclination of the trunnion axis.   (Ans.   3′ 17″)

8 An object is 6.096 m from a convex lens of focal length 150 mm. On the far side of this lens a concave lens of focal length 75 mm is placed. Their principal axes are on the line of the object, and 75 mm apart. Determine the position, magnification and nature of the image formed.

(Ans.   Virtual image 1.555 away from the concave
lens towards object; magnification 0.5)

9 A compound lens consists of two thin lenses, one convex, the other concave, each of focal length 150 mm and placed 75 mm apart with their principal axes common. Find the position of the principal focus of the combination when the light is incident first on (a) the convex lens and (b) the concave lens.

(Ans. (a) real image 150 mm from concave lens away from object;
(b) real image 450 mm from convex lens away from object)

10 (a)   Explain the function of a vernier.
(b)   Construct a vernier reading 114.25 mm on a main scale divided to 2.5 mm.
(c)   A theodolite is fitted with a vernier in which 30 vernier divisions are equal to 14° 30′ on a main scale divided to 30 minutes. Is the vernier direct or retrograde, and what is its least count?   (TP Ans.   direct; 1 min.)

11 The following readings were taken with a theodolite equipped with a compensator for vertical circle readings:

Inst. stn.	Obs. stn.	Horizontal circle readings FL	FR	Vertical circle readings Mean
B	*A*	1° 02′ 20″	181° 02′ 40″	25° 30′ 15″
	C	131° 03′ 30″	311° 04′ 00″	−05° 20′ 00″

Without disturbing the vertical circle and with a random setting of the theodolite telescope the following vertical angles were recorded:

Horizontal FL readings	271°	91°	41°	221°
Vertical circle readings (*A*)	10° 15′ 20″	10° 14′ 20″		
(*C*)			10° 15′ 00″	10° 15′ 30″

Calculate the 'true' horizontal angle *ABC*.

(Ans.   130° 01′ 02″)

Bibliography

BANNISTER, A., and RAYMOND, S., *Surveying*, 4th edition. Pitman (1977)

CLARK, D., *Plane and Geodetic Surveying for Engineers*, 1, 6th edition. Constable (1972)

CLENDINNING, J., and OLLIVER, J. G., *Principles and Use of Surveying Instruments*, 3rd edition. Van Nostrand Reinholt (1972)

CURTIN, W. G., and LANE, R. F., *Concise Practical Surveying*, 2nd edition. English Universities Press (1970)

Wild 'T2', *Theodolite Handbook*. Wild

6

Levelling

6.1 Definitions

Levelling is the process concerned with the determination of the differences in elevation of two or more points between each other or relative to some given datum.

A **datum** may be purely arbitrary but for many purposes it is taken as the mean sea level (MSL) or Ordnance Datum (OD).

A **level surface** can be defined as a plane, tangential to the earth's surface at any given point. The plane is assumed to be perpendicular to the direction of gravity which for most practical purposes is taken as the direction assumed by a plumb-bob.

A **level line**, Fig. 6.1, is a line on which all points are equidistant from the centre of gravity. Therefore, it is curved and (assuming the earth to be a sphere) it is circular. For more precise determinations the geoidal shape of the earth must be taken into consideration.

A **horizontal line**, Fig. 6.1, is tangential to a level line and is taken, neglecting refraction, as the line of collimation of a perfectly adjusted levelling instrument. (As the lengths of sights in levelling are usually less than 150 m, level and horizontal lines are assumed to be the same—see Section 6.6.)

The **line of collimation** is the imaginary line joining the intersection of the main lines of the diaphragm to the optical centre of the object-glass.

Mean sea level. This is the level datum line taken as the reference plane. In the British Isles the Ordnance Survey originally accepted the derived mean sea level value for Liverpool. This has been superseded by a value based upon Newlyn in Cornwall.

Bench mark (⊼) (BM). This is a mark fixed by the Ordnance Survey and cut in stable constructions such as houses or walls. The reduced level of the horizontal bar of the mark is recorded on OS maps and plans.

Temporary bench mark (TBM) Any mark fixed by the observer for reference purposes.

Backsight (BS) is the first sight taken after the setting up of the instrument. Initially it is usually made to some form of bench mark.

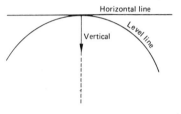

Fig. 6.1

193

Foresight (FS) is the last sight taken before moving the instrument.

Intermediate sight (IS) is any other sight taken.

NB: During the process of levelling the instrument and staff are never moved together, i.e. whilst the instrument is set the staff may be moved, but when the observations at one setting are completed the staff is held at a selected stable point and the instrument is moved forward. The staff station here is known as a **change point** (CP).

6.2   Principles

Let the staff readings be a, b, c etc. In Fig. 6.2,

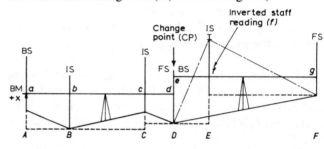

Fig. 6.2

difference in level A to B $= a - b$ $a < b$ therefore $-$ ve i.e. fall

B to C $= b - c$ $b > c$ therefore $+$ ve i.e. rise

C to D $= c - d$ $c < d$ therefore $-$ ve i.e. fall

D to E $= e - (-f)$ inverted staff therefore $+$ ve i.e. rise

E to F $= -f - g$ $-$ ve i.e. fall

$$A \text{ to } F = (a-b) + (b-c) + (c-d) + (e+f) + (-f-g)$$

$$= a - d + e - g$$

$$= (a+e) - (d+g)$$

$$= \Sigma \, BS - \Sigma \, FS$$

$$\Sigma \text{ rises} = (b-c) + (e+f)$$

$$\Sigma \text{ falls} = (b-a) + (d-c) + (f+g)$$

$$\Sigma \text{ rises} - \Sigma \text{ falls} = (a+e) - (d+g) = \Sigma \, BS - \Sigma FS \qquad (6.1)$$

The difference in level between the start and finish

$$= \Sigma \, BS - \Sigma \, FS = \Sigma \text{ rises} - \Sigma \text{ falls} \qquad (6.2)$$

Notes

(1)   Intermediate values have no effect on the final results, and thus reading errors at intermediate points are not shown up.
(2)   Where the staff is inverted the readings are treated as negative values and indicated in bookings by brackets or by an asterisk.

6.3 Booking of readings

6.3.1 Method 1, rise and fall

BS	IS	FS	Rise	Fall	Reduced level
a					x
	b			$b-a$	$x-(b-a)$
	c		$b-c$		$x-(b-a)+(b-c)$
e		d		$d-c$	$x-(b-a)+(b-c)$ $-(d-c)$
	(f)		$e-(-f)$		$x-(b-a)+(b-c)$ $-(d-c)+(e+f)$
		g		$f+g$	$x-(b-a)+(b-c)$ $-(d-c)+(e+f)$ $-(f+g)$
$a+e$ $-(d+g)$		$d+g$	$(b-c)+(e+f)$	$(b-a)+(d-c)$ $+(f+g)$	

Example 6.1 Given the following readings:

$$a = 0.628 \quad e = 2.259$$
$$b = 1.564 \quad f = -2.085$$
$$c = 1.000 \quad g = 0.991$$
$$d = 1.210$$

Notes

(1) The difference between adjacent readings from the same instrument position gives rise or fall according to the sign $+$ or $-$.

(2) At the change point BS and FS are recorded on the same line.

(3) Check $\Sigma\,\mathrm{BS} - \Sigma\,\mathrm{FS} = \Sigma\,\mathrm{rise} - \Sigma\,\mathrm{fall}$ before working out reduced levels. Difference between reduced levels at start and finish must equal $\Sigma\,\mathrm{BS} - \Sigma\,\mathrm{FS}$.

Explanation of bookings (Fig. 6.3)

(1) Sum of BS = 2.887

(2) Sum of FS = 2.201
    Difference = $+0.686$ (rise)

(3) Reduced level of Station A = 30.480
    Difference (as above)   = $+0.686$
    Reduced level of Station F = 31.166
    Known level of Station F = 31.170
    Field error   =   0.004

These calculations must be made in the field immediately after completion of the last FS to verify the acceptance or otherwise of the observations.

TRENT
POLYTECHNIC

LEVELLING BOOKING FORM

Instrument _N°1_

Location _'X' CONSTRUCTION SITE_

Page I

Observer _I. C. Well_

Booker _R. U. SHORE_

Date _7 - 5 - 83_

BS	IS	FS	R	F	RL	REMARKS
0·628					30·480	St A BM 30·480m
	1·564			0·936	29·544	St B
	1·000	0·564			30·108	St C
2·259	1·210			0·210	29·898	St D
	(2·085)	4·344			34·242	St E Inverted staff on girder
	0·991		3·076		31·166	TBM 31·170 m
2·897	2·201	4·908	4·222		30·480	(NB Field error 0·004
2·201		4·222				assuming TBM value
0·686		0·686			0·686	correct)

Fig. 6.3

(4)   The bookings may now be completed as follows:
Rise and Fall columns.

Station B $0.628 - 1.564$ $= -0.936$, i.e. a fall from Station A
       C $1.564 - 1.000$ $= +0.564$, i.e. a rise from Station B
       D $1.000 - 1.210$ $= -0.210$, i.e. a fall from Station C
       E $2.259 - (-2.085)$ $= +4.344$, i.e. a rise from Station D
       F $-2.085 - 0.991$ $= -3.076$, i.e. a fall from Station E

(5)   Sum of rises $= 4.908$
(6)   Sum of falls $= 4.222$
       Difference $= +0.686$ (rise)

This agrees with the difference above and thus proves that the arithmetical processes have been carried out correctly in reducing the rises and falls.

(7)   Reduced level column

Station B $30.480 - 0.936 = 29.544$
       C $29.544 + 0.564 = 30.108$
       D $30.108 - 0.210 = 29.898$
       E $29.898 + 4.344 = 34.242$
       F $34.242 - 3.076 = 31.166$ (should be 31.170, thus
       A $= 30.480$ field error 0.004)

                   difference $=$ 0.686 as previously shown

These checks are vital to prove the accuracy of the reduced values.

6.3.2 Method 2, height of collimation

BS	IS	FS	Height of collimation	Reduced level
a			$x+a$	x
	b			$x+a-b$
	c			$x+a-c$
e		d	$x+a-d+e$	$x+a-d$
	(f)			$x+a-d+e-(-f)$
		g		$x+a-d+e-g$
$a+e$	$b+c-f$	$d+g$		$6x+5a-b-c-3d+2e+f-g$

Arithmetical check

Σ Height of each collimation $\times$ no. of applications

$$= 3(x+a)+2(x+a-d+e)$$
$$= 5x+5a-2d+2e$$

$$\begin{aligned}
\Sigma \text{ Reduced levels} - \text{first} &= 5x+5a-b-c-3d+2e+f-g \\
\Sigma \text{ IS} &= \qquad\quad +b+c \qquad\qquad -f \\
\Sigma \text{ FS} &= \qquad\qquad\qquad +d \qquad\quad +g \\
&\quad 5x+5a \qquad\quad -2d+2e
\end{aligned}$$

Thus the full arithmetical check is given as:

Σ Reduced levels less the first $+\Sigma\,\text{IS}+\Sigma\,\text{FS}$ should equal
Σ Height of each collimation $\times$ no. of applications (6.3)

Using the values in Example 6.1,

BS	IS	FS	HC		RL	REMARKS
0·628			31·108		30·480	St A BM 30·480 m
	1·564		[x 3]		29·544	St B
	1·000			93·324	30·108	St C
2·259		1·210	32·157		29·898	St D
	(2·085)		[x 2]		34·242	St E Inverted staff on girder
		0·991		64·314	31·166	TBM 31·170 m
2·887	2·564	2·201		157·638	154·958	(NB Field error 0·004
2·201	−2·085	0·479				assuming TBM value
0·686	0·479	154·958			0·686	correct)
		157·638				

Fig. 6.4

Notes
(1)   The height of collimation = the reduced level of the BS + the BS reading.
(2)   The reduced level of any station = height of collimation − reading at that station.
(3)   Whilst Σ BS − Σ FS = the difference in the reduced level of start and finish this does not give a complete check on the intermediate values; an arithmetical error can be made without being noticed.
(4)   The full arithmetical check is needed to ensure there is no arithmetical error.

Explanation of bookings (Fig. 6.4). BS, IS, and FS columns as before, with the same field check made by summation of the columns.

(1)   Height of collimation column and the reduced level column are now completed together.

1st set up: 30.480 + 0.628 (BS)	= 31.108	Height of collimation
Reduced level of B = 31.108 − 1.564	= 29.544	
Reduced level of C = 31.108 − 1.000	= 30.108	
Reduced level of D = 30.108 − 1.210	= 29.898	
2nd set up: 29.898 + 2.259 (BS)	= 32.157	Height of collimation
Reduced level of E = 32.157 − (− 2.085)	= 34.242	
Reduced level of F = 32.157 − 0.991	= 31.166	

(2)   This again proves that a field error of − 0.004 has occurred compared with the apparent known value of station F and that the arithmetic is correct for all BS and FS stations.
(3)   In order to ensure that the reduced levels of the intermediate sights are correct, the following procedure *must* be adopted *at all times*.

Sum of IS	= 0.479
Sum of FS	= 2.201
Sum of reduced levels excluding the first	= 154.958
	157.638
Sum of collimation values used to obtain reduced levels	= 157.638

It is only by this complete arithmetical check that the accuracy of the reduced levels can be verified.

Example 6.2   Using the height of collimation calculate the respective levels of floor and roof at each staff station relative to the floor level at A which is 20.00 m above an assumed datum. It is important that a complete arithmetical check on the results should be shown. Note that the staff readings enclosed by brackets thus (1.05) were taken with the staff reversed.

BS	IS	FS	Height of collimation	Reduced level	Remarks
0.75			20.75	20.00	Floor at A
	(1.05)		[× 7]	21.80	Roof at A
	1.21		145.25	19.54	Floor at B
	(0.63)			21.38	Roof at B
	1.27			19.48	Floor at C
	(0.37)			21.12	Roof at C
	1.08			19.67	Floor at D
(1.25)		(0.83)	20.33	21.58	Roof at D
	0.60		[× 6]	19.73	Floor at E
	(1.62)		121.98	21.95	Roof at E
	0.87			19.46	Floor at F
	(0.94)			21.27	Roof at F
	1.40			18.93	Floor at G
(1.09)		(0.35)	19.59	20.68	Roof at G
	0.68		[× 4]	18.91	Floor at H
	(1.42)		78.36	21.01	Roof at H
	(1.85)			21.44	Floor at I
		0.35		19.24	Roof at I (19.25)
+0.75	+7.11	+0.35	345.59	347.19	
−2.34	−7.88	−1.18			
−1.59	−0.77	−0.83			
−0.83		−0.77			
−0.76		−1.60		−0.76	
		347.19			
		345.59			

NB: The checking process should at all times be part of the booking system and not something apart.

Inverted staff readings must always be treated as negative values.

Example 6.3 The following readings were taken with a level and a 4 m staff. Draw up a level book page and reduce the levels by:

(a) the rise and fall method;
(b) the height of collimation method.

0.683, 1.109, 1.838, 3.399, (3.877 and 0.451) CP, 1.405, 1.896, 2.676 BM (31.126 AOD), 3.478, (3.999 and 1.834) CP, 0.649, 1.706.

What error would occur in the final level if the staff had been wrongly extended and a plain gap of 0.012 had occurred at the 2 m section joint? (LU)

This question highlights three fundamental levelling mistakes:

(1) The staff has not been correctly assembled.
(2) The most important sight on to the bench mark should not be an intermediate sight as this cannot be checked.
(3) There is no check on the field work as there is no circuit closure.

(a)

	BS	IS	FS	Rise	Fall	Reduced level	Remarks
	0.683					36.545	
		1.109			0.426	36.119	
		1.838			0.729	35.390	
		3.399			1.561	33.829	
	0.451		3.877		0.478	33.351	
		1.405			0.954	32.397	
		1.896			0.491	31.906	
		2.676			0.780	31.126	BM 31.126 m
		3.478			0.802	30.324	AOD
	1.834		3.999		0.521	29.803	
		0.649		1.185		30.988	
			1.706		1.057	29.931	
	2.968		9.582	1.185	7.799	36.545	
	9.582			7.799			
	−6.614			−6.614		−6.614	

(b)

BS	IS	FS	Height of collimation	Reduced level	Remarks
0.683			37.228	36.545	
	1.109		[× 4]	36.119	
	1.838		148.912	35.390	
	3.399			33.829	
0.451		3.877	33.802	33.351	
	1.405		[× 5]	32.397	
	1.896		169.010	31.906	
	2.676			31.126	BM 31.126 m
	3.478			30.324	AOD
1.834		3.999	31.637	29.803	
	0.649		[× 2]	30.988	
		1.706	63.274	29.931	
2.968	16.450	9.582	381.196	355.164	
9.582		16.450			
−6.614		355.164		−6.614	
		381.196			

Booking is completed as follows:

(a)   Complete booking for rise and fall.

Starting at the BM reduce levels below by normal methods and above by reversing falls for rises.

(b)   In the HC method starting at BM the height of collimation = 31.126 + 2.676 (IS) = 33.802.

A combined booking is shown for convenience.

If a 0.012 mm gap occurred at the 2 m section all readings > 2 m will be 0.012 mm too small.

The final level value will only be affected by the BS and FS readings after the reduced level of the datum, i.e. 31.126, although the IS 2.676 would need to be treated as a BS for booking purposes.

BS (IS) 2.676 should be 2.664 FS 3.999 should be 3.987
$$\underline{1.834} \qquad\qquad \underline{1.706}$$
$$4.498 \qquad\qquad 5.693$$
$$\underline{5.693}$$
$$\text{difference} = -1.195$$

Reduced level of the last point = $31.126 - 1.195 = 29.931$.

Thus the BS and FS are affected in the same manner and the final value is not altered.

Example 6.4 The following readings were observed with a level:

1.143 (BM 34.223), 1.765, 2.566, 3.819, CP, 1.390, 2.262, 0.664 0.433, CP, 3.722, 2.886, 1.618, 0.616, TBM (value thought to be 35.290 m)

(a) Reduce the levels by the rise and fall method.
(b) Calculate the level of the TBM if the line of collimation was tilted upwards at an angle of 6 min and each backsight length was 90 m and the foresight length 30 m.
(c) Calculate the level of the TBM if the staff was not held upright but leaning backwards at 5° to the vertical in all cases.

(LU)

(a)

BS	IS	FS	Rise	Fall	Reduced level	Remarks
1.143					34.223	BM 34.223
	1.765			0.622	33.601	
	2.566			0.801	32.800	
1.390		3.819		1.253	31.547	CP
	2.262			0.872	30.675	
	0.664		1.598		32.273	
3.722		0.433	0.231		32.504	CP
	2.866		0.836		33.340	
	1.618		1.268		34.608	
		0.616	1.002		35.610	TBM 35.290
6.255		4.868	4.935	3.548	34.223	
4.868			3.548			
1.387			1.387		1.387	

(b) In Fig. 6.5,
$$\text{true difference in level per set up} = (a - 3e) - (b - e)$$
$$= (a - b) - 2e$$
$$\text{where } e = 30.000 \times 06' \text{ radians}$$
$$= 0.053 \text{ m per } 30 \text{ m}$$
$$\text{total length of BS's} = 3 \times 90 = 270 \text{ m}$$
$$\text{of foresight} = 3 \times 30 = 90 \text{ m}$$

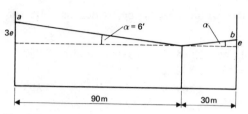

Fig. 6.5

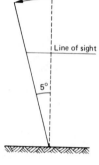

Fig. 6.6

effective difference in length = 3 × 60 = 180 m

therefore

error = 0.053 × 180/30

= 0.318 m

i.e. sum of BS is effectively too large by 0.318.

Therefore true difference in level = 1.387 − 0.318 = 1.069

reduced level of TBM = 34.223 + 1.069 = 35.292

Check   35.610 − 0.318 = 35.292

(c) If the staff was not held vertical (Fig. 6.6) the readings would be too large, the value depending on the staff reading.

true reading = observed reading × cos 5°

apparent difference in level Σ BS − Σ FS = 4.55

true difference in level = Σ BS cos 5° − Σ FS cos 5°

= (Σ BS − Σ FS) cos 5°

= 1.387 cos 50° = 1.382

Therefore level of TBM = 34.223 + 1.382 = 35.605

Example 6.5 *Missing values in booking.* It has been found necessary to consult the notes of a dumpy levelling carried out some years ago.

Whilst various staff readings, rises and falls and reduced levels, are undecipherable, sufficient data remain from which all the missing values can be calculated.

BS	IS	FS	Rise	Fall	Reduced level	Remarks
0.719					36.990	0 m BM on house
				0.591		100 m
1.234		2.222				200 m
				1.359		300 m
	1.314					400 m
	2.112					500 m
					34.540	600 m
		2.374				715 m Peg 36
	0.981		0.481			800 m
					34.141	900 m
	1.990					1000 m
					34.603	1100 m
		1.786				1200 m
				0.945		1236 m BM on wall
4.560						

Calculate the missing values and show the conventional arithmetical checks on your results.

BS	IS	FS	Rise	Fall	Reduced level	Remarks
0.719					36.990	0 m BM on house
	(a) *1.310*			0.591	*36.399*	100
1.234		2.222		(b) *0.912*	*35.487*	200
	(c) *2.593*			1.359	*34.128*	300
	1.314		(d) *1.279*		*35.407*	400
	2.112			(e) *0.798*	*34.609*	500
	(f) *2.181*			0.069	*34.540*	600
(g) *1.462*		2.374		0.193	*34.347*	715 Peg 36
	0.981		0.481		*34.828*	800
	(h) *1.668*			0.687	34.141	900
	1.990			(j) *0.322*	*33.819*	1000
	(k) *1.206*		0.784		34.603	1100
(m) *1.145*		1.786		(l) *0.580*	*34.023*	1200
		(n) *2.090*		0.945	*33.078*	1236 BM on wall
4.560		*8.472*	2.544	6.456		
8.472			6.456		*36.990*	
−3.912		*−3.912*			*−3.912*	

Notes

(a) 1.310 IS is deduced from fall 0.591.

(b) Fall 0.912 is obtained from IS–FS, 1.310–2.222.

(c) 2.593 as (a).

(d) Rise 1.279 as (b).

(e) Fall 0.798 as (b).

(f) Reduced levels $34.609 - 34.540$ give fall of 0.069, which gives staff reading 2.181.

(g) 1.462 BS must occur on line of FS deduced value from rise 0.481.

(h) 1.668 as (f).

(j) 0.322 as normal.

(k) 1.206 as (f).

(l) Fall 0.580 as normal.

(m) 1.145 BS must occur opposite FS 1.786, value from sum of BS 4.560.

(n) 2.090 from fall 0.945.

Checks as normal.

Exercises 6.1 (Booking)

1 The undernoted staff readings were taken successively with a level along an underground roadway.

Staff readings	Distances from A	Remarks
1.759	0	BS to A
0.652	40 m	IS
0.091	80 m	FS
1.689		BS
0.430	120 m	IS
0.917	160 m	FS
0.680		BS
0.671	200 m	IS
0.494	240 m	FS
2.243		BS
1.682	280 m	IS
0.216	320 m	FS
1.520		BS
0.686	360 m	FS to B

Using the Height of Collimation method, calculate the reduced level of each staff station relative to the level of A, which is 1216.039 m below OD.

Thereafter check your results by application of the appropriate method used for verifying levelling calculations derived from heights of collimation.

(MQB/S Ans. Reduced level of $B - 1210.552$)

2 Reduce the following notes of a levelling made along a railway affected by subsidence. Points A and B are outside the affected area, and the grade was originally constant between them.

Find the original grade of the track, the amount of subsidence at each 20 m length and the maximum grade in any 20 m length.

BS	IS	FS	Distance (m)	Remarks
3.094			0	Pt A (105.632 m AOD)
	2.694		20	
2.310		2.344	40	
	2.027		60	
1.283		1.676	80	
	0.829		100	
2.502		1.692		Not on track
	1.146		120	
	0.725		140	
		0.332	160	Pt B

(Ans.   Grade 1 in 50.87; subsidence $+ 0.007$, 0.036, 0.146, 0.188, 0.127, 0.027, $+ 0.001$; max. gradient 1 in 40.57 between 100 and 120 m)

3   The undernoted readings, in metres on a levelling staff, were taken along a roadway AB with a dumpy level, the staff being held in the first case at a starting point A and then at 20 m intervals: 0.765, 1.064, [0.616], 1.835, 1.524.

The level was then moved forward to another position and further readings taken. These were as follows, the last reading being at B: 2.356, 1.378, [2.063], 0.677, 2.027.

The level of A is 41.819 m AOD.

Set out the staff readings and complete the bookings.

Calculate the gradient from A to B.

(Figures in brackets denote inverted staff readings.)

(RICS Ans.   1 in 372)

4   An extract from a level book is given below, in which various bookings are missing. Fill in the missing bookings and re-book and complete the figures by the Height of Collimation method.

BS	IS	FS	Rise	Fall	Reduced level
					47.201
	2.322			1.128	
	0.707		1.615		
2.167					45.025
		1.692	0.475		45.500
	2.246		0.610		46.110
	2.657				45.699
				0.491	
		1.292	1.856		47.064

All the figures are assumed correct.   (ICE)

5   The following figures are the staff readings taken in order on a particular scheme, the backsights being shown in italics:

0.813, 2.170, 2.908, 2.630, 3.133, *3.752*, 3.277, 1.899, 2.390, *2.810*, 1.542, 1.274, 0.643

The first reading was taken on a bench mark 39.563 OD.

Enter the readings in level book form, check the entries and find the reduced level of the last point.

Comment on your completed reduction.   (LU/E)

6.4   Field testing of the level

Methods available are (1) by reciprocal levelling, (2) the two-peg methods.

6.4.1   Reciprocal levelling method

In Fig. 6.7, the instrument is first set at A of height a. The line of sight is assumed to be inclined at an angle of elevation $+\alpha$ giving an error e in the length AB. The reading on the staff at B is b.

difference in level $A - B = a - (b - e)$ (6.4)

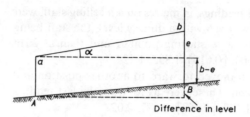

Fig. 6.7

In Fig. 6.8, the instrument is set at B of height b_1 and the reading on the staff at A is a_1.

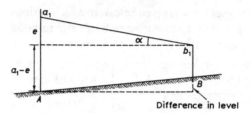

Fig. 6.8

difference in level $A - B = (a_1 - e) - b_1$ (6.5)

Thus, from equations 6.4 and 6.5,

difference in level $= a - b + e$ (Equation 6.4)

$\qquad\qquad\qquad = a_1 - b_1 - e$ (Equation 6.5)

Adding equations 6.4 and 6.5,

$$= \tfrac{1}{2}[(a - b) + (a_1 - b_1)] \qquad (6.6)$$

By subtracting equations 6.4 and 6.5, the *error in collimation* is

$$e = \tfrac{1}{2}[(a_1 - b_1) - (a - b)] \qquad (6.7)$$

Example 6.6 A dumpy level is set up with the eyepiece vertically over a peg A. The height from the top of A to the centre of the eyepiece is measured and found to be 1.408 m. A level staff is then held on a distant peg B and read. This reading is 0.646 m. The level is then set over B. The height of the eyepiece above B is 1.362 m and a reading on A is 2.009 m.

(1)   What is the difference in level between A and B?

(2)   Is the collimation of the telescope in adjustment?

(3)   If out of adjustment, can the collimation be corrected without moving the level from its position at B?   (ICE)

(1)   From equation 6.6,

difference in level $A - B = \tfrac{1}{2}[(1.408 - 0.646) + (2.009 - 1.362)]$

$\qquad\qquad\qquad\qquad = \tfrac{1}{2}[0.762 + 0.647]$

$\qquad\qquad\qquad\qquad = +0.704(5)\,\text{m}$

(2)   From equation 6.7,

error in collimation $e = \frac{1}{2}[0.647 - 0.762]$
$$= -0.0575 \text{ m per length } AB$$

(i.e. the line of sight is depressed.)

(3)   True staff reading at A (instrument at B) should be

$$2.009 - (-0.0575)$$
$$= 2.009 + 0.058 = 2.067 \text{ m}$$

The cross-hairs must be adjusted to provide this reading.

6.4.2   Two-peg method

In the following field tests the true difference in level is ensured by making backsight and foresight of equal length (Fig. 6.9).

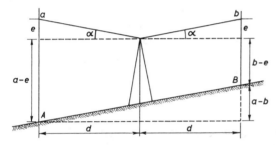

Fig. 6.9

Assuming the line of collimation is elevated by α,

the displacement vertically $= d \tan \alpha$

Thus, if BS $=$ FS,

$$d \tan \alpha = e \text{ in each case}$$

Therefore

true difference in level $AB = (a - e) - (b - e)$
$$= a - b$$

Method (a).   Pegs are inserted at A and B so that the staff reading $a = b$ when the instrument is midway between A and B. The instrument may now be moved to A or B.

In Fig. 6.10, if the height of the instrument at B is b_1 above peg

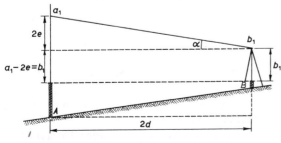

Fig. 6.10

B, the staff reading at peg A should be b_1 if there is no error, i.e. if $\alpha = 0$.

If the reading is a_1 and the distance $AB = 2d$, then the true reading at A should be

$$b_1 = a_1 - 2e$$

therefore $\quad e = \tfrac{1}{2}[a_1 - b_1] \quad\quad\quad (6.8)$

Method (b). The instrument is placed midway between staff positions A and B, Fig. 6.11. Readings taken give the true difference in level $= a - b$.

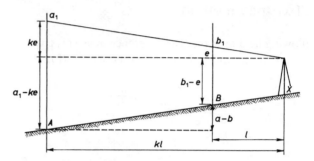

Fig. 6.11

The instrument is now placed at X so that

$$XA = kXB$$

where k is the multiplying factor depending on the ratio of AX/BX.

If $AX = kBX$, then the error at $A = ke$. Then

true difference in level $= (a_1 - ke) - (b_1 - e) = a - b$

therefore $\quad\quad\quad\quad a_1 - b_1 - (k-1)e = a - b$

therefore

$$\text{error per length } BX = e = \frac{(a_1 - b_1) - (a - b)}{k - 1} \quad\quad (6.9)$$

Notes

(1)   If the instrument is placed nearer to A than B, k will be less than 1 and $k - 1$ will be negative (see Example 6.8).

(2)   If the instrument is placed at station B, then equation 6.9 is modified as follows:

$$(a_1 - E) - b_1 = a - b$$

therefore $\quad\quad E = (a_1 - b_1) - (a - b) \quad\quad\quad (6.10)$

where E is the error in the length AB (see Example 6.9).

Example 6.7 (a)   When checking a dumpy level, the following readings were obtained in the two-peg test:

Level set up midway between two staff stations A and B 100 m apart, staff readings on A 1.753 and on B 1.314 m.

Level set up 10 m behind B and in line AB; staff reading on B 1.039 m and on A 1.509 m.

Complete the calculation and state the amount of instrumental error.

(b)   Describe the necessary adjustments to the following types of level, making use of the above readings in each case:

(i)   Dumpy level fitted with diaphragm screws and level tube screws.

(ii)   A level fitted with level screws and tilting screw.

(a)   By equation 6.9,

$$e = \frac{(1.509 - 1.039) - (1.753 - 1.314)}{11 - 1}$$

$$= \frac{0.470 - 0.439}{10} = 0.0031 \text{ m per } 10 \text{ m}$$

Check

With instrument 10 m beyond B,

staff reading at A should be $1.509 - (11 \times 0.0031)$

$$= 1.509 - 0.034 = 1.475$$

staff reading at B should be $1.039 - 0.003 = \underline{1.036}$

$$\text{difference in level} = \overline{0.439}$$

This agrees with the first readings $1.753 - 1.314 = 0.439$.

(b) (i)   With a dumpy level the main spirit should be first adjusted.

The collimation error is then adjusted by means of the diaphragm screws until the staff reading at A from the second setting is 1.475 m and this should check with the staff reading at B of 1.036 m.

(ii)   With the tilting level, the circular (pill-box) level should be first adjusted.

The line of sight should be set by the tilting screw until the calculated readings above are obtained.

The main bubble will now be off centre and must be centralised by the level tube screws.

Example 6.8 (a)   Describe with the aid of a diagram the basic principles of a tilting level, and state the advantages and disadvantages of this type of level compared with the dumpy level.

(b)   The following readings were obtained with a tilting level to two staves A and B 63 m apart.

Position of instrument	Reading at A (m)	Reading at B (m)
Midway between A and B	1.655	1.865
3 m from A and 60 m from B	1.881	2.033

What is the error in the line of sight per 20 m of distance and how would you adjust the instrument?   (RICS)

Part (b) illustrates the testing of a level involving a negative angle of inclination of the line of collimation and a fractional k value.

Using equation 6.9,

$$e = \frac{(a_1 - b_1) - (a - b)}{k - 1} \quad \text{where} \quad k = \frac{AX}{BX} = \frac{3}{60}$$

$$= \frac{(1.881 - 2.033) - (1.655 - 1.865)}{0.05 - 1}$$

$$= \frac{-0.152 - (-0.210)}{-0.95} = -0.061 \text{ m per 60 m}$$

$$= -0.020 \text{ m per 20 m and } 0.003 \text{ m per 3 m}$$

Check

At X, reading on A should be $1.881 + 0.003 = \quad 1.884$

reading on B should be $2.033 + 0.061 = \quad \underline{2.094}$

difference in level $= -0.210$

Alternative solution from first principles (Fig. 6.12)

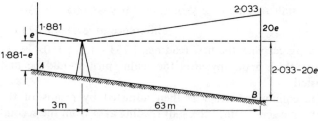

Fig. 6.12

True difference in level

$$= (1.881 - e) - (2.033 - 20e) = (1.655 - 1.865)$$

i.e. $19e = 0.210 + 0.152$

$$e = \frac{0.058}{19}$$

$$= -0.003 \text{ m per 3 m}$$

Example 6.9   The following readings taken on to two stations A and B were obtained during a field test of a dumpy level. Suggest what type of error exists in the level and give the magnitude of the error as a percentage. How would you correct it in the field?

BS	FS	Remarks
1.893		Staff at station A $A - B$ 60 m apart
	1.664	Staff at station B Instrument midway
1.521		Staff at station A Instrument v. near to B
	1.311	Staff at station B

(RICS)

By equation 6.10,

$$E = (a_1 - b_1) - (a - b)$$
$$= (1.521 - 1.311) - (1.893 - 1.664)$$
$$= 0.210 - 0.229$$
$$= -0.019 \text{ m per } 60 \text{ m}$$
$$= -0.006 \text{ m per } 20 \text{ m}$$

Check

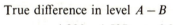

$$1.521 + 0.019 = 1.540$$
$$1.311 + 0.0 \quad = \underline{1.311}$$
$$= 0.229 \text{ m}$$

True difference in level $= 1.893 - 1.664 = 0.229$ m.

Example 6.10   In levelling up a hillside to establish a TBM the average lengths of ten backsights and ten foresights were 30 m and 15 m respectively.

As the reduced level of the TBM of 25.146 m AOD was in doubt, the level was set up midway between two pegs A and B 75 m apart, the reading on A being 1.390 and that on B 1.597. When the instrument was moved 15 m beyond B on the line AB produced, the reading on A was 1.628 and on B 1.792 m. Calculate the true value of the reduced level.

True difference in level $A - B$

$$= 1.390 - 1.597 = -0.207$$

When set up 15 m beyond B,

$$(1.628 - 6e) - (1.792 - e) = -0.207$$
$$1.628 - 1.792 - 5e = -0.207$$
$$5e = 0.043$$
$$e = 0.0086 \text{ m per } 15 \text{ m}$$

Check
True readings should have been:

$$\text{at } A \quad 1.628 - 0.052 = \quad 1.576$$
$$\text{at } B \quad 1.792 - 0.009 = \quad \underline{1.783}$$
$$-0.207$$

Error in levelling $= 0.0086$ m per 15 m
Difference in length between BS and FS per set $= 30 - 15 = 15$ m

$$\text{per 10 sets} = 150 \text{ m}$$

Therefore error $= 10 \times 0.0086 \quad = \quad 0.086$
Therefore true value of TBM $= 25.146 - 0.086$
$$= 25.060 \text{ m AOD}$$

Exercises 6.2 (Adjustment)

1 Describe how you would adjust a level fitted with tribrach screws graduated tilting screw and bubble-tube screws, introducing into your answer the following readings which were taken in a 2 peg test:

Staff stations at *A* and *B* 100 m apart.

Level set up halfway between *A* and *B*: staff readings on *A* 1.283 m, on *B* 0.860 m.

Level set up 10 m behind *B* in line *AB*: staff readings on *A* 1.612 m, on *B* 1.219 m.

Complete the calculation and show how the result would be used to adjust the level.

(LU Ans. Error −0.003 m per 10 m)

2 A modern dumpy level was set up at a position equidistant from two pegs *A* and *B*. The bubble was adjusted to its central position for each reading, as it did not remain quite central when the telescope was moved from *A* to *B*. The readings on *A* and *B* were 1.481 m and 1.591 m respectively. The instrument was then moved to *D*, so that the distance *DB* was about five times the distance *DA*, and the readings with the bubble central were 1.560 m and 1.655 m respectively. Was the instrument in adjustment? (ICE Ans. Error = 0.004 m per distance *AD*)

3 A level was set up on the line of two pegs *A* and *B* and readings were taken to a staff with the bubble central. If *A* and *B* were 150 metres apart, and the readings were 2.763 m and 1.792 m respectively, compute the collimation error. The reduced levels were known to be 27.002 m and 27.995 m respectively.

The level was subsequently used, without adjustment, to level between two points *X* and *Y* situated 1 km apart. The average length of the backsights was 45 m and of the foresights 55 m. What is the error in the difference in level between *X* and *Y*?

(TP Ans. 30.5″ depressed; error +0.0145 m)

4 A level set up in a position 30 m from peg *A* and 60 m from peg *B* reads 1.914 m on a staff held on *A* and 2.237 m on a staff held on *B*, the bubble having been carefully brought to the centre of its run before each reading. It is known that the reduced levels of the tops of the pegs *A* and *B* are 87.575 m and 87.279 OD respectively. Find
(a) the collimation error, and
(b) the readings that would have been obtained had there been no collimation error.

(LU Ans. (a) −0.027 m per 30 m (b) 1.887 m; 2.183 m)

5 *P* and *Q* are two points on opposite banks of a river about 100 m wide. A level with an anallatic telescope with a constant 100 is set up at *A* on the line *QP* produced, then at *B* on the line

PQ produced, and the following readings taken on to a graduated staff held vertically at *P* and *Q*.

From	To	Staff readings in m		
		Upper stadia	Collimation	Lower stadia
A	*P*	1.567	1.423	1.280
	Q	0.997	0.369	below ground
B	*P*	3.240	2.594	1.948
	Q	1.603	1.442	1.280

What is the true difference in level between *P* and *Q* and what is the collimation error of the level expressed in seconds of arc, there being 206 265 seconds in a radian?

(ICE Ans.   1.103 m; 105″ above horizon)

6.5 Sensitivity of the bubble tube

The sensitivity of the bubble tube depends on the radius of curvature (R) and is usually expressed as an angle (θ) per unit division (d) of the bubble scale.

6.5.1 Field test

Staff readings may be recorded as the position of the bubble is changed by a footscrew or tilting screw. Readings at the eye and objective ends of the bubble may be recorded or alternatively the bubble may be set to the exact scale division.

In Fig. 6.13,

$$\tan(n\theta) = s/l \tag{6.11}$$

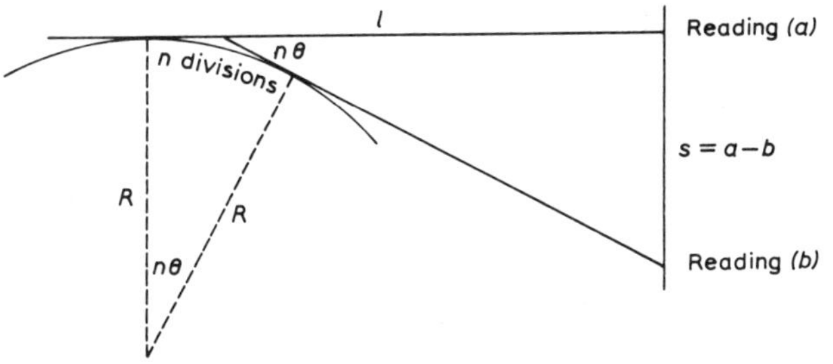

Fig. 6.13

but θ is very small (usually 1 to 60 seconds). Therefore

$$n\theta_{rad} = s/l$$

$$\theta_{rad} = s/nl \tag{6.12}$$

$$\theta_{sec} = 206\,265\, s/nl \tag{6.13}$$

where s = difference in staff readings a and b
n = number of divisions the bubble is displaced between readings
l = distance from staff to instrument

If d = length of 1 division on the bubble tube, then
$$d = R\theta_{rad}$$
i.e. $R = d/\theta$ (6.14)
$$= ndl/s$$ (6.15)

6.5.2   O–E correction

If the bubble tube is graduated from the centre then an accurate reading is possible, particularly when seen through a prismatic reader, Fig. 6.14.

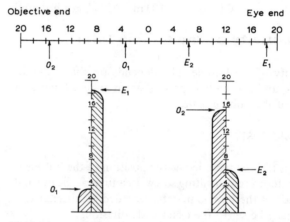

Fig. 6.14

If the readings at the objective end are O_1 and O_2 and those at the eye end E_1 and E_2, then the movement of the bubble in n divisions will equal

$$\frac{(O_2 - E_2) - (O_1 - E_1)}{2}$$ (6.16)

The length of the bubble will be
$$O + E$$ (6.17)
The displacement of the bubble will be
$$\frac{O - E}{2}$$ (6.18)

If $O > E$ the telescope is elevated;
    $O < E$ the telescope is depressed.

6.5.3   Bubble scale correction

With a geodetic level, the bubble is generally very sensitive, say 1 division = 1 second.

Instead of attempting to line up the prismatically viewed ends of the bubble, their relative positions are read on the scale provided and observed in the eyepiece at the time of the staff reading.

The correction to the middle levelling hair is thus required.

By equation 6.13,

$$\theta'' = 205\,265\,s/nl$$

Transposing gives

$$e = \frac{nl\theta''}{206\,265} \tag{6.19}$$

where e = the error in the staff reading
θ = the sensitivity of the bubble tube in seconds
n = the number of divisions displaced
l = the length of sight.

Example 6.11 Find the radius of curvature of the bubble tube attached to a level and the angular value of each 2 mm division from the following readings taken to a staff 60.96 m from the instrument.

Staff readings		1.070	1.141
Bubble readings	Eye end	18.3	6.4
	Objective end	3.4	15.3

By equation 6.16,

$$n = \tfrac{1}{2}[(15.3 - 6.4) - (3.4 - 18.3)]$$
$$= 11.9 \text{ divisions}$$

Then, by equation 6.13,

$$\theta'' = \frac{206\,265 \times (1.141 - 1.070)}{11.9 \times 60.96} = 20\,\text{sec}$$

By equation 6.15,

$$R = \frac{0.002 \times 60.96 \times 11.9}{0.071} = 20.43\,\text{m}$$

Example 6.12 The following readings were taken through the eyepiece during precise levelling. What should be the true middle hair reading of the bubble value if 1 division is 1 second. The stadia constant of the level is $\times\,100$.

Stadia readings			Bubble scale readings	
Top	Middle	Bottom	E	O
1.9421	1.6787	1.4152	10.6	8.48

By equation 6.18,

$$n = \frac{O - E}{2}$$

$$= \frac{8.4 - 10.6}{2} = -1.1$$

Then by equation 6.19,

$$e = \frac{nl\theta''}{206\,265}$$

$$= \frac{-1.1 \times 100\,(1.9421 - 1.4152) \times 1''}{206\,265}$$

$$= -0.000\,28$$

Therefore true middle reading should be $1.6787 + 0.0003$
$= 1.6790$.

Exercises 6.3 (Sensitivity)

1 A level is set with the telescope perpendicular to two footscrews at a distance of 30 m from a staff.

The graduations on the bubble were found to be 2.5 mm apart and after moving the bubble through 3 divisions the staff readings differed by 0.0088 m.

Find the sensitivity of the spirit bubble tube and its radius of curvature.   (Ans. $\theta \simeq 20$ seconds; $R = 25.6$ m)

2 State what is meant by the term 'sensitivity' when applied to a spirit level, and discuss briefly the factors which influence the choice of spirit level of sensitivity appropriate for the levelling instrument of specified precision.

The spirit level attached to a levelling instrument contains a bubble which moves 2.5 mm per 20 seconds change in the inclination of the axis of the spirit level tube. Calculate the radius of curvature of the spirit level tube.   (Ans.   25.8 m)

6.5.4   Gradient screws (tilting mechanism)

On some instruments the tilting screw is graduated as shown in Fig. 6.15.

The vertical scale indicates the number of complete revolutions whilst the horizontal scale indicates the fraction of a revolution.

The positive and negative tilts of the telescope are usually shown in black and red respectively and these must be correlated with similar colours on the horizontal scale.

The gradient of the line of sight is given as 1 in x, where

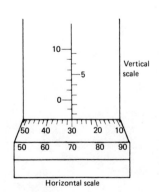

Fig. 6.15

$$1/x = nr \tag{6.20}$$

and n = number of revs

$\quad r$ = the ratio of 1 rev (frequently 1/1000)

Using the gradient screw, it is also possible to obtain the approximate distance by taking staff readings. If

$$\text{gradient} = s/L = nr$$

then $\qquad L = s/nr \qquad\qquad\qquad\qquad$ (6.21)

where s = staff intercept

$\qquad L$ = length of sight

Example 6.13   Staff reading (a) = 1.926 m

$\qquad\qquad\qquad\qquad\qquad$ (b) = 2.085 m

$\qquad\qquad$ number of revs (n) = 6.35

$\qquad\qquad$ gradient ratio (r) = 1/1000

Then $\qquad L = \dfrac{(2.085 - 1.926) \times 1000}{6.35}$

$\qquad\qquad = 25.039$ m

6.6   The effect of the earth's curvature and atmospheric refraction

6.6.1   The earth's curvature

Over long distances the effect of the earth's curvature becomes significant.

Let the error due to the earth's curvature $E = AC$ (see Fig. 6.16).

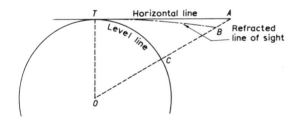

Fig. 6.16

In Fig. 6.16, by Pythagoras,

$$AO^2 = OT^2 + AT^2$$

i.e. $\qquad (E + R)^2 = R^2 + L^2$

$$E^2 + 2RE + R^2 = R^2 + L^2$$

$$E = \frac{L^2}{2R + E}$$

i.e. $\qquad\qquad \simeq \dfrac{L^2}{2R} \qquad\qquad\qquad\qquad$ (6.22)

As R, the radius of the earth, is $\simeq 6380$ km,

$$E \simeq \frac{L^2 \times 10^3}{2 \times 6380} = 7.84 \times 10^{-8} L^2 \text{ m} \tag{6.23}$$

6.6.2 Atmospheric refraction

Due to variation in the density of the earth's atmosphere, affected by atmospheric pressure and temperature, a horizontal ray of light TA is refracted to give the bent line TB.

If the coefficient of refraction K is defined as the multiplying factor applied to the angle TOA (subtended at the centre) to give the angle ATB,

angle of refraction $ATB = K\hat{TO}A = 2KA\hat{T}C$

As the angles are small,

$$AB:AC :: A\hat{T}B : A\hat{T}C$$

Then

$$AB = \frac{2KA\hat{T}C}{A\hat{T}C} \times AC$$

$$= 2KAC \tag{6.24}$$

The value of K varies with time, geographical position, atmospheric pressure and temperature. A mean value is frequently taken as 0.07. Therefore

$$AB = 0.14 \, AC$$

error due to refraction $\simeq \frac{1}{7}AC$ \tag{6.25}

$$\simeq 1.10 \times 10^{-8} L^2 \tag{6.26}$$

6.6.3 The combined effect of curvature and refraction

The net effect $\quad e = BC$

$$= AC - AB$$

$$= \frac{L^2}{2R} - 2K\frac{L^2}{2R}$$

$$= \frac{L^2}{2R}[1 - 2K] \tag{6.27}$$

If K is taken as 0.07 then

$$e = 7.84 \times 10^{-2} L^2 (1 - 0.14)$$
$$= 6.74 \times 10^{-2} L^2 \text{ m} \tag{6.28}$$

NB: *When the line of sight is close to the ground, refraction is very variable and may even be negative.*

Example 6.14 What will be the combined effect of curvature and refraction at the following distances: (a) 200 m; (b) 1 km; (c) 5 km; (d) 160 km?

(a) $E_a = 6.74 \times 10^{-8} \times 200^2$ $=$ 0.0027   say   0.003 m.

(b) $E_b = 6.74 \times 10^{-8} \times 1000^2 =$ 0.0674   say   0.067 m.

(c) $E_c = E_b \times 5^2$ $=$ 1.6850   say   1.685 m.

(d) $E_d = E_b \times 160^2$ $= 1725.4400$ say 1725.440 m.

In ordinary precise levelling it is essential that the lengths of the backsight and foresight be equal to eliminate instrumental error. This is also required to counteract the error due to curvature and refraction, as this error should be the same in both directions providing the climatic conditions remain constant. To minimise the effect of climatic change the length of sights should be kept below 50 m.

In precise surveys, where the length of sight is greater than this value and climatic change is possible, e.g. crossing a river or ravine, 'reciprocal levelling' is employed.

Exercise 6.4 (Curvature and refraction)

1   Derive the expression for the combined curvature and refraction correction used in levelling practice.

If the sensitivity of the bubble tube of a level is 20 seconds of arc per division, at what distance does the combined curvature and refraction correction become numerically equal to the error induced by dislevelment of one division of the level tube?

(RICS Ans.   1445.5 m)

6.6.4   Intervisibility

The earth's curvature and the effect of atmospheric refraction affect the maximum length of sight, Fig. 6.17.

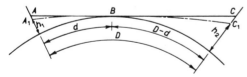

Fig. 6.17

$$h_1 = Kd^2$$
$$h_2 = K(D-d)^2$$

where $K = 6.74 \times 10^{-8}$

This will give the minimum height h_2 at C which can be observed from A height h_1 assuming the ray grazes the surface of the earth or sea.

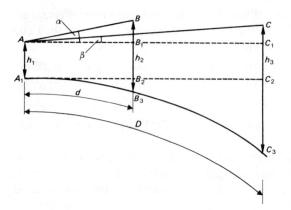

Fig. 6.18

With intervening ground at B. In Fig. 6.18,

let the height of $AA_1 = h_1$

of $BB_3 = h_2 = BB_1 + B_1B_2 + B_2B_3$

$$= d \tan \alpha + h_1 + Kd^2 \qquad (6.29)$$

of $CC_3 = h_3$

$$= D \tan \beta + h_1 + KD^2 \qquad (6.30)$$

If C is to be visible from A then $\alpha \leqslant \beta$.
If $\alpha = \beta$, then

$$\tan \alpha = \frac{h_2 - h_1 - Kd^2}{d} = \frac{h_3 - h_1 - KD^2}{D} \qquad (6.31)$$

Thus the minimum height is

$$h_2 = \frac{d}{D}(h_3 - h_1 - KD^2) + h_1 + Kd^2$$

$$= \frac{dh_3}{D} + (D-d)\left(\frac{h_1}{D} - Kd\right) \qquad (6.32)$$

$$h_3 = \left[h_2 - (D-d)\left(\frac{h_1}{D} - Kd\right)\right]\frac{D}{d} \qquad (6.33)$$

Clendinning quotes the formula as

$$h_2 = \frac{dh_3}{D} + (D-d)\left(\frac{h_1}{D} - Kd \operatorname{cosec}^2 Z\right) \qquad (6.34)$$

where Z is the zenith angle of observation.
Over large distances $Z \simeq 90°$, therefore $\operatorname{cosec}^2 Z \simeq 1$.

Example 6.15   If $h_1 = 700\,\mathrm{m}$ (at A), $d = 74\,\mathrm{km}$
$h_2 = 320\,\mathrm{m}$ (at B), $D = 135\,\mathrm{km}$
$h_3 = 550\,\mathrm{m}$ (at C),

can C be seen from A?

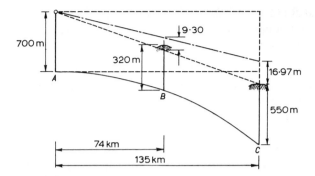

Fig. 6.19

See Fig. 6.19. By equation 6.32,

$$h_2 = \frac{74 \times 550}{135} + (135 - 74)\left(\frac{700}{135} - 6.74 \times 10^{-2} \times 74\right)$$

$$= 313.6\,\text{m}$$

The station C cannot be seen from A as h_2 is greater than 313.6 m.

If the line of sight is not to be nearer than 3 m to the surface at B, then it would be necessary to erect a tower at C of such a height that the line of sight would be 3 m above B, i.e. so that

$$\text{height } h = (320.0 + 3.0 - 313.6) \times 135/74$$

$$= 17.2\,\text{m, say } 17\,\text{m}$$

Alternatively, by equation 6.33,

$$h_3 = \left[h_2 - (D - d)\left(\frac{h_1}{D} - Kd\right) \right]\frac{D}{d}$$

$$= \left[323.0 - 61\left(\frac{700}{135} - 6.74 \times 10^{-2} \times 74\right) \right]\frac{135}{74}$$

$$= 567.3\,\text{m}$$

Existing value of $h_3 = 550$, thus the tower must have

$$\text{height} = 17.3\,\text{m, say } 17\,\text{m}.$$

Exercises 6.5

1 Two ships A and B are 32.2 km apart. If the observer at A is 6.10 m above sea level, what should be the height of the mast of B above the sea for it to be seen at A? (Ans. 34.7 m)

2 As part of a minor triangulation a station A was selected at 215.99 m AOD. Resection has been difficult in the area and as an additional check it is required to observe a triangulation station C 56.3 km away (reduced level 99.29 m AOD). If there is an intermediate hill at B, 24.1 km from A (spot height shown on map near B 112.8 m AOD), will it be possible to observe station C from A assuming that the ray should be 3 m above B?

(TP Ans.   Instrument + target should be $\simeq$ 6 m)

3 (a)   Discuss the effects of curvature and refraction on long sights as met with in triangulation, deriving a compounded equation for their correction.

(b)   A colliery headgear at A, ground level 137.92 m AOD, is 44.2 m to the observing platform.

It is required to observe a triangulation station C, reduced level 125.78 m AOD, which is 24.1 km from A, but it is thought that intervening ground at B approximately 152.4 m AOD and 8.0 km from A will prevent the line of sight.

Assuming that the ray should not be nearer than 3 m to the ground at any point, will the observation be possible?

If not, what height should the target be at C?

(RICS/M Ans.   Height of target greater than 2 m)

4   Describe the effect of the earth's curvature and refraction on long sights. Show how these effects can be cancelled by taking reciprocal observations.

Two beacons A and B are 96.0 km apart and are respectively 36.5 m and 365.0 km above mean sea level. At C, which is in the line AB and is 24.1 km from B, the ground level is 167.0 m above mean sea level.

Find by how much, if at all, B should be raised so that the line of sight from A to B should pass 3 m above the ground at C. The mean radius of the earth may be taken as 6371.6 km.

(LU Ans.   +4.6 m)

6.6.5   Trigonometrical levelling

For plane surveying purposes where the length of sight is limited to say 15 km the foregoing principles can be applied, Fig. 6.20.

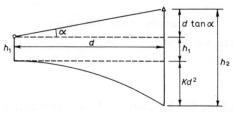

Fig. 6.20

The difference in elevation is given by
$$\Delta h = h_2 - h_1 = d \tan \alpha + 6.74 \times 10^{-8} d^2 \qquad (6.35)$$
If the instrument height is given as h_i and the target height is h_t, then
$$\Delta h = d \tan \alpha + 6.74 \times 10^{-8} d^2 + (h_i - h_t) \qquad (6.36)$$

Alternatively the vertical angle α may be adjusted for (Eye $-$ Object), i.e. for $h_i - h_t$.

$$\delta\alpha° = 57.2958\,\delta h\cos\alpha/d \tag{6.37}$$

NB: It is considered advisable in trigonometrical levelling, and in normal geometrical levelling over long distances, to observe in both directions, simultaneously where possible, in order to eliminate the effects of curvature and refraction, as well as instrumental errors. This is known as reciprocal levelling. (See Shepherd, *Advanced Engineering Surveying*, Edward Arnold (1980).)

Example 6.16   The reduced level of the observation station A is 106.790 m AOD. From A, instrument height 1.314 m, the angle of elevation is $5° 30''$ to station B, target height 1.963 m. If the computed distance AB is 1087.53 m, what is the reduced level of B.

By equation 6.36,

$$\Delta h = 1087.53\tan 5.5000° + 6.74\times10^{-8}\times1087.53^2 + 1.314 - 1.963$$
$$= 104.147\,\text{m}$$

Alternatively, by equation 6.37,

$$\delta\alpha = 57.2958\times -0.649\cos 5.5000°/1087.53$$
$$= -0.0340$$

therefore,

$$\alpha = 5.4660$$
$$\Delta h = 1087.53\tan 5.4660° + 6.74\times10^{-8}\times1087.53^2$$
$$= 104.145\,\text{m}$$

Therefore reduced level of $B = 106.790 + 104.147$
$$= 210.937\text{ say }210.94\,\text{m}$$

6.7   Reciprocal levelling

Corrections for curvature and refraction are only approximations as they depend on the observer's position, the shape of the geoid and atmospheric conditions.

To eliminate the need for corrections a system of reciprocal levelling is adopted for long sights.

In Fig. 6.21, difference in level

$$d = BX_1 = a_1 + c + e - r - b_1$$

from A
$$= (a_1 - b_1) + (c - r) + e$$

Also from B $d = AX_2 = -(b_2 + c + e - r - a_2)$
$$= (a_2 - b_2) - (c - r) - e$$

By adding, $2d = (a_1 - b_1) + (a_2 - b_2)$
$$d = \tfrac{1}{2}[(a_1 - b_1) + (a_2 - b_2)] \tag{6.38}$$

Subtracting, $2(c - r + e) = [(a_2 - b_2) - (a_1 - b_1)]$

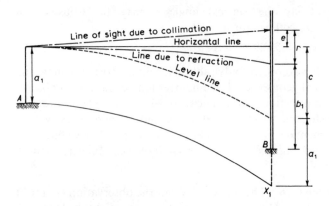

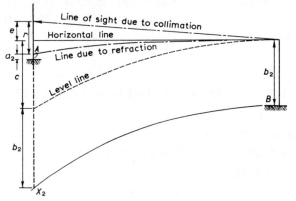

Fig. 6.21

Therefore total error

$$(c-r)+e = \tfrac{1}{2}[(a_2-b_2)-(a_1-b_1)] \qquad (6.39)$$

By calculating the error due to refraction and curvature, equation 6.28, for $(c-r)$ the collimation error e may be derived (see Section 6.4).

Example 6.17 (a)   Obtain from first principles an expression giving the combined correction for earth's curvature and atmospheric refraction in levelling, assuming that the earth is a sphere of 12 740 km diameter.

(b)   Reciprocal levelling between two points Y and Z 731.5 m apart on opposite sides of a river gave the following results:

Instrument at	Height of instrument	Staff at	Staff reading
Y	1.463	Z	1.689
Z	1.436	Y	0.991

Determine the difference in level between Y and Z and the amount of any collimation error in the instrument.   (ICE)

(a)   By equation 6.28, $c \simeq 6.74 \times 10^{-8} d^2$

(b)   By equation 6.36,

$$\text{difference in level} = \tfrac{1}{2}[(a_1 - b_1) + (a_2 - b_2)]$$
$$= \tfrac{1}{2}[(1.463 - 1.689) + (0.991 - 1.436)]$$
$$= \tfrac{1}{2}[-0.226 - 0.445]$$
$$= -0.335\,\text{m}$$

i.e. Z is 0.335 m below Y.

By equation 6.37,

$$\text{total error } (c - r) + e = \tfrac{1}{2}[(a_2 - b_2) - (a_1 - b_1)]$$
$$= \tfrac{1}{2}[-0.445 + 0.226]$$
$$= -0.219/2 = -0.109\,\text{m}$$

By equation 6.28,

$$(c - r) \simeq 6.74 \times 10^{-8} d^2$$
$$= 6.74 \times 10^{-8} \times 731.5^2$$
$$= 0.036\,\text{m}$$

therefore

$$e = -0.145\,\text{m per }731.5\,\text{m}$$
$$= -1.98 \times 10^{-4}\,\text{m per m}$$
$$\text{(collimation depressed)}$$

Check   Difference in level $= 1.463 - 1.689 - 0.145 + 0.036 = -0.335.$

Also $\qquad\qquad\qquad 0.991 - 1.436 + 0.145 - 0.036 = -0.336.$

6.7.1   The use of two instruments

To improve the observations by removing the likelihood of climatic change two instruments should be used, as in the following example.

Example 6.18

Staff at A	Mean	Staff at B	Mean	Apparent difference in level	Remarks
2.1135		2.8724			Inst. I on same
2.0678	2.0678	2.4603	2.4605	−0.3927	side as A
2.0220		2.0489			
1.9309		1.9586			Inst. II on same
1.5194	1.5192	1.9129	1.9129	−0.3937	side as B
1.1073		1.8672			
1.9666		1.9855			Inst. I on same
1.5539	1.5542	1.9397	1.9397	−0.3855	side as B
1.1421		1.8940			
2.0672		2.8191			Inst. II on same
2.0214	2.0214	2.4058	2.4064	−0.3850	side as A
1.9757		1.9943			

True difference in level   4) −1.5569

−0.3892

Thus B is 0.3892 m below A.

Exercises 6.6 (Reciprocal levelling)

1 The results of reciprocal levelling between stations A and B 250 m apart on opposite sides of a wide river were as follows:

Level at	Height of eyepiece (m)	Staff readings
A	1.399	2.518 on B
B	1.332	0.524 on A

Find
(a) the true difference in level between the stations;
(b) the error due to imperfect adjustment of the instrument assuming the mean radius of the earth 6365 km.
(LU/E Ans. (a) -0.964 m; (b) $+0.060$ m per 100 m)

2 In levelling across a wide river the following readings were taken:

Instrument at	Staff reading at A	Staff reading at B
A	1.823 m	2.481 m
B	2.499 m	3.182 m

If the reduced level at A is 31.282 m above datum, what is the reduced level of B? (Ans. 30.612 m)

6.8 Levelling for construction

6.8.1 Grading of constructions

The gradient of the proposed construction will be expressed as 1 in x, i.e. 1 vertical to x horizontal.

The reduced formation level is then computed from the reduced level of a point on the formation, e.g. the starting point, and the proposed gradient.

By comparing the existing reduced levels with the proposed reduced levels the amount of cut and fill is obtained.

If formation > existing, fill is required.

If formation < existing, cut is required.

Example 6.19 The following notes of a sectional levelling were taken along a line of a proposed road on the surface.

BS	IS	FS	Height of collimation	Reduced level	Horizontal distance	Remarks
3.121				31.858		BM
	1.411				20	Station 1
	0.448				40	Station 2
2.597		0.125			60	Station 3
	1.594				80	Station 4
3.853		1.027			100	Station 5
0.364		1.789			120	Station 6
0.051		3.727				
		3.312				BM

Calculate the reduced level of each station and apply the conventional arithmetical checks. Thereafter calculate the depth of cutting and filling necessary at each station to form an even gradient rising at 1 in 15 and starting at a level of 32.000 above datum at station 1. (MQB/M)

BS	IS	FS	Height of collimation	Reduced level	Remarks
3.121			34.979	31.858	BM
	1.411		[× 3]	33.568	1
	0.448		104.937	34.531	2
2.597		0.125	37.451	34.854	3
	1.594		[× 2]	35.857	4
3.853		1.027	74.902	36.424	5
0.364		1.789	40.277	38.488	6
0.051		3.727	38.852	35.125	
		3.312	35.176	31.864	BM
9.986	3.453	9.980	294.144	280.711	
9.980		3.453			
0.006		280.711			
		294.144		0.006	Field error

To calculate the cut and fill: (see Fig. 6.22)

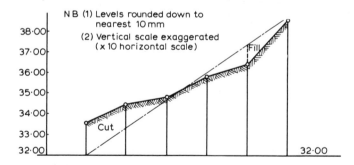

NB (1) Levels rounded down to nearest 10 mm
(2) Vertical scale exaggerated (× 10 horizontal scale)

20	40	60	80	100	120	Chainage
33·57	34·53	34·85	35·86	36·42	38·49	Existing level
32·00	33·33	34·67	36·00	37·33	38·67	Formation level
1·57	1·20	0·18				Cut
			0·14	0·91	0·18	Fill

Fig. 6.22

Reduced level	Formation level	Cut	Fill	Station
31.858				BM
33.568	32.000	1.568		1
34.531	33.333	1.198		2
34.854	34.667	0.187		3
35.857	36.000		0.143	4
36.424	37.333		0.909	5
38.488	38.667		0.179	6

6.8.2 The use of sight rails and boning (or travelling) rods

Sight rails and boning rods are used for excavation purposes associated with the grading of drains and sewers.

The sight rails are established at fixed points along the excavation line, at a height above the formation level equal to the length of the boning rod. The formation level compared with the surface level gives the depth of excavation, Fig. 6.23.

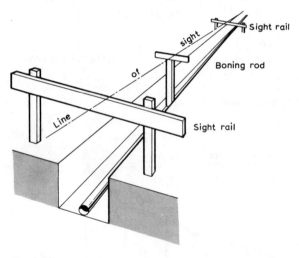

Fig. 6.23

When the boning rod is in line with the sight rails the excavation is at the correct depth, Fig. 6.24. Where the trench has to be dug by machine, the sight rails have to be set to the side

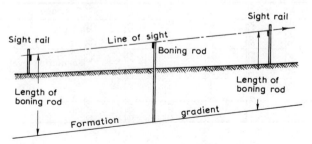

Fig. 6.24

of the trench to allow for the passage of the machine. The same mathematical principles apply.

The sight rails have thus to be set at a specific height above the invert level and this necessitates the need for a reversal of the booking principles. *If the full arithmetical checks are applied (as they must be at all times), there is no advantage of one method over the other.*

6.8.3   Booking of levels for setting out

Here the level of the fixed points are decided upon before the staff readings are taken, but there is no reason why the same booking process should not be observed and it is just as important to close the levelling circuit and to complete the checks on the arithmetic as it would be if an initial survey were taking place.

6.8.4   The setting of reduced levels on to pegs and profiles (Fig. 6.25)

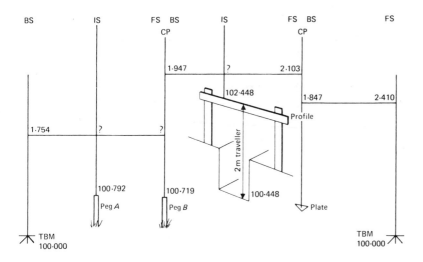

Fig. 6.25

1) At the TBM (reduced level 100.000 m) a staff reading of 1.754 was taken.
2) Peg *A* is to be set at the reduced level of 100.792.
   Calculate the staff reading required.
3) Peg *B* is to be set at the reduced level of 100.719.
   Calculate the staff reading required. This point is then used as a change point with a backsight reading of 1.947.
4) The trench base has a reduced level of 100.448. A 2 m traveller is to be used with the profile. Calculate the staff reading on the profile.
5) To close the levelling, further readings were taken to tie

back on to the **TBM** FS on to a levelling plate of 2.103, BS 1.847; FS on **TBM** 2.410.

6) Complete the field bookings required for the setting of these points and the subsequent checking of both the field work and the arithmetic.

Booking of levels for setting out by Rise and Fall method.
Given values in italic:

BS	IS	FS	Rise	Fall	Reduced level	Remarks
1.754					*100.000*	TBM value 100.00
	0.962		0.792		100.792	Peg *A*
1.947		1.035		0.073	100.719	Peg *B* CP
	0.218		1.729		102.448	Profile
1.847		2.103		1.885	100.563	CP
		2.410		0.563	100.000	TBM value 100.000
5.548		5.548	2.521	2.521		
5.548			2.521			
0.000			0.000		0.000	

Order of computation

1) From given reduced levels, compute the rise or fall:
e.g. $100.792 - 100.000 = +0.792$ rise.
2) Compute the staff reading to give a rise or fall as above:
e.g. $1.754 - 0.792 = 0.962$.
3) Complete the table in the normal way.
4) Apply the arithmetical checks.

Booking of levels for above by Height of Collimation method.

BS	IS	FS	Reduced level		Remarks
1.754			101.754	*100.000*	TBM value 100.000
	0.962		(101.754)	*100.792*	Peg *A*
1.947		1.035	102.666	*100.719*	Peg *B* CP
	0.218		(102.666)	*102.448*	Profile
1.847		*2.103*	102.410	*100.563*	CP
		2.410		100.000	TBM value 100.000
5.548	1.180	5.548	511.250	504.522	
5.548		1.180			
		504.522			
0.000		511.250		0.000	

Order of computation

1) From first reduced level, compute the height of collimation from staff reading.
2) From HC and reduced level of peg, compute the staff reading on peg.
3) Complete the table in the normal way.
4) Apply the full arithmetical checks. $\Sigma IS + \Sigma FS + \Sigma$ (reduced levels − first) should equal each HC × number of times applied.

Example 6.20 In preparing the fixing of sight rails, the following consecutive staff readings were taken.

Bench mark (50.490 AOD)	0.832
Ground level at A	1.804
Invert of sewer at A	3.240
Ground level at B	1.301
Ground level at C	1.079
Foresight at change point	1.234
Backsight at change point	3.210
Foresight at bench mark	2.808

If the sewer is to rise at 1 in 300 from the invert at A and the distances $AB = 32.00\,\text{m}$ and $BC = 46.63\,\text{m}$, what will be the staff readings on the sight rails at A, B and C for use with a 3 m boning rod?

See Fig. 6.26.

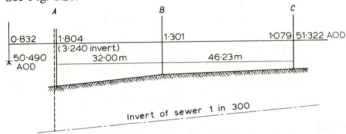

Fig. 6.26

Level of invert at $A = 50.490 + 0.832 - 3.240 = 48.082$

Rise from A to $B = 32.00/300$ $= 0.107$

Level of invert at B $= 48.189$

Rise from B to $C = 46.63/300$ $= 0.155$

Level of invert at C $= 48.344$

The remaining calculations should be made in the level book.

BS	FS	IS	Height of collimation	Reduced level	Remarks
0.832			51.322	50.490	BM 50.490
	1.804		(×9) 461.898	49.518	Ground at A
	3.240			48.082	Invert at A
	0.240			51.082	*Sight rail at A*
	1.301			50.021	Ground at B
	3.133			48.189	Invert at B
	0.133			51.189	*Sight rail at B*
	1.079			50.243	Ground at C
	2.978			48.344	Invert at C
3.210		1.234	53.298	50.088	CP
	1.954		(×2) 106.596	51.344	*Sight rail at C*
		2.808		50.490	BM
4.042	15.862	4.042	568.494	548.590	
4.042		15.862			
		548.590			
0.000		568.494	*Check*	0.000	(No field error)

Alternatively, by the Rise and Fall method of booking:

BS	IS	FS	Rise	Fall	Reduced level	Remarks
0.832					50.490	BM 50.490
	1.804			0.972	49.518	Ground at *A*
	3.240			1.436	48.082	Invert at *A*
	0.240		3.000		51.082	*Sight rail at A*
	1.301			1.061	50.021	Ground at *B*
	3.133			1.832	48.189	Invert at *B*
	0.133		3.000		51.189	*Sight rail at B*
	1.079			0.946	50.243	Ground at *C*
	2.978			1.899	48.344	Invert at *C*
3.210		1.234	1.744		50.088	CP
	1.954		1.256		51.344	*Sight rail at C*
		2.808		0.854	50.490	BM
4.042		4.042	9.000	9.000		
4.042			9.000			
0.000			0.000		0.000	

It can be seen by comparing these two sets of bookings that there is little to choose between them and that there is no basis for the widely held view that height of collimation booking for setting out is better than the rise and fall method.

6.8.5   The setting of slope stakes

A slope stake is set in the ground at the intersection of the ground and the formation slope of the cutting or embankment.

The position of the slope stake relative to the centre line of the formation may be obtained:

(a)   by scaling from the development plan; or
(b)   by calculation involving the cross-slope of the ground and the formation slope, using the rate of approach method; or
(c)   by a trial and error method.

By the rate of approach method, in Fig. 6.27,

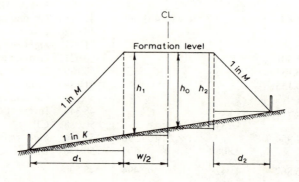

Fig. 6.27

$$h_1 = h_0 + \frac{w}{2K} \tag{6.40}$$

$$h_2 = h_0 - \frac{w}{2K} \tag{6.41}$$

$$d_1 = \frac{h_1}{\dfrac{1}{M} - \dfrac{1}{K}} \tag{6.42}$$

Similarly, $$d_2 = \frac{h_2}{\dfrac{1}{M} + \dfrac{1}{K}} \tag{6.43}$$

Example 6.21   To determine the position of slope stakes, staff readings were taken at ground level as follows:

Point *A* — Centre line of proposed road (Reduced level 31.614 m AOD)   1.716 m

Point *B* — 15.2 m from centre line and at right angles to it   1.868 m

If the reduced level of the formation at the centre line is to be 37.783 AOD, the formation width 6.1 m, and the batter is to be 1 in 2, what will be the staff reading, from the same instrument height at the slope stake and how far will the peg be from the centre line point *A*?

Gradient of $AB = (1.868 - 1.716)$ in 15.2 m

i.e. 0.152 m in 15.2 m

i.e. 1 in 100

In Fig. 6.28,

$$AA_1 = 37.783 - 31.614 \qquad = 6.169$$

By equation 6.38,

$$XX_1 = h = 6.169 + \frac{6.1}{2 \times 100} = 6.200$$

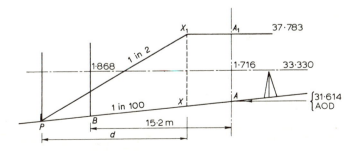

Fig. 6.28

By equation 6.42, the horizontal distance d, i.e. XP, is

$$d = \frac{h}{1/M - 1/K} = \frac{6.2 \times 100}{50 - 1}$$

$$= 12.653$$

Therefore distance from the centre line point A
$$= 12.653 + 6.1/2 = 15.703$$

the inclined length

$$XP = \frac{12.653 \times \sqrt{(100^2 + 1)}}{100} \qquad = \quad 12.654$$

level of A $\qquad = \quad 31.614$

difference in level $AP = \dfrac{12.653 + 3.050}{100} \qquad = \quad -0.157$

level of P $\qquad = \quad 31.457$

height of collimation $= 31.614 + 1.716 \quad = \quad 33.330$

Therefore staff reading at P $\qquad = \quad 1.873$

Trial and error method. Assuming the ground is level, the side width peg would be located by the general equation:

$$W = w/2 + mh \qquad\qquad (6.44)$$

where $W =$ the distance from the centre line to the trial point P
$\quad\; w =$ the width of formation
$\quad\; h =$ the difference in level between the ground at P and the formation level
$\quad\; m =$ the gradient of the batter

As the ground is often sloping the position of P will be incorrect and the second trial point, P_1, may be found by the modified equation

$$W' = W + mh' \qquad\qquad (6.45)$$

where $h' =$ the difference in level between W and W'
or $\delta W = mh'$.

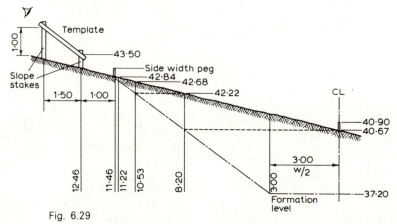

Fig. 6.29

Example 6.22 Given that $w = 6\,\text{m}$, formation level $37.20\,\text{m}$, the existing ground level at the centre line $40.67\,\text{m}$ and the batter 1 in 1.5. Levels are taken to a TBM, a peg on the centre line $(40.90\,\text{m})$ and distances step chained (if necessary) as through chainages from the centre line conforming to equation 6.45 (see Fig. 6.29).

The booking of the observations and computations is shown below:

Initial calculations
$$W = w/2 + mh$$
$$= 3.00 + 1.5\,(40.67 - 37.20) = 8.20$$

Booking of levels (Rise and Fall)

BS	IS	FS	Rise	Fall	Reduced level	Distance	Remarks
3.59					40.90	0	TBM (40.90 m) Formation 37.20
	3.82			0.23	40.67	0	CL $W = 3.00 + 1.5\,(3.47) = 8.20$
	2.27		1.55		42.22	8.20	$\delta W = 1.5\,(1.55) = 2.33$
	1.81		0.46		42.68	10.53	$\delta W = 1.5\,(0.46) = 0.69$
	1.65		0.16		42.84	11.22	$\delta W = 1.5\,(0.16) = 0.24$
						11.46	Side width peg
	0.99		0.66		43.50	12.46	Batter rail ($h = W/1.5 = 0.66$)
		3.59		2.60	40.90		TBM (40.90)
3.59		3.59	2.83	2.83			
3.59			2.83				
0.00			0.00		0.00		

Booking of levels (Height of Collimation)

BS	IS	FS	Height of collimation	Reduced level	Distance	Remarks
3.59			44.49	40.90	0	TBM (40.90) Formation 37.20
	3.82			40.67	0	CL $W = 3.00 + 1.5\,(3.47) = 8.20$
	2.27			42.22	8.20	$\Delta h = 1.55;\ \delta W = 1.5\,(1.55) = 2.33$
	1.81			42.68	10.53	$\Delta h = 0.46;\ \delta W = 1.5\,(0.46) = 0.69$
	1.65			42.84	11.22	$\Delta h = 0.16;\ \delta W = 1.5\,(0.16) = 0.24$
					11.46	Side width peg
	0.99			43.50	12.46	Batter rail ($h = W/1.5 = 0.66$)
		3.59		40.90		TBM (40.90)
3.59	10.54	3.59		252.81		
3.59		10.54				
		252.81	($\times 6$)			
0.00		266.94	266.94	0.00		

Exercises 6.7 (Construction levelling)

1 Sight rails are to be fixed at A and B $106.68\,\text{m}$ apart for the setting out of a sewer at an inclination of 1 in 200 rising towards B.

If the levels of the surface are *A* 32.379 and *B* 31.839, and the invert level at *A* is 30.706, at what height above ground should the sight rails be set for use with boning rods 3 m long?

(Ans.   1.327 m at *A*; 2.400 m at *B*)

2   A sewer is to be laid at a uniform gradient of 1 in 200 between two points *X* and *Y*, 243.84 m apart. The reduced level of the invert at the outfall *X* is 150.821 m.

In order to fix sight rails at *X* and *Y*, readings are taken with a level in the following order:

	Reading (m)	Staff station
BS	0.811	TBM (near *X*) Reduced level 153.814 m
IS	*a*	Top of sight rail at *X*
IS	1.073	Peg at *X*
FS	0.549	TP between *X* and *Y*
BS	2.146	TP between *X* and *Y*
IS	*b*	Top of sight at *Y*
FS	1.875	Peg at *Y*

(i)   Draw up a level book and find the reduced levels of the pegs.

(ii)   If a boning rod of length 3 m is to be used, find the readings *a* and *b*.

(iii)   Find the height of the sight rails above the pegs at *X* and *Y*.   (Ans.   (ii) 0.804, 1.182; (iii) 0.269, 0.693)

3   The levelling shown on the field sheet given below was undertaken during the laying out of a sewer line. Determine the height of the ground at each observed point along the sewer line and calculate the depth of the trench at points *X* and *Y* if the sewer is to have a gradient of 1 in 200 downwards from *A* to *B* and is to be 1.280 m below the surface at *A*.

BS	IS	FS	Distance (m)	Remarks
3.417				BM 98.002 m
1.390		1.774	0	
	1.152		20	
3.551		1.116	40	Point *X*
0.732		1.088	60	
2.384		3.295	80	
	1.801		100	
	1.999		120	Point *Y*
1.936		2.637	140	Point *B*
		1.161		BM 100.324

(RICS/ML Ans.   1.754 m, 2.439 m)

Exercises 6.8 (General)

1   The following staff readings in fact were taken successively with a level, the instrument having been moved forward after the second, fourth and eighth readings. 0.469, 2.207, 1.228, 0.351, 2.627, 2.597, 1.953, 0.344, 2.228, 0.838 and 1.649. (First reading on TBM 29.619.)

The last reading was taken with the staff on a bench mark having an elevation of 31.620 m.

Enter the readings in level book form, complete the reduced levels and apply the normal arithmetical checks. How accurate is the TBM value?   (Ans.   Apparent error 3 mm)

2   The following readings were taken using a dumpy level on a slightly undulating underground roadway.

BS	IS	FS	Reduced level	Distance	Remarks
1.622			− 380.329 m OD	0	Point *A* 380.329 m
	1.960			20 m	below OD
	1.594			40 m	
0.933		1.256		60 m	
	1.310			80 m	
	0.680			100 m	
0.920		0.332		120 m	
	1.222			140 m	
	1.561			160 m	
		2.033		180 m	Point *B* 380.480 m
					below OD

Work out the reduced levels of the points along the roadway and state the amount of excavation necessary at point *B* to form an even gradient dipping 1 in 300 from *A* to *B*, the levels given for *A* and *B* being assumed correct.   (Ans.   0.449 m)

3   A levelling party ran a line of levels from point *A* at elevation 41.279 m to point *B* for which the reduced level was found to be 26.563 m. A series of flying levels (as below) was taken back to the starting point *A*.

BS	FS	Remarks
2.947		*B* 26.563 m AOD
3.517	0.421	
2.505	1.466	
2.420	1.021	
3.219	0.631	
3.024	1.624	
2.707	0.317	
	0.128	*A* 41.279 m AOD

Find the misclosure on the starting point.   (Ans.   0.015 m)

4   The level book refers to a grid of levels taken at 20 m intervals on 4 parallel lines 20 m apart.
(a)   Reduce and check the level book.
(b)   Draw a grid to a scale of 1/500 and plot the contours for a 0.5 m vertical interval.

BS	IS	FS	Reduced level	Distance	Remarks
0.37			27.17	0	Line A TBM
	0.84			20	
	1.07			40	
	1.30			60	
	3.76			60	Line B
	2.76			40	
	2.07			20	
	1.27			0	
1.26		2.12		0	Line C
	1.68			20	
	2.40			40	
	3.19			60	
	2.72			60	
	2.19			40	
	1.63			20	
	1.40			0	
2.55		0.80			CP
		1.26			TBM

5   The record of a levelling made some years ago has become of current importance. Some of the data are undecipherable but sufficient remain to enable all the missing values to be calculated. Reproduce the following levelling notes and calculate and insert the missing values.

BS	IS	FS	Rise	Fall	Reduced level	Remarks
0.719					36.991	BM at No. 1 shaft
				0.591		
1.234		2.222				
				1.359		
	1.314					
	2.112					
					34.540	
		2.374		0.192		BM on school
	0.981		0.482			
					34.141	
	1.990					
					34.604	
		1.786				
				0.945		BM on church
4.560		____	____	6.455		

If the **BM** values on the school and church are 34.352 m and 33.090 m respectively, what are the field errors at these points?

(Ans.   4 mm and 10 mm)

6   In order to check the underground levellings of a colliery it was decided to remeasure the depth of the shaft and connect the levelling to a recently established Ordnance Survey Bench Mark A, 83.043 AOD.

The following levels were taken with a dumpy level starting at
A to the mouth of the shaft at D.

BS	FS	Reduced level	Remarks
0.661		83.043	BM at A
1.024	3.450		Mark B
1.764	2.417		Mark C
	0.000		Mark D on rails

The vertical depth of the shaft was then measured from D to E
at the pit bottom and found to be 532.092 m.

A backsight underground to E was found to be 1.213 and a
foresight to the colliery Bench Mark F on a wall near the pit
bottom was 0.832 m.

Tabulate the above readings and find the value of the
underground BM at F expressing this as a depth below
Ordnance Datum.   (Ans.   −451.086 m)

7   In levelling up a hillside, the sight lengths were observed
with stadia lines, the average length of the ten backsights and
foresights being 22 m and 11 m respectively.

Since the observed difference of the reduced level of 23.896 m
was disputed, the level was set up midway between two pegs
A and B 100 m apart, and the reading on A was 1.402 m and on
B 1.558 m, and when set up in line AB, 10 m behind B, the
reading on A was 1.576 m and on B 1.719 m.

Calculate the true difference of reduced level.
(LU Ans.   23.882 m; 1.3×10^{-3} m per 10 m)

8 A, B, C, D, E and F are the sites of manholes, 100 m apart on
a straight sewer. The natural ground can be considered as a
plane surface rising uniformly from A to F at a gradient of 1
vertically in 500 horizontally, the ground level at A being
31.394 m. The level of the sewer invert is to be 28.956 m at A, the
invert then rising uniformly at 1 in 200 to F. Site rails are to be
set up at A, B, C, D, E and F so that a 3 m boning rod or traveller
can be used. The backsights and foresights were made approximately equal and a peg at ground level at A was used as
datum.

Draw up a level book showing the readings.   (LU)

9   The following staff readings were obtained when running a
line of levels between two bench marks A and B:

1.085 (A) 2.036, 2.231, 3.014 change point, 0.613, 2.003, 2.335,
CP 1.622, 1.283, 0.543, CP 1.426, 1.795, 0.911

Enter and reduce the readings in an accepted form of field
book. The reduced levels of the bench marks at A and B were
known to be 43.650 m and 41.672 m respectively.

It is found after the readings have been taken with the staff supposedly vertical, as indicated by a level on the staff, that the level is 5° in error in the plane of the staff and instrument.

Is the collimation error of the instrument elevated or depressed and what is its value in seconds if the backsights and foresights averaged 30 m and 60 m respectively?

(LU Ans. True difference in level 2.046 m; collimation elevated 117″)

10 The centre line of a section of a proposed road in cutting is indicated by pegs at equal intervals and the corresponding longitudinal section gives the existing ground level and the proposed formation level at each peg, but no cross-sections have been taken, or sidelong slopes observed.

Given the proposed formation width (*d*) and the batter of the sides (*S* horizontal to 1 vertical) how would you set out the batter pegs marking the tops of the slopes at each centre line peg, without taking and plotting the usual cross-sections?

Any alternative method would be acceptable.   (ICE)

11 (a) Determine from first principles the approximate distance at which correction for curvature and refraction in levelling amounts to 3 mm, assuming that the effect of refraction is one seventh that of the earth's curvature and that the earth is a sphere of 12 746 km miles diameter.

(b) Two survey stations *A* and *B* on opposite sides of a river are 250 m apart, and reciprocal levels have been taken between them with the following results:

Instrument at	Height of instrument	Staff at	Staff reading
A	1.47 m	*B*	1.83 m
B	1.50 m	*A*	1.21 m

Compute the ratio of refraction correction to curvature correction, and the difference in level between *A* and *B*.

(ICE Ans. (a) 211 m; (b) *A* is 0.325 m below *B*.
Ratio ≃ 0.14 to 1)

Bibliography

BANNISTER, A., and RAYMOND, S., *Surveying*, 4th edition. Pitman (1977)

BRIGHTY, S. G., *Setting Out—Guide for Site Engineers*. Crosby Lockwood Staples (1975)

CURTIN, W. G., and LANE, R. F., *Concise Practical Surveying*, 2nd edition. English Universities Press (1970)

HOLLAND, J. L., WARDELL, K., and WEBSTER, A. G., Surveying, *Coal Mining Series*. Virtue

IRVINE, W., *Surveying for Construction*, 2nd edition. McGraw-Hill (1980)

SCHOFIELD, W., *Engineering Surveying*, 3rd edition. Butterworth (1978)

7

Dip and fault problems

Problems on gradients take a number of different forms and may be solved graphically or trigonometrically according to the accuracy required.

7.1 Definitions

Let *ABCD* represent a plane inclined at δ to the horizontal (Fig. 7.1).

Dip of a bed, seam or road in any direction (ϕ) may be given as the angle of inclination (δ) or as a gradient of 1 in x, i.e. 1 vertical to x horizontal. Then

$$\delta = \cot^{-1} x \tag{7.1}$$

NB: The term rise denotes the opposite of dip.

Full dip (or true dip) is the maximum inclination (δ) and its direction is at right angles to the level line or *strike* ($\alpha = 0$).

Apparent dip (α) is the inclination in any direction other than full dip and lies at an angle λ to the direction of full dip.

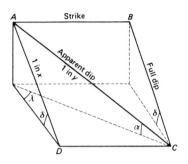

Fig. 7.1

7.2 Dip problems

7.2.1 Given the rate and direction of full dip to find the apparent dip in any other direction

Let Fig. 7.2 represent the plan of the inclined plane *ABCD* with *AB* and *CD* as strike lines.

If the rate of full dip is 1 in x then for a fall of 1 unit these strike lines may be drawn as x units apart, i.e. $AD = x$ units.

For a fall of 1 in y then the length *AC* will be y units.

In triangle *ACD*,

$$AC = y = x/\cos \lambda \tag{7.2}$$

or $\quad AD = x = y \cos \lambda \tag{7.3}$

i.e. the gradient value of full dip (x) equals the gradient value of apparent dip multiplied by the cosine of the angle between them.

Also

$$\lambda = \cos^{-1} x/y \tag{7.4}$$

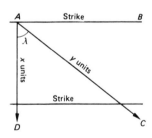

Fig. 7.2

Example 7.1   The full dip of a seam is 1 in 4 on a bearing 030° 00′.

Find the gradients of roadways driven in the seam (a) due N, (b) 075° 00′, (c) due E.

Graphically,

$$AB = 4.0 \text{ units}$$
$$AC = 4.6 \text{ units}$$
$$AD = 5.7 \text{ units}$$
$$AE = 8.0 \text{ units}$$

In Fig. 7.3, AB represents full dip 1 in 4 (4 units) 030° 00′. Then

$$CAB = \lambda_c = 30° 00' \quad \text{and} \quad AC = 4/\cos 30° 00' = 4.62$$
$$BAD = \lambda_d = 45° 00' \quad\quad\quad AD = 4/\cos 45° 00' = 5.66$$
$$BAE = \lambda_e = 60° 00' \quad\quad\quad AE = 4/\cos 60° 00' = 8.00$$

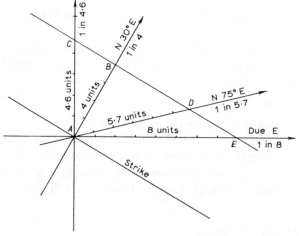

Fig. 7.3

Thus gradient due N is 1 in 4.62; gradient 075° 00′ is 1 in 5.66 and gradient due E is 1 in 8.00.

Example 7.2   A seam dips at 1 in 5 in a direction 208° 00′. In what direction will a roadway in the seam dip 1 in 8?

By equation 7.4,

$$\lambda = \cos^{-1} 5/8 = 51° 19'$$

then $\phi_{AD} = \phi_{FD} \pm \lambda$

$$= 208° 00' + 51° 19' = 259° 19'$$

or $\quad = 208° 00' - 51° 19' = 156° 41'$

7.2.2   Given two apparent dips to find the rate and direction of full dip

Graphically (Fig. 7.4).   Plot the directions of apparent dips AC and AD of length y and z respectively. Join AB at right angles to CD. Measure $AB(x)$ in the same units as y and z.

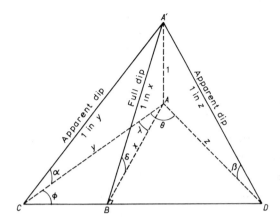

Fig. 7.4

Trigonometrically (Fig. 7.5). In triangle ADC,

$$AC = y, \ AD = z, \ DAC = \theta, \ AB = x \ \text{(full dip)}$$

Then $\quad \dfrac{C-D}{2} = \tan^{-1} \dfrac{z-y}{z+y} \tan \dfrac{C+D}{2}$

$$\dfrac{C+D}{2} = \dfrac{180-\theta}{2}$$

Solve for angle C; then

$$\lambda = 90 - C$$

In triangle ABC,

$$AB = x = y \cos \lambda = z \cos (\theta - \lambda)$$

$$\phi_{AB} = \phi_{AC} - \lambda$$

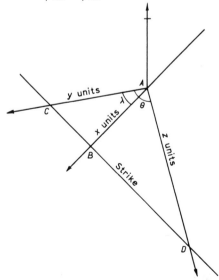

Fig. 7.5

Alternatively,

$$x = y \cos \lambda = z(\cos \theta \cos \lambda + \sin \theta \sin \lambda)$$

then $y = z(\cos \theta + \sin \theta \tan \lambda)$

or $\tan \lambda = y \operatorname{cosec} \theta / z - \cot \theta$ (7.5)

Or as

$$y = \cot \alpha, \ z = \cot \beta \ \text{and} \ x = \cot \delta$$

then $\lambda = \tan^{-1}(\tan \beta \operatorname{cosec} \theta \cot \alpha - \cot \theta)$ (7.6)

and as $x = \cot \alpha \cos \lambda$

$$\cot \delta = \cot \alpha \cos \lambda$$

or $\tan \alpha = \tan \delta \cos \lambda$ (7.7)

i.e. tan apparent dip = tan full dip × cos angle between.

Example 7.3   A roadway dips 1 in 4 in a direction 085° 30′ and intersects another dipping 1 in 6 at 354° 30′. Find the rate and direction of full dip.

In Fig. 7.6,

$$DAC = \phi_{AC} - \phi_{AD} = 91° \ 00′$$

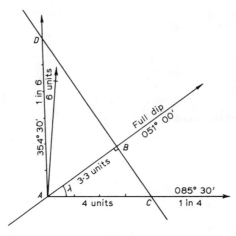

Fig. 7.6

Let $AD = 6$; $AC = 4$

$$\tan \frac{C - D}{2} = \frac{6 - 4}{6 + 4} \tan \frac{89° \ 00′}{2}$$

Then

$$(C - D)/2 = 11.119$$
$$(C + D)/2 = 44.500$$
$$C = 55.619$$
$$\lambda = 90 - C = 34.381$$
$$\phi_{AB} = \phi_{AC} - \lambda$$

$$= 85.500 - 34.381 = 51.119 \ (51° 07')$$

$$AB = AC \sin C$$

$$= 4 \sin 55.619 \qquad = 3.30$$

Thus full dip is 1 in 3.3, bearing 051° 07'.

Alternatively, by equation 7.6,

$$\lambda = \tan^{-1} (\tan \beta \operatorname{cosec} \theta \cot \alpha - \cot \theta)$$

$$= \tan^{-1} (1/6 \times 1/4 \times 1/\sin 91.000 - \cot 91.000)$$

$$= 34.381 \quad \text{as above.}$$

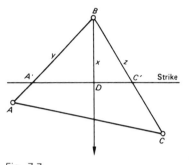

Fig. 7.7

7.2.3 Given the levels and relative positions of three points in a plane (bed or seam) to find the rate and direction of full dip

Graphical solution (Fig. 7.7). Select the highest point, say B, and derive the gradients from this point.

Gradient BA is $(L_A - L_B)$ in S_{BA}, say 1 in y.

Gradient BC is $(L_C - L_B)$ in S_{BC}, say 1 in z.

Set off y units on the line BA and z units on the line BC.

Join $A'C'$ as a strike line.

Draw BD at 90° to $A'C'$ and measure BD as x units as before, i.e. two apparent dips to find full dip.

Trigonometrical solution (Fig. 7.8). This follows the graphical solution.

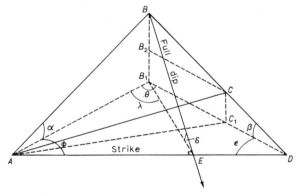

Fig. 7.8

Example 7.4 The following data were recorded at three boreholes A, B and C.

	E (m)	N (m)	Surface level (AOD)	Seam depth (m)
A	10 000.00	10 000.00	112.78	320.04
B	9485.37	12 791.21	68.58	123.44
C	8408.57	9832.73	77.72	56.39

Calculate the rate and direction of full dip.

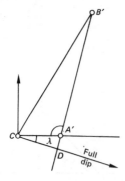

Fig. 7.9

Levels in the seam:

At A $112.78 - 320.04 = -207.26$ m

B $68.58 - 123.44 = -54.86$

C $77.72 - 56.39 = 21.33$ (C is the highest point)

$\phi_{CA} = 084.000$ $S_{CA} = 1600.20$

$\phi_{CB} = 020.000$ $S_{CB} = 3148.35$

$\theta = 64.000$

Gradient $CA = (-207.26 - 21.33)$ in 1600.20, i.e. 1 in 7.00

$CB = (-54.86 - 21.33)$ in 3148.35, i.e. 1 in 41.32

In Fig. 7.9,

$$\tan\frac{A' - B'}{2} = \frac{41.32 - 7.00}{41.32 + 7.00}\tan(116.000/2)$$

$(A' - B')/2 = 48.660$

$(A' + B')/2 = 58.000$

then $A' = 106.660$

$\lambda = 90 - (180 - 106.660) = 16.660$

$CD = 7.00\cos 16.660 = 6.71$

Full dip is then 1 in 6.71.

$\phi_{CD} = \phi_{CA} + 16.660 = 100.660 = 100°\,40'$

7.3   The rate of approach method for convergent lines

7.3.1   When both lines converge to a common point

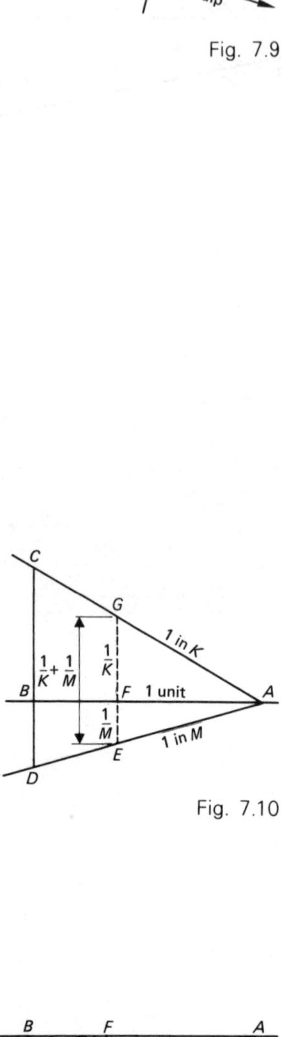

In Fig. 7.10, let AC rise from A at 1 in K

AD dip from A at 1 in M

AB be a horizontal line with $AF = 1$ unit

Then $GF = 1/K$ and $FE = 1/M$

Comparing the similar triangles ADC and AEG,

$CD/BA = GE/FA = 1/K + 1/M$

$CD = BA(1/K + 1/M)$ (7.8)

$BA = CD/(1/K + 1/M)$ (7.9)

Thus if two convergent lines CA and DA are CD vertically apart, then the horizontal length BA when they meet is given as $CD/(1/K + 1/M)$.

7.3.2   When both lines dip (or rise) from a common point

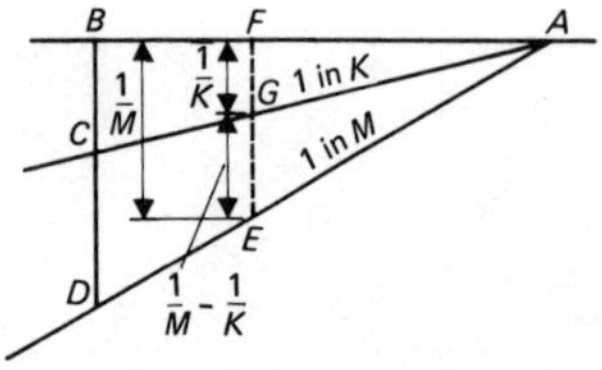

In Fig. 7.11, let AC dip from A at 1 in K

AD dip from A at 1 in M

Fig. 7.10

Fig. 7.11

As before,

$$CD/BA = GE/FA = 1/M - 1/K$$

$$CD = BA(1/M - 1/K) \tag{7.10}$$

$$BA = CD/(1/M - 1/K) \tag{7.11}$$

Thus if two lines CA and DA dip or rise in the same direction they will be CD vertically apart after a horizontal distance of BA given as $CD/(1/M - 1/K)$.

Example 7.5 Two seams of coal, 30.00 m vertically apart, dip at 1 in 6. Find the length of a roadway (drift) driven between the seams:
(a) at a rise of 1 in 4 from the lower to the upper seam;
(b) at a dip of 1 in 2 from the upper to the lower seam.

Referring to Fig. 7.12,

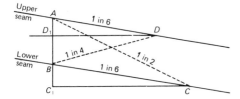

Fig. 7.12

(a) In triangle ADB, AD falls at 1 in 6
BD rises at 1 in 4
$AB = 30.00$ m

Then by equation 7.9,

$$D_1D = AB/(1/M + 1/K)$$
$$= 30.00/(1/4 + 1/6) = 72.00 \text{ m}$$

Thus the inclined length of the drift $= 72.00 \times \sqrt{(4^2 + 1^2)/4}$
$$= 74.22 \text{ m}.$$

(b) Similarly, by equation 7.11,

$$CC_1 = 30.00/(1/2 - 1/6) = 90.00$$

Thus the inclined length of the drift $= 90.00 \times \sqrt{5}/2$
$$= 100.62 \text{ m}.$$

Example 7.6 Two parallel levels, 200.00 m apart, run due East/West in a seam which dips due North at 1 in 12. At a point A in the lower level a cross measures drift rising at 1 in 6 and bearing 030° 00′ is driven to intersect another seam situated 200.00 m vertically above the first seam at C. From a point B in the upper level due South from A another cross measures drift rising at 1 in 3 and bearing 030° 00′ is also driven to intersect the upper seam at a point D. Calculate the inclined length and the bearing of the line CD.

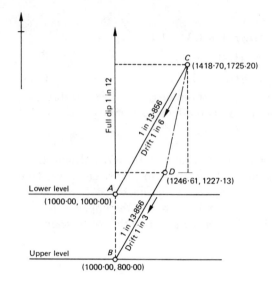

Fig. 7.13

In Fig. 7.13, by equation 7.7,

$$\tan \alpha = \tan \delta \cos \lambda$$
$$= (\cos 30)/12 = 1/13.856$$

length of the drift $AC = 200.00/(1/6 + 1/13.856) = 837.39$ m
length of the drift $BD = 200.00/(1/3 + 1/13.856) = 493.21$ m
Assuming the two levels are 200.00 m apart in plan, the relative positions of B, C and D may be coordinated relative to A (1000.00, 1000.00).

Line AC $\phi =$ 030° 00′ $S =$ 837.39
$\Delta E =$ 418.70 $E_C =$ 1418.70
$\Delta N =$ 725.20 $N_C =$ 1725.20
Line AB $\phi =$ 180° 00′ $S =$ 200.00
$\Delta E =$ 0.00 $E_B =$ 1000.00
$\Delta N = -200.00$ $N_B =$ 800.00
Line BD $\phi =$ 030° 00′ $S =$ 493.21
$\Delta E =$ 246.61 $E_D =$ 1246.61
$\Delta N =$ 427.13 $N_D =$ 1227.13
Line CD $\Delta E = -172.09$ $\Delta N = -498.07$
Then $\phi_{CD} = 199° 03′ 40″$ and $S_{CD} = 526.96$ m.

Assuming the level of $A = 0$
AC rises at 1 in 6   then $\Delta L = 837.39/6 = 139.56$
BD rises at 1 in 3 $\Delta L = 493.21/3 = 164.40$
AB rises at 1 in 12 $\Delta L = 200.00/12 = 16.67$

Then

$$\Delta L_{DC} = 164.40 + 16.67 - 139.56 = 41.51$$

The inclined length CD is then $= \sqrt{(526.96^2 + 41.51^2)}$
$$= 528.59.$$

Exercises 7.1

1   The full dip of a seam is 1 in 9. Calculate the angle included between full dip and an apparent dip of 1 in 12.

(Ans. $41° 24'$)

2   The angle included between full dip and an apparent dip is $60° 00'$. If the inclination of apparent dip is $9° 10'$ calculate the gradient of full dip.   (Ans.   1 in 3.1)

3   On a hillside sloping at $18°$ runs a track at an angle of $50°$ with the line of greatest slope. Calculate the inclination of the track.

(Ans. $11° 48'$)

4   The full dip of a seam is 1 in 3 in a direction $274° 46'$. A roadway is driven in the seam in a southerly direction dipping at 1 in 10. Calculate the bearing of the roadway.

(Ans. $202° 14'$)

5   The full dip of a seam is 1 in 5, $356° 00'$. A roadway is to be set off rising at 1 in 8. Calculate the alternative bearings.

(Ans. $227° 19'$; $124° 41'$)

6   A seam dips at 1 in 12.75, $197° 00'$ and at 1 in 12.41, $159° 45'$. Calculate the rate and direction of full dip.

7   The coordinates and levels of points A, B and C are as follows:

	E (m)	N (m)	Levels (m) relative to datum
A	1119.0	1074.0	− 128.0
B	750.0	1787.5	− 297.0
C	1812.0	2011.0	− 195.0

Calculate the amount and direction of full dip.

(Ans.   1 in 4.67, $322° 04'$)

8   Two parallel seams, 60 m vertically apart, dip due W at 1 in 6. A drift with a falling gradient of 1 in 12 is driven from the upper to the lower seam in a direction due E. Calculate the length of the drift.   (Ans.   240.82 m)

9   Three boreholes A, B and C intersect a seam at depths of 540 m, 624 m, and 990 m respectively. A is 1800 m North of C and 2400 m East of B. Calculate the rate and direction of full dip.   (Ans.   1 in 3.96, $187° 58'$)

10   Two parallel roadways AB and CD advancing due North on the strike of a seam are connected by a roadway BD in the

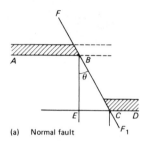

(a) Normal fault

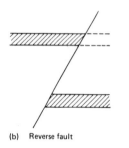

(b) Reverse fault

Fig. 7.14

seam, on a bearing 060° 00′. The plan length of *BD* is 150.00 m and the rate of dip of the seam is 1 in 5 in the direction *BD*.

Another roadway is to be driven on the bearing 045° 00′ to connect the two roadways commencing at a point 200.00 m out from *B* on the roadway *AB*, the first 20.00 m to be level.

Calculate the total length of the new roadway and the gradient of the inclined portion. (Ans. 186.44 m; 1 in 5.46)

7.4 Fault problems

7.4.1 Definitions

Fig. 7.14 illustrates the end view of a fault and no indication is given of movement in any other direction.

FF_1 is known as the fault plane.

B is the upthrow side of the fault.

C is the downthrow side of the fault.

θ is the angle of hade of the fault (measured from the vertical).

BE is the vertical displacement or throw of the fault.

EC is the horizontal displacement causing an area of 'want' or barren ground in a normal fault.

Where the direction and amount of full dip remain the same on both sides of the fault, a simple type of fault is indicated and the lines of contact between the seam and the fault on both sides of the fault are parallel. Where the direction and/or the amount of dip changes, rotation of the strata has taken place and the lines of contact will converge or diverge. The vertical throw diminishes towards the convergence until there is a change in direction of the throw which then increases (Fig. 7.15).

NB: The strike of the fault is its true bearing which will differ from the bearings of the line of contact between the seam and the fault unless the seam is level.

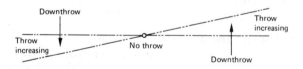

Fig. 7.15

Example 7.7 A vertical shaft, which is being sunk with an excavated diameter of 7.160 m, passes through a well defined fault of uniform direction and hade. Depths of the fault plane below a convenient horizontal plane are taken vertically at the extremities of two diameters *AB* and *CD* which bear N/S and E/W respectively. The undernoted depths were measured as:

at *A* (N) 3.07 m, at *B* (S) 8.00 m

at *C* (E) 1.22 m, at *D* (W) 9.86 m

Calculate the direction and throw of the fault and the amount of hade.

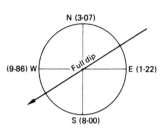

Fig. 7.16

Gradient of N/S line is $(8.00 - 3.07)$ in 7.160,

i.e. 1 in 1.4523 $\quad \beta = 34.5499$

E/W line is $(9.86 - 1.22)$ in 7.160,

i.e. 1 in 0.8287 $\quad \alpha = 50.3513$

In Fig. 7.16, by equation 7.6,

$$\tan \lambda = \tan \beta / (\tan \alpha \sin \theta) - \cot \theta$$

but $\qquad \theta = 90°$

then $\qquad \lambda = \tan^{-1} (\tan \beta / \tan \alpha)$

$\qquad = \tan^{-1} (\tan 34.5499 / \tan 50.3513) \ = 29.7098$

$\qquad \delta = \tan^{-1} (\tan \alpha / \cos \lambda)$

$\qquad = \tan^{-1} (\tan 50.3513 / \cos 29.7098) = 54.2548$

$\qquad \phi_{FD} = 270 - 29.7098 = 240.2902$, i.e. $240° \, 17' \, 25''$

angle of hade

$$\theta = 90 - 54.2548 = 35.7452, \text{ i.e. } 35° \, 44' \, 43'', \text{ say } 35° \, 45'$$

Example 7.8 A roadway dips at 1 in 8 in the direction of full dip, strikes an upthrow fault, bearing at right angles thereto. Following the fault plane a distance of 35.50 m the seam is again located and the hade of the fault is proved to be 30°. Calculate the length of a cross measures drift to win the seam, commencing at the lower side of the fault and rising at 1 in 6 in the same direction as the roadway.

In Fig. 7.17,

vertical throw of the fault $\quad FB = 35.50 \cos 30 = 30.74$

lateral displacement $\qquad\qquad FC = 35.50 \sin 30 \ = 17.75$

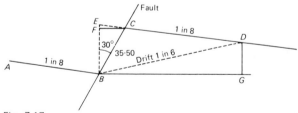

Fig. 7.17

In triangle EFC,

$$EF = EC/8 = 17.75/8 \qquad\qquad = 2.22$$

$$EB = FB + EF = 30.74 + 2.22 \ = 32.96$$

To find the plan length of the drift by the rate of approach method:

$$BG = 32.96/(\tfrac{1}{8} + \tfrac{1}{6}) \qquad = 113.01$$

the inclined length

$$BD = 113.01 \times \sqrt{37/6} = 114.56$$

Example 7.9   A roadway, advancing due East in the direction of full dip of 1 in 8, meets a downthrow fault bearing 145° 00′, with a throw of 20.50 m and a hade of 30°. At a distance of 51.25 m along the roadway, back from the fault, a drift is to be driven in the same direction as the roadway in such a way that it meets the point of contact of seam and fault on the downthrow side. Calculate the inclined length and gradient of the drift, assuming the seam is constant on both sides of the fault.

In plan (Fig. 7.18):

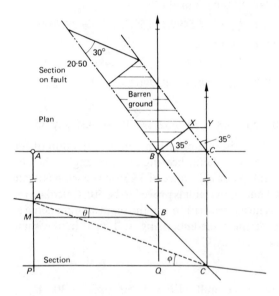

Fig. 7.18

width of the barren ground
$$BX = 20.50 \tan 30 = 11.84$$

on the line of the roadway
$$BC = 11.84/\cos 35 = 14.45 = QC$$
$$XC = BC \sin 35$$
$$XY = XC \sin 35 = BC \sin^2 35$$
$$= 14.45 \sin^2 35 = 4.75$$

As the dip of XY is 1 in 8, the level of Y relative to X is given as
$$-4.75/8 = -0.59$$

YC is the line of the strike; then
$$\text{level of } Y = \text{level of } C = 0.59 \text{ m below } X$$
$$= 20.50 \text{ m} + 0.59 \text{ m below } A$$
$$= 21.09 \text{ m below } A \ (BQ = MP)$$

In section

gradient of roadway AB $= 1$ in 8; $\theta = \cot^{-1} 8 = 7.125$

$$AM = 51.25 \sin 7.125 = 6.36$$

$$MB = 51.25 \cos 7.125 = 50.85 = PQ$$

$$\Delta L_{AC} = AP = AM + MP$$

$$= 6.36 + 21.09 = 27.45$$

plan length of drift $AC = PC = PQ + QC$

$$= 50.85 + 14.45 = 65.30$$

gradient of drift $= 27.45$ in 65.30

then $\alpha = \tan^{-1}(27.45/65.30) = 22.800$

length of drift $= 65.30/\cos 22.800 = 70.84\,\text{m}$

7.5   To find the relationship between the true and apparent bearings of a fault

The true bearing of a fault is the bearing of its strike. The apparent bearings of the fault are the bearings of the lines of contact between the seam and the fault.

The bearings of the lines of contact will be equal provided the dip is constant on both sides of the fault, i.e. the throw is constant.

Two general cases are considered:

(1)   when the throw is in the same direction as the dip;
(2)   when the throw is in the opposite direction to the dip.

Let the full dip on the upthrow side of the fault be 1 in x
      the full dip on the downthrow side of the fault be 1 in y
      the angle between full dip and the line of contact be α
      the angle between the contact line and the true bearing of
         the fault be β
      the angle of hade be θ
      the throw of the fault be t
      the angle between the full dip (x) and the true bearing of the
         fault be ψ
      the angle between the full dip (y) and the true bearing of the
         fault be δ

7.5.1   To find the true bearing of a fault when the downthrow of the fault is in the same direction as the dip of the seam (Fig. 7.19)

If the throw of the fault is t then D will be t below A and for a dip of 1 in x then $BC = tx$.

C will be t below A and B and thus

$$AD = t \tan \theta$$

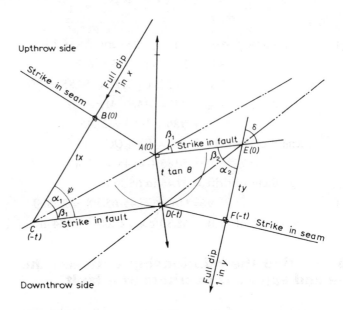

Fig. 7.19

The strike of the fault is shown as CD, and $AE : EF = ty$ and F is t below E. Then

$$AC = BC \sec \alpha_1 = tx \sec \alpha_1 \qquad (7.12)$$

$$\sin \beta_1 = (t \tan \theta)/(tx \sec \alpha_1) = \tan \theta \cos \alpha_1 / x \qquad (7.13)$$

But $\qquad \psi = \alpha_1 + \beta_1$

then $\qquad \alpha_1 = \psi - \beta_1$

so $\quad \sin \beta_1 = \tan \theta (\cos \psi \cos \beta_1 + \sin \psi \sin \beta_1)/x$

$\qquad x \cot \theta = \cos \psi \cot \beta_1 + \sin \psi$

$\qquad \cot \beta_1 = (x \cot \theta - \sin \psi) \sec \psi \qquad (7.14)$

Similarly

$$\sin \beta_2 = (t \tan \theta)/(ty \sec \alpha_2) = \tan \theta \cos \alpha_2 / y$$

and $\qquad \alpha_2 = \delta - \beta_2$

then $\quad \cot \beta_2 = (y \cot \theta - \sin \delta) \sec \delta \qquad (7.15)$

The true bearing of the fault $= \phi_{CA} + \beta_1 = \phi_{DE} + \beta_2$

7.5.2 To find the true bearing of a fault when the downthrow of the fault opposes the dip of the seam (Fig. 7.20)

As before:

$$AC = AB \sec \alpha_2 = ty \sec \alpha_2 \qquad (7.16)$$

$$\sin \beta_2 = \tan \theta \cos \alpha_2 / y \qquad (7.17)$$

and $\qquad \sin \beta_1 = \tan \theta \cos \alpha_1 / x \qquad (7.18)$

Here $\qquad \alpha_2 = \delta + \beta_2 \quad$ and $\quad \alpha_1 = \psi + \beta_1$

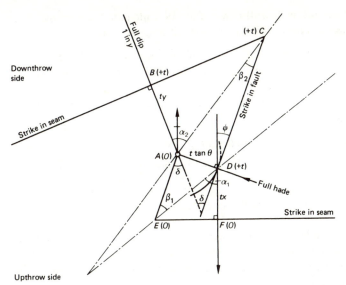

Fig. 7.20

then $\cot \beta_2 = (y \cot \theta + \sin \delta) \sec \delta$ (7.19)

and $\cot \beta_1 = (x \cot \theta + \sin \psi) \sec \psi$ (7.20)

The true bearing of the fault $= \phi_{AC} - \beta_2 = \phi_{ED} - \beta_1$

7.5.3   General relationship between true and apparent bearings

$$\sin \beta_1 = \tan \theta \cos \alpha_1 / x \qquad \text{(Equation 7.18)}$$
$$\sin \beta_2 = \tan \theta \cos \alpha_2 / y \qquad \text{(Equation 7.17)}$$
$$\cot \beta_1 = (x \cot \theta \pm \sin \psi) \sec \psi \qquad (7.21)$$
$$\cot \beta_2 = (y \cot \theta \pm \sin \delta) \sec \delta \qquad (7.22)$$

The sign is $+$ when the downthrow is in the opposite direction
to the dip

and $-$ when the downthrow is in the same direction as
the dip.

Example 7.10   A plan of workings in a seam dipping 1 in 3 in a
direction 150° 00′ shows a fault bearing 225° 00′ in the seam
which throws the measures down to the NW. The hade of the
fault is 30° to the vertical. Calculate the true bearing of the fault.

In Fig. 7.21, by equation 7.17,

$$\sin \beta = \tan \theta \cos \alpha / y$$
$$\beta = \sin^{-1} (\tan 30 \cos 75)/3 = 2° 51′$$
$$\phi_{CD} = \phi_{CA} - \beta$$
$$= 225° 00′ - 2° 51′ = 222° 09′$$

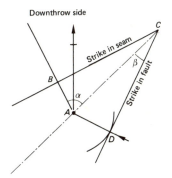

Fig. 7.21

Example 7.11   A seam dipping 1 in 5 on a bearing 120° 00′ is
intersected by a fault the hade of which is 30° and bears 325° 00′.

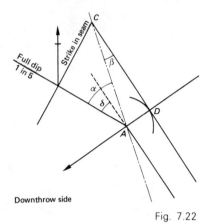

Fig. 7.22

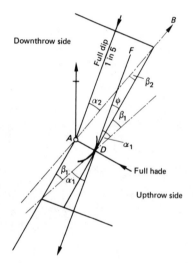

Fig. 7.23

The fault is downthrow to the SW. Calculate the bearing of the line of contact between the fault and the seam.

The amount and direction of full dip is assumed the same on both sides of the fault and thus the contact lines are parallel (Fig. 7.22).

As the full dip and the hade oppose one another then, by equation 7.21,

$$\cot \beta = (x \cot \theta + \sin \psi) \sec \psi$$
$$\beta = \cot^{-1} (5 \cot 30° + \sin 25° \, 00') \sec 25° \, 00'$$
$$= 5° \, 42'$$

The bearing of the line of contact $= 325° \, 00' + 5° \, 42' = 333° \, 42'$

Example 7.12   Headings in a seam at A and B have made contact with a previously unlocated fault which throws the measures up 30.5 m to the SE with a true hade of 40° to the vertical.

Full dip is known to be of constant bearing 202° 30′ but the amount of dip changes from 1 in 5 on the north side to 1 in 3 on the south side of the fault. Given the coordinates of A and B as follows:

	E (m)	N (m)
A	1119.37	1790.27
B	1513.41	2217.23

Calculate:
(a)   the true bearing of the fault and
(b)   the bearings of the lines of contact between the fault and the seam.

In Fig. 7.23, for the line AB
$$\Delta E = 394.04; \quad \Delta N = 426.96$$
then $\phi_{AB} = 042.704°$
$$\alpha_2 = \phi_{BA} - \phi_{FD} = 222.704 - 202.500 = 20.204°$$
By equation 7.17,
$$\beta_2 = \sin^{-1} (\cos 20.204 \tan 40)/5 \qquad = 9.061°$$
To find the bearing of the line of strike in the fault, i.e.
true bearing of the fault $= 042.704 - 9.061 = \underline{33.643°}$
$$\psi = \alpha_1 - \beta_1$$
$$= 20.204 - 9.061 = 11.143°$$
By equation 7.21, $\cot \beta_1 = (3 \cot 40 + \sin 11.143)/\cos 11.143$
$$\beta_1 = 14.592°$$
Bearing of the line of contact on the upthrow side of the fault
$$= 033.643 + 14.592 = 48.236°$$
$$= \underline{048° \, 14'}$$

Exercises 7.2

1 A roadway in a level seam advancing due N meets a normal fault with a hade of 30° from the vertical and bearing at right angles to the roadway.

An exploring drift is set off due N and rising at 1 in 1. At a distance of 12.50 m, as measured on the slope of the drift, the seam on the north side of the fault is again intersected.

(a) Calculate the throw of the fault and the width of the barren ground.

(b) If the drift had been driven at 1 in 4 (instead of 1 in 1) what would be the throw of the fault and the width of the barren ground?   (Ans.   8.84 m, 5.09 m; 3.02 m, 1.74 m)

2 A roadway AB, driven on the full rise of a seam at a gradient of 1 in 10, is intersected at B by an upthrow fault, the bearing of which is parallel with the direction of the level course of the seam, with a hade of 30° from the vertical. From B a cross measures drift has been driven in the line AB produced, to intersect the seam at a point 57.91 m above the level of B and 117.65 m from the upper side of the fault as measured in the seam. Calculate the amount of vertical displacement of the fault and the length and gradient of the cross measures drift. Assume that the direction and rate of dip of the seam are the same on each side of the fault.   (Ans.   46.21 m, 154.77 m, 1 in 2.5)

3 A roadway advancing due East in a level seam meets a fault bearing N/S, which hades at 30°. A drift, driven up the fault plane in the same direction as the roadway, meets the seam again at a distance of 36.58 m. Calculate the length and gradient of a drift rising from a point on the road 121.92 m to the west of the fault which intersects the seam 30.48 m East of the fault.
                (Ans.   173.74 m, 1 in 5.4)

4 A roadway driven to the full dip of 1 in 12 in a coal seam meets a 80 m downthrow fault. A cross measures drift is set off in the direction of the roadway at a gradient of 1 in 6. In what distance will it strike the seam again on the downthrow side if
(a) the hade is 8° from the vertical or
(b) the hade is vertical?   (Ans.   961.9 m; 973.3 m)

5 A roadway in a seam dipping 1 in 7 on the line of the roadway meets a downthrow fault of 30 m with a hade of 2° vertical to 1 horizontal. Calculate the length of the drift, dipping at 1 in 4 in the line of the roadway to win the seam, a plan distance of 50.00 m from the dip side of the fault; also the plan distance from the rise side of the fault from which the drift must be set off.

Assume the gradient of the seam to be uniform and the line of the fault at right angles to the roadway.
                (Ans.   267.97 m, 194.97 m)

6   A roadway, bearing due East in a seam which dips due South at 1 in 11, has struck a fault at a point A. The fault which, on this side, runs in the seam at $350°\ 00'$ is found to hade at $20°$ to the vertical and to throw the seam down 30 m from the point A. The dip of the seam on the lower side of the fault is in the same direction as the upper side but the dip is 1 in 6.

From a point B in the roadway 420.0 m West of A a slant road is driven in the seam on a bearing $050°\ 00'$ and is continued in the same direction and at the same gradient until the seam on the East side of the fault is intersected at a point C. Draw to a scale of 1/1250 a plan of the roads and the fault and mark the point C. State the length of the slant road BC.

(Ans.   792.0 m)

7   A seam dips at 1 in 4, $208°\ 30'$. Headings A and B have proved the bearing of the contact line AB to be $075°\ 00'$. If the hade of the fault is $30°$, what is the true bearing of the fault if:
(a)   it is a downthrow to the South or
(b)   it is a downthrow to the North?

(Ans. $080°\ 42'$; $069°\ 18'$)

8 (a)   Define the true and apparent azimuth of a fault.
(b)   A fault exposed in a certain seam has an azimuth of $086°\ 10'$, and a hade of $33°$. It throws down to the NW. The full dip of the seam is 1 in 6.5 at $236°\ 15'$. Calculate the true azimuth of the fault. Check by plotting.
(c)   Two seams, separated by 80.00 m of strata, dip at 1 in 13 in a direction $216°\ 00'$. They are to be connected by a drift falling at 1 in 5, $074°\ 30'$. Calculate the plan and slope length of the drift.

(TP Ans. $081°\ 12'$; 330.51 m, 337.07 m)

8

Traverse surveys

A traverse is a series of related points or stations which, when connected together with angular and linear values, forms a framework.

The purpose of the traverse is to control subsequent detail, i.e. the fixing of specific points to which detail can be related. The accuracy of the control survey must be superior to that of the subsidiary survey. The introduction of electromagnetic distance measurement (EDM) has revolutionised traversing and its application is extended as a main method of providing plan control.

8.1 General principles

All traverses should start on a fixed point and finish on another (or the same) fixed point. The initial and final bearings should be known. The accuracy of the fixed data should be better than that required within the traverse.

8.2 Types of traverse

(a) A *link traverse* joining two fixed points
(b) A *loop traverse* starting and finishing on the same point
(c) An *open traverse* which does not close on to a fixed point

Generally a link traverse is preferred to a loop traverse as the latter will not disclose any scale error (see p. 282).

The open traverse should not be used unless there is no alternative as the only check possible is by comparison with known bearings.

8.3 Traverse conditions

8.3.1 Link traverse (Fig. 8.1)

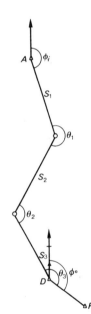

Fig. 8.1 Link traverse

Angular
$$\phi^o = \phi_i + \Sigma\theta \pm n.180 \qquad (8.1)$$
where ϕ^o = the final control bearing observed (ϕ^c – computed)
ϕ_i = the initial control bearing
$\Sigma\theta$ = the sum of the clockwise angles
$n.180$ = the number of applications of 180
(depends upon the figures used)

the angular misclosure $\quad \delta\phi = \phi^o - \phi^c$ $\qquad$ (8.2)

$\qquad\qquad\qquad\qquad$ (observed or fixed − computed)

Coordinates $\quad$ The final station coordinates are:

$$E_n = E_i + \Sigma\Delta E \qquad (8.3)$$

and $\qquad\qquad N_n = N_i + \Sigma\Delta N \qquad (8.4)$

$\quad$ misclosure $\qquad \delta E = E^o - E^c \qquad (8.5)$

$$\delta N = N^o - N^c \qquad (8.6)$$

$$\delta S = \sqrt{(\delta E^2 + \delta N^2)} \qquad (8.7)$$

where E_n, N_n = the final computed coordinates

$\qquad\quad E_i$, N_i = the initial fixed coordinates

$\qquad\quad E^o$, N^o = the observed or known coordinates of the final station

$\qquad\quad E^c$, N^c = the computed traverse coordinates of the final station

$\qquad\qquad \delta S$ = the closing linear vector

8.3.2 Loop traverse (Fig. 8.2)

Angular

$\qquad$ internal $\Sigma\theta' = (2n - 4) \times 90 \qquad (8.8)$

$\qquad$ external $\Sigma\theta = (2n + 4) \times 90 \qquad (8.9)$

$\quad$ misclosure $\quad \delta\phi = \Sigma\theta - ((2n \pm 4) \times 90) \qquad (8.10)$

Coordinates $\quad$ As $E_i = E_n$ and $N_i = N_n$, then

$$\Sigma\Delta E = 0 \qquad (8.11)$$

$$\Sigma\Delta N = 0 \qquad (8.12)$$

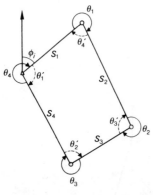

Fig. 8.2 Loop traverse

8.3.3 Accuracy

The misclosure in both cases of δS is the vector length and is compared with the total length of the traverse $= \Sigma S$. The relative accuracy (error ratio) is expressed as the ratio $\delta S / \Sigma S$, i.e. 1 in X where X will vary from 5000 to 50 000 dependent upon the equipment used.

8.4 Methods of traversing

The method is dependent upon the accuracy required and the equipment available. The following are alternative methods.

(1) *Compass traversing* using one of the following:
$\qquad$ (a) a prismatic compass
$\qquad$ (b) a miners' dial
$\qquad$ (c) a tubular or trough compass fitted to a theodolite
$\qquad$ (d) a special compass theodolite.
(2) *Continuous azimuth* (fixed needle traversing) using either
$\qquad$ (a) a miners' dial or
$\qquad$ (b) a theodolite.

(3) *Separate angular measurement* using any angular measuring equipment.

8.4.1 Compass traversing (loose needle traversing) (Fig. 8.3)

Application. Reconnaissance or exploratory surveys.

Advantages.
(1) Rapid surveys.
(2) Each line is independent—errors tend to compensate.
(3) The bearing of a line can be observed at any point along the line.
(4) Only every second station needs to be occupied (this is not recommended because of the possibility of local attraction).

Disadvantages.
(1) Lack of accuracy.
(2) Local attraction.

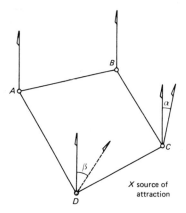

Fig. 8.3 Compass traversing

Accuracy of survey. Due to magnetic variations, instrument and observation errors, the maximum accuracy is probably limited to ± 10 min, i.e. linear equivalent 1 in 300.

Detection of effects of local attraction. Forward and back bearings should differ by 180° assuming no instrumental or personal errors exist.

Elimination of the effect of local attraction. The effect of local attraction is that all bearings from a given station will be in error by a constant value, the angle between adjacent bearings being correct.

Where forward and back bearings of a line agree this indicates that the terminal stations are both free of local attraction.

Thus, starting from bearings which are unaffected, a comparison of forward and back bearings will show the correction factors to be applied.

In Fig. 8.3 the bearings at A and B are correct. The back bearing of CB will be in error by α compared with the forward bearing BC. The forward bearing CD can thus be corrected by α. Comparison of the corrected forward bearing CD with the observed back bearing DC will show the error β by which the forward bearing DA must be corrected. This should finally check with back bearing AD.

Example 8.1

Line	Forward bearing	Back bearing
AB	120° 10′	300° 10′
BC	124° 08′	306° 15′
CD	137° 10′	310° 08′
DE	159° 08′	349° 08′
EF	138° 15′	313° 10′

Station	Back bearing	Forward bearing	Correction	Corrected forward bearing	± 180°
A	—	120° 10′	—		
B	300° 10′	124° 08′	—	124° 08′	304° 08′
C	306° 15′	137° 10′	− 2° 07′	135° 03′	315° 03′
D	310° 08′	159° 08′	+ 4° 55′	164° 03′	344° 03′
E	349° 08′	138° 15′	− 5° 05′	133° 10′	313° 10′
F	313° 10′				

Thus stations A, B, and F are free from local attraction.

Notes

(1)   Line AB. Forward and back bearings agree; therefore stations A and B are free from attraction.

(2)   Corrected forward bearing at B + 180° compared with back bearing at C shows an error of 2° 07′, i.e. 304° 08′ − 306° 15′ = − 2° 07′.

(3)   Corrected forward bearing CD 137° 10′ − 2° 07′ = 135° 03′.

(4)   Comparison of corrected forward bearing EF + 180° agrees with back bearing FE. Therefore station F is also free from local attraction.

8.4.2   Continuous azimuth method (Fig. 8.4)

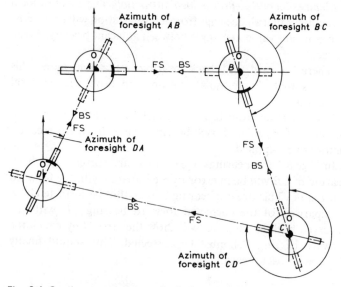

Fig. 8.4 Continuous azimuth method of traversing

This method was ideally suited to the old type of miners' dial with open-vane sights which could be used in either direction.

The instrument is orientated at each station by observing the backsight, with the reader clamped, from the reverse end of the 'dial' sights.

The recorded value of each foresight is thus the bearing of each line relative to the original orientation. For mining purposes this was the magnetic meridian and hence the method was known as 'the fixed needle method'.

The method may be modified for use with a theodolite by changing face between backsight and foresight observations.

8.4.3 Separate angular measurement

This involves mainly horizontal angles but with the introduction of EDM, vertical angle measurement becomes more widely needed.

Horizontal angles are always measured clockwise and represent the difference between adjacent pointings. After extracting the mean values these are converted into bearings for coordinate or plotting purposes.

The angles may be measured using:

(a) separate independent targets, or
(b) targets as part of a forced centre system of traversing (3 tripod system).

In either case the observations should be booked in a systematic way and the following method is recommended (Fig. 8.5). The booking system minimises the instrumental errors (see Chapter 5).

Sta	F	Reading	Angle	MEAN	Diagram
A	L	304° 18' 14"			
C	L	92 42 16	148 24 02		
C	R	272 42 23	148 24 17		
A	R	124 18 06			
A	R	52 36 34		148 24 09	
C	R	201 00 45^{6}	148 24 12		
C	L	21 00 41	148 24 03		
A	L	232 36 38			

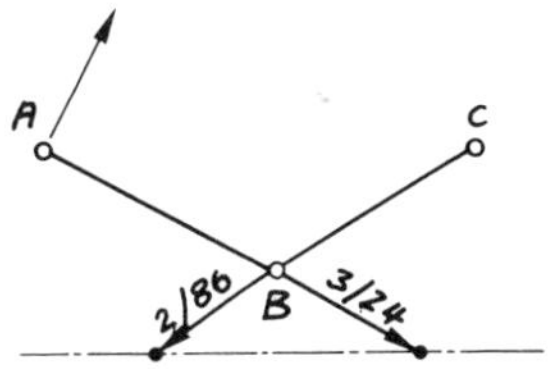

Fig. 8.5

(1)   Graduation errors—the readings should be repeated.
(2)   Eccentricity—by changing face on the forward station as shown the readings become diametrically opposed, i.e. $304°\,(F_L)$, $124°\,(F_R)$, etc.
(3)   Collimation error—a change of face.
(4)   Trunnion axis error—a change of face.

The three tripod system of traversing involves the interchange of theodolite for targets in the tribrach of a tripod previously plumbed over the station marker. This has the effect of reducing the angular closing error due to eccentricity between target and theodolite positions relative to the station marker.

Vertical angles are related to the horizon so it is essential that the index error is minimised and if present constant. The mean of face left and right eliminates this error. (NB: In modern instruments the vertical circle is graduated $0-360$ from the zenith, and thus zenith angles are read and if necessary converted to vertical angles.)
Two instrument types exist:

(1)   Manually operated setting of the index bubble before the vertical angle is read.
(2)   Automatic indexing when the automatic compensator requires no preliminary operation.

As with the horizontal angles a systematic booking method is necessary and the following is recommended:

Station set at . . *A* . .   Station observed . *B* . . .

		VA		
FL	89° 30′ 20″	+ 0° 29′ 40″		
FR	270° 29′ 50″	+ 0° 29′ 50″		
	360° 00′ 10″			
Δ/2)	180° 59′ 30″			
	90° 29′ 45″	+ 0° 29′ 45″	Mean value.	

Station observed . *C* . .

FL	91° 27′ 40″	− 1° 27′ 40″	
FR	268° 32′ 30″	− 1° 27′ 30″	
	360° 00′ 10″		
Δ/2)	177° 04′ 50″		
	88° 32′ 25″	− 1° 27′ 35″	Mean value.

Notes
(1)   The value of the vertical angle is computed as follows:
$$\text{VA} = 90° - Z_{FL} = Z_{FR} - 270°$$
(2)   The sum of the zenith angles should always $= 360°$ if observations are perfect.
(3)   The mean VA may be obtained by taking half the difference of the zenith angles and then $\text{VA}_m = \Delta/2 - 90°$.

(4) It is probably more common to extract the vertical angles from each zenith reading.

(5) If the vertical angles need to be repeated in order to increase the accuracy and to eliminate mistakes it is good practice to read the circle as the three cross hairs are set on the target, e.g.

	Z	VA
U_L	86° 18′ 54″	3° 41′ 06″
M_L	86° 01′ 45″	3° 58′ 15″
L_L	85° 44′ 32″	4° 15′ 28″
U_R	274° 15′ 38″	4° 15′ 38″
M_R	273° 58′ 26″	3° 58′ 26″
L_R	273° 41′ 17″	3° 41′ 17″

$$6)23° 50′ 10″$$
$$3° 58′ 21.7″$$

8.5 Designation of traverse stations

It is essential that the stations are numbered or lettered sequentially. In the case of a link traverse the sequence ensures that the correct clockwise angle is recorded. In the case of the loop traverse the clockwise angle becomes the internal angle if the work proceeds anticlockwise and the external angle if the work proceeds clockwise (see Fig. 8.2).

The angular error $\delta\theta$ at each station is assumed equal unless there is evidence to the contrary. If any gross error (mistake) exists, remeasurement is necessary. If there is no gross error the misclosure is distributed as

$$\delta\theta = \delta\phi/n$$

8.6 The accuracy of the traverse

The propagation of the errors is dependent upon three factors:

(1) the angular accuracy,
(2) the linear accuracy, and
(3) the shape of the traverse.

Empirical formulae may be applied (disregarding the shape which is beyond the scope of this book; see Clark, Vol. 1). Acceptable limits are given as:

(a) *Angular misclosure*

$$\delta\phi'' = x \sqrt{(2n/K)} \tag{8.13}$$

where x = the least estimated count of the theodolite (secs)
$\quad n$ = the number of settings of the theodolite
$\quad K$ = the number of arcs per station

For example, given $x = \pm 10''$; $n = 9$; $K = 1$, then

$$\delta\phi'' = 10\sqrt{(18/1)} = 42.4''$$
$$\delta\theta'' = 42.4/9 \qquad = 4.7''$$

The standard error in the measurement would be $\delta\phi/\sqrt{n} = \pm 14''$.

Given the number of arcs measured was 3, then

$$\delta\phi'' = 24.5''; \quad \delta\theta'' = 2.7''$$

the standard error in the measurement would be $\pm 8''$.

(b) The error ratio quoted as $\delta S/S$ is largely dependent upon the accuracy of the linear measurements. The following is a rough guide to the accuracies expected from various types of work.

Tacheometric surveys	1/500
Subsidiary and detail surveys	1/5000
Control surveys (steel tape)	1/15 000
(EDM)	1/25 000 to 1/50 000 +

NB: With more advanced triangulation and trilateration surveys using EDM the accuracy would be in excess of 1/50 000.

Exercises 8.1

1   A colliery plan has been laid down on the national grid of the Ordnance Survey and the coordinates of the two stations A and B have been converted into metres and reduced to A as a local origin.

	Departure (m)	Latitude (m)
Station A	0	0
Station B	East 109.2	South 991.7

Calculate the Grid bearing of the line AB. The mean magnetic bearing of the line AB is S 3° 54′ W and the mean magnetic bearing of an underground line CD is N 17° 55′ W.

State the Grid bearing of the line CD.

(MQB/M Ans.   331° 54′)

2   The following angles were measured in a clockwise direction, from the National Grid North lines on a colliery plan:

(a) 156° 15′   (b) 181° 30′   (c) 354° 00′   (d) 17° 45′

At the present time in this locality, the magnetic north is found to be 10° 30′ W of the Grid North.

Express the above directions as quadrant bearings to be set off using the magnetic needle.

(MQB/UM Ans.   (a) S 13° 15′ E; (b) S 12° 00′ W;
(c) N 4° 30′ E; (d) N 28° 15′ E)

3   The following underground survey was made with a miners' dial in the presence of iron rails. Assuming that station *A* was free from local attraction calculate the correct magnetic bearing of each line.

Station	BS	FS
A		352° 00′
B	358° 30′	12° 20′
C	14° 35′	282° 15′
D	280° 00′	164° 24′
E	168° 42′	200° 22′

(Ans.   352° 00′; 05° 50′; 273° 30′; 157° 54′; 189° 34′)
(NB: A miners' dial has vane sights, i.e. BS = FS, not reciprocal bearings.)

4   The geographical azimuth of a church spire is observed from a triangulation station as 346° 20′. At a certain time of the day a magnetic bearing was taken of this same line as 003° 23′. On the following day at the same time an underground survey line was magnetically observed as 195° 20′ with the same instrument. Calculate
(a)   the magnetic declination,
(b)   the true bearing of the underground line.
(Ans.   17° 03′ W; 178° 17′)

5   Describe and sketch a prismatic compass. What precautions are taken when using the compass for field observations?
   The following readings were obtained in a short traverse *ABCA*. Adjust the readings and calculate the coordinates of *B* and *C* if the coordinates of *A* are 250 m E, 75 m N.

Line	Compass bearing	Length (m)
AC	00° 00′	195.5
AB	44° 59′	169.5
BA	225° 01′	169.5
BC	302° 10′	141.7
CB	122° 10′	141.7
CA	180° 00′	195.5

(RICS Ans.   *B* 370 E, 195 N; *C* 250 E, 270 N)

6   The following notes were obtained during a compass survey made to determine the approximate area covered by an old dirt-tip.
   Correct the compass readings for local attraction. Plot the survey to a scale of 1/2000 and adjust graphically by Bowditch's rule.

Thereafter find the area enclosed by equalising to a triangle.

Line	Forward bearing	Back bearing	Length (m)
AB	N 57° 10′ E	S 58° 20′ W	750
BC	N 81° 40′ E	S 78° 00′ W	828
CD	S 15° 30′ E	N 15° 30′ W	764
DE	S 10° 20′ W	N 12° 00′ E	405
EF	S 78° 50′ W	N 76° 00′ E	540
FG	N 69° 30′ W	S 68° 30′ W	950
GA	N 22° 10′ W	S 19° 30′ E	383

(TP Ans.   AB 54° 40′; BC 78° 00′; CD 164° 30′; DE 190° 20′;
EF 257° 10′; FG 291° 40′; GA 338° 00′; area 13.20 ha)

7   A and B are two reference stations in an underground roadway, and it is required to extend the survey through a drift from station B to a third station C. The observations at B were as follows:

Horizontal angle ABC	271° 05′ 20″
Vertical angle to staff at C	+ 10° 15′ 00″
Staff reading at C	0.457 m
Instrument height at B	1.701 m
Measured distance BC	86.825 m

The bearing of AB was 349° 56′ 10″ and the coordinates of B 1429.274 m E, 1790.243 m N. Calculate the true slope of BC to the nearest 10 seconds, the horizontal length of BC, its bearing, and the coordinates of C.

(TP Ans.   11° 03′ 28″; 85.213 m; 081° 01′ 30″;
1513.444 m E, 1803.537 m N)

8.7   Office tests for locating mistakes in traversing

8.7.1   A mistake in the linear value of one line

If the figure is closed the coordinates can be computed and the closing error found.

Let the computed coordinates give values for $ABC_1D_1A_1$, Fig. 8.6.

The length and bearing of AA_1 suggests that the mistake lies in this direction, and if it is parallel to any given line of the traverse this is where the mistake has been made.

The amount AA_1 is therefore the linear mistake, and a correction to the line BC gives the new station values of C_1D and thus closes on A.

If the closing error is parallel to a number of lines then a repetition of their measurements is suggested.

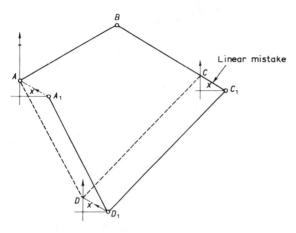

Fig. 8.6 Location of a linear mistake

8.7.2   A mistake in the angular value at one station

Let the traverse be plotted as $ABCD_1A_1$, Fig. 8.7.

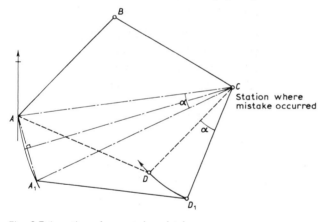

Fig. 8.7 Location of an angular mistake

The closing error AA_1 is not parallel to any line but the perpendicular bisector of AA_1 when produced passes through station C. Here an angular mistake exists.

Proof. AA_1 represents a chord of a circle of radius AC, the perpendicular bisector of the chord passing through the centre of the circle of centre C.

The line CD_1 must be turned through the angle $\alpha = ACA_1$.

8.7.3   When a traverse is closed on to fixed points and a mistake in the bearing is known to exist

The survey should be plotted or computed from each end in turn, i.e. $ABCDE - E_1D_1C_1BA$, Fig. 8.8.

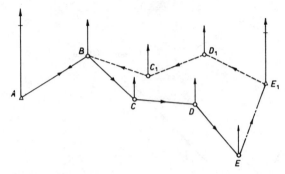

Fig. 8.8

The station which is common to both systems will suggest where the mistake has been made.

If there are two or more mistakes, either in length or bearing, then it is impossible to locate their positions but they may be localised.

Example 8.2 *Location of a linear mistake.* The field results of a subsidiary loop traverse are as follows:

Line	WCB	Length (m)
AB	00° 00′	50.60
BC	063° 49′	91.08
CD	089° 13′	67.06
DE	160° 55′	61.57
EF	264° 02′	41.15
FA	258° 18′	121.62

The observed angles are assumed to be correct but it is thought that there is a mistake in the length of one of the lines. Which line was measured incorrectly and by how much?

	ϕ	S	ΔE	ΔN
AB	0° 00′	50.60	0.000	50.600
BC	063° 49′	91.08	81.734	40.189
CD	089° 13′	67.06	67.054	0.917
DE	160° 55′	61.57	20.130	− 58.186
EF	264° 02′	41.15	− 40.927	− 4.278
FA	258° 18′	121.62	− 119.093	− 24.663
		Errors	8.898	4.579

Vector of misclosure

$$\delta S = \sqrt{(8.898^2 + 4.579^2)} = 10.007$$

Bearing of vector

$$= \tan^{-1}(8.898/4.579) = 062° 46′$$

Line *BC* is most likely to contain the mistake of 10.000 m, i.e.

$BC = 81.08$ m. Then

$$\Delta E_{BC} = 72.760 \qquad \Delta N_{BC} = 35.776$$
$$\Sigma \Delta E = -0.076 \qquad \Sigma \Delta N = 0.166$$
$$\delta S = 0.183 \qquad \delta S / \Sigma S = 1/2311$$

8.8   The adjustment of closed traverses

8.8.1   The general traverse

This is represented by the link traverse (Fig. 8.9).

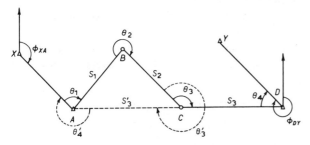

NB If traverse is closed back on to A
    angles measured are θ'_3 and θ'_4
    line measured is S'_3

Fig. 8.9

Given: the fixed points $A(E_A, N_A)$ and $D(E_D, N_D)$
         the fixed bearings ϕ_{XA} and ϕ_{DY}
Observations made: angles $\theta_1, \theta_2, \theta_3$ and θ_4
                     distances S_1, S_2 and S_3

Notes

(1)   If a loop traverse $ABCA$ exists then this is a special case and
the observations made are then $\theta_1, \theta_2, \theta'_3$ and θ'_4; S_1, S_2 and S'_3.
(2)   When station X does not exist then ϕ_{AB} is assumed and
$\theta_1 = 0$.

Conditions to be fulfilled

bearing $\phi_{DY} = \phi_{XA} + \Sigma\theta$ The misclosure is usually distributed equally
                                  between the angles measured.
coordinates: $E_D = E_A + \Sigma\Delta E$ The misclosures are adjusted in order
               $N_D = N_A + \Sigma\Delta N$ that the coordinates are compatible.

8.8.2   Coordinate adjustment

NB: *Good adjustment does not improve bad field work. If the
field work is good the method of adjustment becomes less
important.*
      Only simple adjustment is considered here.

*Where a subsidiary traverse is adjusted to fit a superior survey
by transposition of coordinates:*
The start and finish of a traverse are fixed; this assumes that all
the traverse errors are in the distances and that the superior
survey coordinates are correct.

8.8.3   Where the start and finish of a traverse are fixed

The length and bearing of the line joining these points are
known and must be in agreement with the length and bearing of
the closing line of the traverse.

Where the traverse is not orientated to the fixed line an angle
of swing (α) has to be applied.

Where there is discrepancy between the closing lengths, a
scale factor k must be applied to all the traverse lengths (see
Section 4.8):

$$k = \frac{\text{length between the fixed points}}{\text{closing length of traverse}}$$

In Fig. 8.10, the traverse is turned through angle α so that
traverse $ABCD$ becomes AB_1C_1D and AD_1 is orientated on to
line XY. The scale factor $k = \dfrac{XY}{AD}$ must be applied to the traverse
lines.

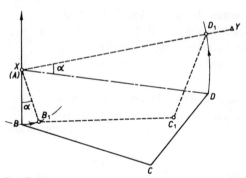

Fig. 8.10

Traverses are often orientated originally on their first line.
Coordinates are then computed, and from these the length and
bearing of the closing line (AD). The latter is then compared
with the length and bearing XY.

Coordinates of the traverse can now be adjusted by

(1)   recomputing the traverse by adjusting the bearings by the
angle α and the length by multiplying by k, or
(2)   transposing the coordinates by changing the grid
(equations 4.40 and 4.41) and also applying the scale factor
k, or

(3) applying one of the following adjustment methods, e.g. Bowditch.

NB: The factor k can be a compounded value involving:
(a) traverse error,
(b) local scale factor—(ground distance to national grid),
(c) change of units, e.g. feet to metres.

Example 8.3 A traverse XaY is made between two survey stations X (1000.00 m E, 1000.00 m N) and Y(1424.50 m E, 754.90 m N). Based upon an assumed meridian, the following partial coordinates are computed:

	ΔE	ΔN
Xa	69.54 m	− 393.92 m
aY	199.32 m	− 17.46 m

Adjust the traverse so that the coordinates conform to those of the fixed stations X and Y.

	E	N
X	1000.00	1000.00
Y	1424.50	754.90
$\Delta E'$	424.50	$\Delta N'$ − 245.10

then $\phi'_{XY} = 120° 00' 05''$

$S'_{XY} = 490.18$

From the partial coordinates,

$\Delta E_{XY} = 69.54 + 199.32 = 268.86$

$\Delta N_{XY} = -393.92 - 17.46 = -411.38$

$\phi_{XY} = 146° 50' 00''$

$S_{XY} = 491.45$

Then angle of swing

$\alpha = 146° 50' 00'' - 120° 00' 05''$

$= 26° 49' 55''$

scale factor $k = 490.18/491.45$

$= 0.997416$

Alternatively, by equations 4.41 and 4.43,

$\Delta E' = m\Delta E - n\Delta N$

$\Delta N' = m\Delta N + n\Delta E$

i.e. $424.50 = 268.86 m + 411.38 n$

$-245.10 = -411.38 m + 268.86 n$

Solving for m and n,

$m = k \cos \alpha = 0.890034$

$n = k \sin \alpha = 0.450205$

Then by equation 3.48,

$$\alpha = \tan^{-1} n/m = 26°\, 49'\, 54'' \quad \text{(check)}$$

and by equation 3.49,

$$k = \sqrt{(m^2 + n^2)} = 0.997\,419 \quad \text{(check)}$$

As the subsequent calculations are entirely based upon m and n, the latter solution is considered better as no trigonometrical values are required.

To find the new coordinates of 'a':

By equation 4.40,

$$E'_a = E'_o + m E_a - n N_a$$
$$= 1000.00 + (0.890\,034 \times 69.54) - (0.450\,205 \times -393.92)$$
$$= 1000.00 + 61.89 + 177.34$$
$$= 1239.23$$

and, by equation 4.41,

$$N'_a = N'_o + m N_a + n E_a$$
$$= 1000.00 + (0.890\,034 \times -393.92) + (0.450\,205 \times 69.54)$$
$$= 1000.00 - 350.60 + 31.31$$
$$= 680.71$$

Example 8.4 An underground traverse between two wires in shafts A and D based on an assumed meridian gives the following partial coordinates:

	ΔE(ft)	ΔN(ft)
AB	0	− 263.516
BC	+ 523.684	+ 21.743
CD	+ 36.862	+ 421.827

If the grid coordinates of the wires are:

	E (metres)	N (metres)
A	552 361.63	441 372.48
D	552 532.50	441 428.18

Transform the underground partials into grid coordinates.

Bearing of traverse line AD

$$= \tan^{-1} + 560.546/ + 180.054 = \text{N } 72°\, 11'\, 33''\, \text{E}$$

Length of traverse line AD

$$= 560.546/\sin 72°\, 11'\, 33'' = 588.754\, \text{ft}$$

Bearing of grid line AD

$$= \tan^{-1} 170.87/55.70 = \text{N } 71°\, 56'\, 42''\, \text{E}$$

Length of grid line AD

$$= 170.87/\sin 71°\, 56'\, 42'' = 179.719\, \text{m}$$

Angle of swing α

$$= 72°\, 11'\, 33'' - 71°\, 56'\, 42'' = +0°\, 14'\, 51''$$

Scale factor k

$$= 179.719/588.755 = 0.305\,253$$

By equations 4.44 and 4.45,

$$\Delta E' = m\Delta E - n\Delta N$$

$$\Delta N' = m\Delta N + n\Delta E$$

i.e. $170.87 = 560.546\,m - 180.054\,n$

$55.70 = 180.054\,m + 560.546\,n$

Solving simultaneously,

$$0.304\,828 = m - 0.321\,212\,n$$

$$0.309\,352 = m + 3.113\,210\,n$$

then $n = 0.001\,317$

and $m = 0.305\,251$

$$\alpha = \tan^{-1} n/m = 0°\,14'\,50''$$

$$k = \sqrt{(m^2 + n^2)} = 0.305\,254$$

By equations 4.40 and 4.41,

$$E'_p = E'_0 + mE_p - nN_p$$

$$N'_p = N'_0 + mN_p + nE_p$$

Tabulating:

Pt	E	N	mE_p	nN_p	mN_p	nE_p	$\Delta E'_p$	$\Delta N'_p$
A	0.000	0.000	0.000				0.000	0.000
B	0.000	−263.516	0.000	−0.347	−80.439	0.000	0.347	−80.439
C	523.684	−241.773	159.855	−0.318	−73.801	0.689	160.173	−73.112
D	560.546	180.054	171.107	0.237	54.962	0.738	170.870	55.700

Then

$$E_B = 552\,361.98 \quad N_B = 441\,292.04$$

$$E_C = 552\,521.80 \quad N_C = 441\,299.37$$

$$E_D = 552\,532.50 \quad N_D = 441\,428.18$$

8.8.4   The Bowditch method (*compass rule*)
(Fig. 8.11)

This method is more widely used than any other because of its simplicity. It was originally devised for the adjustment of compass traverses.

Bowditch assumed that (a) the linear errors were compensating and thus the probable error (PE) was proportional to the square root of the distance S, i.e.

$$\text{PE} \propto \sqrt{S}$$

and (b) the angular error $\delta\phi$ in the bearing ϕ would produce an equal displacement B_1B_2 at right angles to the line AB.

A resultant BB_2 is thus developed with the total probable

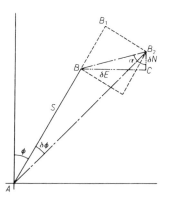

Fig. 8.11

error. The resultant

$$BB_2 = \sqrt{B_1B^2 + B_1B_2^2}$$
$$= \sqrt{2B_1B} \quad (B_1B = B_1B_2)$$

also $\quad = \sqrt{\delta E^2 + \delta N^2}$

where δE and δN are the adjustments to the partial coordinates. It can be shown that

$$\delta E_n = \frac{\Sigma \delta E . S_n}{\Sigma S} = k_1 S_n \tag{8.14}$$

$$\delta N_n = \frac{\Sigma \delta N . S_n}{\Sigma S} = k_2 S_n \tag{8.15}$$

where δE_n, δN_n are the adjustments to the partial coordinates of the line n, δE, δN are the coordinate misclosures, ΣS the sum of the length of the lines, and k_1 and k_2 are the constants $\Sigma \delta E / \Sigma S$, and $\Sigma \delta N / \Sigma S$ respectively. Thus

adjustment of the partial coordinate

$$= \text{total adjustment} \times \frac{\text{length of corresponding side}}{\text{total length of traverse}} \tag{8.16}$$

The effect at any station is that the resultant BB_2 will be equal to the misclosure x, the ratio $S/\Sigma S$ and parallel to the bearing of the closing error, i.e.

$$\phi = \tan^{-1} \delta E_n / \delta N_n = \tan^{-1} \Sigma \delta E / \Sigma \delta N$$

The correction can thus be applied either graphically in the manner originally intended or mathematically to the partial or to the total coordinates.

The total movement of each station is therefore parallel to the closing error and equal to

$$\frac{\Sigma (\text{lengths up to that point})}{\text{total length of traverse}} \times \text{closing error}$$

The correction can thus be applied either graphically in the manner originally intended or mathematically to the coordinates.

Jameson points out that the bearings of all the lines are altered unless they lie in the direction of the closing error and that the maximum alteration in the bearing occurs when the line is at right angles to the closing bearing, when it becomes

$$\delta \theta_{rad} = \frac{(S_n/\Sigma S) \times \text{closing error}}{S_n} = \frac{\text{closing error}}{\Sigma S}$$

The closing error expressed as a fraction of the length of the traverse may vary from 1/1000 to 1/50000, so taking the maximum error as 1/1000.

$$\delta \theta'' = \frac{206\,265}{1000} = 206'' = 03'\,26''$$

a value far in excess of any theodolite station error. A change of bearing of 20″ represents 1/10 000 and this would be excessive even using a 20″ theodolite, whilst 1/50 000 gives $\delta\theta = 4''$ and a 1″ theodolite is then needed.

Graphical solution by the Bowditch method (Fig. 8.12)

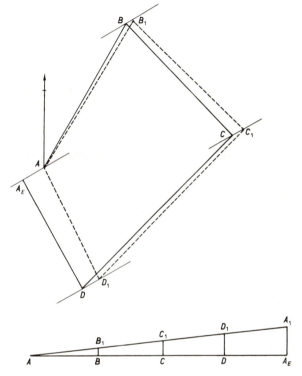

Fig. 8.12

(1) Plot the survey and obtain the closing error AA_E.
(2) Draw a line representing the length of each line of the traverse to any convenient scale.
(3) At A_E draw a perpendicular $A_E A_1$, equal to the closing error and to the same scale as the plan.
(4) Join AA_1, forming a triangle $AA_1 A_E$, and then through B, C and D similarly draw perpendiculars to cut the line AA_1 at B_1, C_1 and D_1.
(5) Draw a line through each station parallel to the closing error and plot lines equal to BB_1, CC_1 and DD_1, giving the new figure $AB_1 C_1 D_1 A$.

8.8.5   The Transit or Wilson method (*theodolite rule*)

This is an empirical method which can only be justified on the basis that (a) it is simple to operate, (b) it has generally less effect on the bearings than the Bowditch method. It can be stated as:

the correction to the partial coordinate

$$= \text{the partial coordinate} \times \frac{\text{closing error in the coordinate}}{\Sigma \, \text{partial coordinates}}$$

(ignoring the signs)

(8.17)

i.e. $\delta E_n = \dfrac{\Delta E_n \Sigma \delta E}{\Sigma \Delta E} = K_1 \Delta E_n$ (8.18)

$$\delta N_n = \frac{\Delta N_n \Sigma \delta N}{\Sigma \Delta N} = K_2 \Delta N_n$$ (8.19)

NB: *Coordinates derived from polar values, i.e. length and bearing, can only be quoted to the same accuracy as the original data. For adjustment purposes extra digits may be used to produce compatibility, but the final values must be reduced.*

8.8.6   Comparison of methods of simple adjustment

Example 8.5   *Link traverse.* Given:

		E(m)	N(m)
Initial coordinates	A	1685.430	1976.845
	B	1299.215	1341.065
Final coordinates	Y	1384.538	780.404
	Z	1514.589	510.058

Angles observed (3 arcs measured with a 20″ theodolite)

ABC	120° 36′ 44″
BCD	215° 21′ 58″
CDY	191° 58′ 14″
DYZ	135° 04′ 30″

Distances measured with a steel tape (expected accuracy 1/15 000)

BC	306.415 m
CD	192.358 m
DY	105.486 m

Line AB $\Delta E = -386.215$, $\Delta N = -635.780$, $\phi = 211° 16′ 38″$

YZ $\Delta E = 130.051$, $\Delta N = -270.346$, $\phi = 154° 18′ 36″$

Bearing misclosure

$$\phi_{YZ} = \phi_{AB} + \theta \pm n.180$$

211° 16′ 38″		
120° 36′ 44″	+8″	36′ 52″
215° 21′ 58″	+8″	22′ 06″
191° 58′ 14″	+8″	58′ 22″
135° 04′ 30″	+8″	04′ 38″

$\overline{ 874° 18′ 04″}$

$n \times 180 = 720°$

$\overline{ 154° 18′ 04″}$

should be $154° 18′ 36″$

$\overline{}$

$\delta\phi = 32″$

Check: by equation 8.13,

$$\delta\phi'' = x \sqrt{(2n/K)} = 20 \sqrt{(2 \times 4/3)} = 32.7''$$

The traverse computation is now tabulated (Fig. 8.13).

TRENT POLYTECHNIC

TRAVERSE COMPUTATION SHEET No. _1_

Date _14 · 1 · 82_

Fixed pts _(B) 1299·215_   E _1341·065_ N   Initial φ _211° 16' 38"_

_(Y) 1384·538_   E _780·404_ N   Final φ _154° 18' 36"_

Computed by _F.A.S._

ΔE _85·323_   ΔN _−560·661_   Δφ _____

Checked by _____

LINE	Sum (φ + α) ±180/−540 WCB (φ) Angle (α)	Length	ΔE / δE / ΔE′	ΔN / δN / ΔN′	Station Coordinates E	N	STN
A					1685·430	1976·845	A
t							
o							
	211° 16' 38"						
B	120 36 52				1299·215	1341·065	B
t	331 53 30		144·3644	−270·2759			
o	180		0·0077	0·0149			
	151 53 30	306·415	144·3721	−270·2610			
C	215 22 06				1443·587	1070·804	C
t	367 15 36		−24·3087	−190·8158			
o	180		0·0048	0·0093			
	187 15 36	192·358	−24·3039	−190·8065			
D	191 58 22				1419·283	879·998	D
t	379 13 58		−34·7478	−99·5986			
o	180		0·0026	0·0051			
	199 13 58	105·486	−34·7452	−99·5935			
Y	135 04 38				1384·538	780·404	Y
t	334 18 36						
o	180						
	154 18 36						
Z					1514·589	510·058	Z
t							
o							
t							
o							
	Σ	604·259	85·3079	−560·6903			
	Sh be		85·3230	−560·6610			
	Errors		− 0·0151	− 0·0293			

$$K_1 = \frac{0·0151}{604·3} = 2·499 \times 10^{-5}$$

$$K_2 = \frac{0·0293}{604·3} = 4·849 \times 10^{-5}$$

Fig. 8.13

Explanation of tabulation

Col. 1   Stations.

2   Adjusted angles are filled in opposite station, e.g. 120° 36′ 52″ (θ).

Initial bearing ϕ, 211° 16′ 38″, applied above station line.

Sum ($\phi + \theta$).

$\pm$ or -540.

3   Lengths are entered opposite computed bearing.

4/5   Partial coordinates ΔE, ΔN
Adjustment δE, δN
Adjusted partials ΔE′, ΔN′

6/7   Total coordinates   E, N

The adjustment is by the Bowditch rule.

By summing columns 3, 4 and 5 the following values are derived:

$$\Sigma S = 604.259$$

$$\Sigma \delta E = -0.0151 \qquad \Sigma \delta N = -0.0293$$

$$K_1 = 0.0151/604.259 \qquad = 2.499 \times 10^{-5}$$

$$K_2 = 0.0293/604.259 \qquad = 4.849 \times 10^{-5}$$

$$\text{Then } \delta E_{BC} = 306.415\,K_1 \qquad = 0.0077$$

$$\delta N_{BC} = 306.415\,K_2 \qquad = 0.0149$$

The relative accuracy of the traverse is then given as $\delta S/\Sigma S$.

$$\delta S = \sqrt{(\delta E^2 + \delta N^2)} = \sqrt{(0.015^2 + 0.029^2)} = 0.033$$

$$\delta S/\Sigma S = 0.033/694.259, \text{ i.e. 1 in } 18\,499 \quad \text{(acceptable)}$$

Example 8.6   *Loop traverse*

Measured data: (1 arc measured with 20″ theodolite)

	Angles		Lengths (m)
ABC	124° 35′ 20″	AB	53.86
BCD	30° 51′ 30″	BC	248.35
CDA	141° 42′ 40″	CD	129.17
DAB	62° 51′ 30″	DA	169.50

Assumed ϕ_{AD} = due North.

Check on angles:	124° 35′ 20″	$-15″$	35′ 05″
	30° 51′ 30″	$-15″$	51′ 15″
	141° 42′ 40″	$-15″$	42′ 25″
	62° 51′ 30″	$-15″$	51′ 15″
	360° 01′ 00″		

$$(2n-4) \times 90 = 360° 00′ 00″$$

$$\text{Error} = \quad 01′ 00″ \quad (\text{Check} \quad 20\sqrt{8} = 56.6″.)$$

Tabulation as before (Fig. 8.14).

TRENT POLYTECHNIC

TRAVERSE COMPUTATION SHEET No. _2_

Date _15-1-82_

Fixed pts *(A)* *1000·00*   E _1000·00_ N   Initial φ *AD 0° 00' 00"*

(A) *1000·00*   E _1000·00_ N   Final φ _____

Computed by ___*F.A.S.*___

Checked by _____

ΔE_____ ΔN_____   Δφ_____

L I N E	Sum (φ + α) / ± 180/−540 / WCB (φ)	Length	ΔE / δE / ΔE′	ΔN / δN / ΔN′	Station Coordinates E	Station Coordinates N	S T N
A	Angle (α)				1000·00	1000·00	A
t	242° 51' 15"		47·927	24·574			
o	180		−0·003	0·004			
	62 51 15	53·86	47·924	24·578			
B	124 35 05				1047·92	1024·58	B
t	187 26 20		32·154	246·260			
o	180		0·015	0·019			
	007 26 20	248·35	32·139	246·279			
C	30 51 15				1080·06	1270·86	C
t	038 17 35		−80·045	−101·379			
o	180		−0·008	0·010			
	218 17 35	129·17	−80·053	−101·369			
D	141 42 25				1000·01	1169·49	D
t	360 00 00		0·000	−169·500			
o	−180		−0·010	0·013			
	180 00 00	169·50	−0·010	−169·487			
A	62 51 15				1000·00	1000·00	A
t	242 51 15						
o							
t							
o							
t							
o							
	Σ	600·88	0·036	−0·046			

$$K_1 = \frac{-0·036}{600·88} = -5·99 \times 10^{-5}$$

$$K_2 = \frac{0·046}{600·88} = 766 \times 10^{-5}$$

Fig. 8.14

Example 8.7   Consider that the tape was subsequently found to have a standardisation error of 0.010 m in its 30.000 m length. The adjusted lengths would then be:

$$AB = \ \ 53.88$$
$$BC = 248.43$$
$$CD = 129.21$$
$$DA = 169.56$$
$$\Sigma S = 601.08$$

The partial coordinates would then be recomputed as:

AB	47.945		24.583
BC	32.164		246.339
CD	−80.069		−101.411
DA	0.000		−169.560
$\Sigma\delta\text{E}$	0.040	$\Sigma\delta\text{N}$	−0.049

The error ratio would then be

$$\delta S/\Sigma S = 0.063/601.080, \text{ i.e. } 1/9502$$

compared with the previous value

$$= 0.058/600.88, \text{ i.e. } 1/10\,360$$

It can be seen that there is little change and the misclosure does not show any scale error.

If the same tape had been used in the link traverse then the lengths would have been:

$$BC = 306.516$$
$$CD = 192.421$$
$$DY = 105.521$$
$$\Sigma S = 604.458$$

The partial coordinates would then have been:

BC	144.412		−270.365
CD	−24.317		−190.878
DY	−34.759		−99.632
then $\Sigma\delta\text{E} =$	0.013	$\Sigma\delta\text{N} =$	0.214

The error ratio is now

$$0.214/604.458, \text{ i.e. } 1/2819$$

It is obvious that something is now wrong.

Example 8.8   Fig. 8.15 shows a short 'dial' traverse connecting two theodolite lines in a mine survey. The coordinates of TM 64 are 13 899.82 m E, 10 884.38 m N and of TM 86 are 14 070.67 m E, 10 794.08 m N. Calculate the coordinates of the traverse and adjust to close on TM 86.   (TP)

This is a subsidiary survey carried out with a 1 minute instrument (miners' dial) and the method of adjustment should be as simple as possible. See Fig. 8.15.

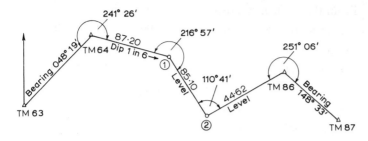

Bearing TM 63 – TM 64   048° 19′
+ angle   241° 26′

          289° 45′   Adjusted bearings
          – 180°

Bearing TM 64 – 1   109° 45′   + 01′   109° 46′
+ angle   216° 57′

          326° 42′
          – 180°

Bearing 1 – 2   146° 42′   + 02′   146° 44′
+ angle   110° 41′

          257° 23′
          – 180°

Bearing 2 – TM 86   077° 23′   + 03′   077° 26′
+ angle   251° 06′

          328° 29′
          – 180°

Bearing TM 86 – TM 87   148° 29′   + 04′   148° 33′
Fixed bearing
     TM 86 – TM 87   148° 33′

          error =   04′

Horizontal length TM 64 – 1 (1 in 6 = 9° 28′)
     = 87.20 cos 9° 28′ = 86.01 m.

Coordinates

Line	Length	Bearing	ΔE	ΔN	E	N
TM 64					13 899.82	10 884.38
to 1	86.01	S 70° 14′ E	80.942	− 29.088		
to 2	85.10	S 33° 16′ E	46.680	− 71.154		
to TM 86	44.62	N 77° 26′ E	43.551	9.708	14 070.67	10 794.08
	215.73		171.173	− 90.534	170.850	− 90.300
					171.173	− 90.534
		Traverse errors			0.323	− 0.234

Method 1
The coordinates may be transposed using the equations 4.40
and 4.41 or alternatively in this simple example by changing the
lengths and bearings.

From the control stations

Bearing $64 - 86 = \tan^{-1} 170.85/-90.30 = 117° 51' 30''$

From the traverse

Bearing $64 - 86 = \tan^{-1} 171.17/-90.53 = 117° 52' 30''$

$$\text{Angle of swing } \alpha = \quad 0° 01' 00''$$

From the control stations

Length $64 - 86 = 193.245$

From the traverse

Length $64 - 86 = 193.640$

Scale factor $k = 193.245/193.940 = 0.997\,963$

The new coordinates may now be computed:

Line	New length	New bearing	ΔE	ΔN	E	N
64	85.834	109° 45'	80.785	− 29.004	13 899.82	10 884.38
1	84.927	146° 43'	46.606	70.986	13 980.61	10 855.38
2	44.529	077° 25'	43.459	9.701	14 027.21	10 784.38
86					14 070.67	10 794.08
			170.850	− 90.299	170.85	− 90.30

Method 2 (Bowditch) using equations 8.14 and 8.15

$k_1 = -0.323/215.73 = -0.001\,48$

$k_2 = 0.234/215.73 = 0.001\,07$

Line	ΔE	δE	$\Delta E'$	ΔN	δN	$\Delta N'$	E	N
64							13 899.82	10 884.38
	80.942	− 0.129	80.813	− 29.088	0.093	− 28.995		
1							13 980.63	10 855.39
	46.680	− 0.127	46.553	− 71.154	0.092	− 71.062		
2							14 027.18	10 784.33
	43.551	− 0.067	43.484	9.708	0.049	9.757		
86							14 070.67	10 794.08
	171.173	− 0.323	170.850	− 90.534	0.234	− 90.300	170.85	− 90.30

Method 3 (Transit) using equations 8.18 and 8.19,

$k_1 = -0.323/171.173 = -0.001\,87$

$k_2 = 0.234/109.950 = 0.002\,128$

Line	ΔE	δE	$\Delta E'$	ΔN	δN	$\Delta N'$	E	N
64							13 899.82	10 884.38
	80.942	− 0.153	80.789	− 29.088	0.062	− 29.026		
1							13 980.61	10 855.35
	46.680	− 0.088	46.592	− 71.154	0.151	− 71.003		
2							14 027.20	10 784.35
	43.551	− 0.082	43.469	9.708	0.021	9.729		
86							14 070.67	10 794.08
	|171.173|	− 0.319	170.850	|109.950|	0.234	− 90.300	170.85	− 90.30

Exercises 8.2   (Traverse adjustment)

1   The mean observed internal angles and measured sides of a closed traverse *ABCDA* (in anticlockwise order) are as follows:

Angle	Observed value	Side	Measured length (m)
DAB	97° 41′	*AB*	22.11
ABC	99° 53′	*BC*	58.34
BCD	72° 23′	*CD*	39.97
CDA	89° 59′	*DA*	52.10

Adjust the angles, compute the partial coordinates assuming that *D* is due N of *A*, adjust the traverse by the Bowditch method, and give the coordinates of *B*, *C* and *D* relative to *A*.

Assess the accuracy of these observations and justify your assessment.

(ICE Ans.   *B* 21.97 m E, − 3.03 m N; *C* 39.75 m E, 52.39 m N; *D* − 0.12 m E, 52.26 m N)

2   A traverse *ACDB* is surveyed by theodolite and chain. The lengths and bearings of the lines *AC*, *CD* and *DB* are given below.

If the coordinates of *A* are $x = 0.00$, $y = 0.00$ and those of *B* are $x = 0.00$, $y = +89.71$, adjust the traverse and determine the coordinates of *C* and *D*.

The coordinates of *A* and *B* must not be altered.

Line	*AC*	*CD*	*DB*
Length	48.060	29.20	44.81
Bearing	025° 19′	037° 53′	301° 00′

(LU Ans.   *C* + 20.52, 43.50; *D* + 38.44, 66.58)

3   The lengths and partial coordinates of the lines of a closed traverse are given below.

In one of the lines it appears that a chainage has been misread by 4 m. Select the line in which the error is most likely to have occurred, correct it and adjust the latitudes and departures by the Bowditch method to the nearest 0.01 m.

Line	Length (m)	ΔN	ΔE
AB	31.05	+ 7.54	+ 30.12
BC	69.58	− 64.25	+ 26.71
CD	49.28	− 39.11	− 29.98
DE	43.17	+ 23.96	− 35.91
EF	34.31	+ 29.60	+ 17.35
FA	40.19	+ 39.89	− 49.0

(LU Ans.   Line *DE* 40 m too long
AB + 30.11 + 7.56 *DE* − 39.25 + 26.20
BC + 26.70 − 64.21 *EF* + 17.34 + 29.62
CD − 29.99 − 39.08 *FA* − 4.91 + 39.91)

4 (a)   Why is the accuracy of angular measurement so important in a traverse for which a theodolite and steel tape are used?
(b)   *A* and *D* are the terminals of traverse *ABCD*. Their plane
rectangular coordinates on the survey grid are:

	Eastings	Northings
A	+ 5861.14 m	+ 3677.90 m
D	+ 6444.46 m	+ 3327.27 m

The bearings adjusted for angular misclosure and the lengths
of the legs are:

AB	111° 53′ 50″	306.57 m
BC	170° 56′ 30″	256.60 m
CD	86° 43′ 10″	303.67 m

Calculate the adjusted coordinates of *B* and *C*.
(NU Ans.   *B* E 6100.70 N 3563.47; *C* E 6141.19 N 3309.99)

5   From an underground traverse between two shaft-wires, *A*
and *D*, the following partial coordinates in feet were obtained:

AB	E 150.632 ft	S 327.958 ft
BC	E 528.314 ft	N 82.115 ft
CD	E 26.075 ft	N 428.862 ft

Transform the above partials to give the total Grid coordinates of station *B* given that the Grid coordinates of *A* and *D*
were:

A	E 520 163.462 metres	N 432 182.684 metres
D	E 520 378.827 metres	N 432 238.359 metres

(TP Ans.   *B* E 520 209.364 N 432 082.480)

6 (a)   A traverse to control the survey of a long straight street
forms an approximate rectangle of which the long sides, on the
pavements, are formed by several legs, each about 100 m long,
and the short sides are about 15 m long; heavy traffic prevents
the measurement of lines obliquely across the road. A theodolite
reading to 20″ and a 30 m tape are used and the coordinates of
the stations are required as accurately as possible.

Explain how the short legs in the traverse can reduce the
accuracy of results and suggest a procedure in measurement and
calculation which will minimise this reduction.
(b)   A traverse *TABP* was run between the fixed stations *T* and
P of which the coordinates are:

	E	N
T	+ 6155.04	+ 9091.73
P	+ 6349.48	+ 9385.14

The coordinate differences for the traverse legs and the data
from which they were calculated are:

	Length	Adjusted bearing	ΔE	ΔN
TA	354.40	210° 41′ 40″	− 180.91	− 304.75
AB	275.82	50° 28′ 30″	+ 212.75	+ 175.54
BP	453.03	20° 59′ 50″	+ 162.33	+ 422.95

Applying the Bowditch rule, calculate the coordinates of *A* and *B*.

(LU Ans.   *A* E 5974.22 N 8786.87; *B* E 6187.04 N 8962.33)

7   The coordinates in metres of survey control stations *A* and *B* in a mine are as follows:

Station *A*   E 843.250   N 698.123
Station *B*   E 935.756   N 414.553

Undernoted are azimuths and distances of a traverse survey between *A* and *B*.

Line	Azimuth	Horizontal distance
A–1	151° 54′ 20″	56.431
1–2	158° 30′ 25″	39.482
2–3	161° 02′ 10″	95.365
3–4	168° 15′ 00″	54.003
4–*B*	170° 03′ 50″	54.890

Adjust the traverse on the assumption that the coordinates of stations *A* and *B* are correct and state the corrected coordinates of the traverse station.

(MQB/S Ans.   *A*, 843.250 698.123; 1, 869.824 648.355
2, 884.291 611.628; 3, 915.283 521.464
4, 926.282 468.607; *B*, 935.756 414.553)

Exercises 8.3   (General)

1   The following are the notes of a theodolite traverse between the two faces of two advancing roadways *BA* and *FG*, which are to be driven until they meet.

Calculate the distance still to be driven in each roadway.

Line	Azimuth	Distance (m)
AB	267° 55′	45.72
BC	355° 01′	106.68
CD	001° 41′	96.07
DE	000° 53′	153.31
EF	086° 01′	323.09
FG	203° 55′	128.02

(Ans.   *BA* produced 107.47 m, *FG* produced 279.35 m)

2   The following measurements were made in a closed traverse, *ABCD*

$$\hat{A} = 70°\,45'; \quad \hat{D} = 39°\,15'$$

$$AB = 121.92\,\text{m}; \quad CD = 213.36\,\text{m}; \quad AD = 310.59\,\text{m}$$

Calculate the missing measurements.

(LU/E Ans. $\hat{B} = 119°\,58'$, $\hat{C} = 130°\,02'$, $BC = 107.02\,\text{m}$)

3   The traverse table below refers to a closed traverse run from station *D*, through *O*, *G* and *H* and closing on *D*. The whole circle bearing of O from D is $06°\ 26'$ and G and H lie to the west of the line OD.

Compute the partial coordinates of O, G and H with reference to D as origin, making any adjustments necessary.

Observed	Internal angles	Length (m)
HDO	79° 47′	DO 54.77
DOG	102° 10′	OG 93.98
OGH	41° 11′	GH 84.02
GHD	136° 56′	HD 42.65

(ICE Ans. $O\ +54.51,\ +6.10;\ G\ +84.61,\ -83.07;$
$H\ +12.16,\ -40.82$)

4   The following traverse was run from station I to station V between which there occur certain obstacles.

Line	Length (m)	Bearing
I–II	107.08	082° 28′
II–III	45.51	030° 41′
III–IV	136.34	098° 17′
IV–V	65.01	093° 39′

It is required to peg the mid-point of I–V.

Calculate the length and bearing of a line from station III to the required point.   (ICE Ans.   52.16 m 137° 32′ E)

5   Two shafts, A and B, have been accurately connected to the National Grid of the Ordnance Survey and the coordinates of the shaft centres, reduced to a local origin, are as follows:

Shaft A E 10 055.02 m   N 9768.32 m
Shaft B E 11 801.90 m   N 8549.68 m

From shaft A, a connection to an underground survey was made by wires and the grid bearing of a base line was established from which the underground survey was calculated. Recently, owing to a holing through between the collieries, an opportunity arose to make an underground traverse survey between the shafts A and B. This survey was based on the grid bearing as established from A by wires, and the coordinates of B in relation to A as origin were computed as

E 5720.8 ft   S 4007.0 ft

Assuming that the underground survey between A and B is correct, state the adjustment required on the underground base line as established from shaft A to conform with the National Grid bearing of that line.   (MQB/S Ans.   00° 06′ 30″)

6   Calculate the coordinate values of the stations B, C, D and E of the traverse $ABCDEA$, the details of which are given below.

Data: coordinates of A 1000.00 m E   1000.00 m N
bearing of line AB 0° 00′ 00″
length of line AB 104.24 m

Interior	Angle	Length (m)
BAE	27° 18′ 00″	AB 104.24
CBA	194° 18′ 40″	BC 125.58
DCB	146° 16′ 00″	CD 125.58
EDC	47° 27′ 20″	DE 180.44
AED	124° 40′ 00″	EA 208.18

(RICS Ans.   B 1000.00 E, 1104.24 N; C 968.97 E, 1225.89 N

D 1010.73 E, 1344.30 N; E 1095.52 E, 1185.01 N)

7   It is proposed to extend a straight road AB in the direction AB produced. The centre line of the extension passes through a small farm and in order to obtain the centre line of the road beyond the farm a traverse is run from B to a point C, where A, B and C lie in the same straight line.

The following angles and distances were recorded, the angles being measured clockwise from the back to the forward station:

$$ABD = 87° 42' \quad BD = 29.02 \text{ m}$$

$$BDE = 282° 36' \quad DE = 77.15 \text{ m}$$

$$DEC = 291° 06'$$

Calculate
(a)   the length of the line EC;
(b)   the angle to be measured at C so that the centre line of the road can be extended beyond C;
(c)   the chainage of C taking the chainage of A as zero and AB = 110.34 m.

(LU Ans.   (a) 17.77 m; (b) 58° 36′; (c) 175.81 m)

8   The following are the notes of a traverse made to ascertain the position if the point F was in line with BA produced.

Line	Azimuth	Distance
AB	355° 30′	182.88 m level
BC	125° 00′	94.49 m rising 1 in 2
CD	210° 18′	115.21 m level
DE	130° 36′	125.58 m level
EF	214° 00′	141.73 m level

Calculate the difference in the azimuths of AF and BA and the extent to which the point F is out of alignment with BA produced.   (TP Ans.   0° 01′; 0.09 m)

9   The following notes were made when running a traverse from a station A to a station E:

Side	WCB	Length (m)
AB	119° 32′	26.48
BC	171° 28′	16.24
CD	223° 36′	18.83
DE	118° 34′	31.65

A series of levels were also taken along the same route as follows:

BS	IS	FS	RL	Remarks
0.684			24.620	BM near A
	0.386			Station A
	1.102			Station B
0.132		1.366		CP1
	0.966			Station C
	1.296			Station D
0.082		1.344		CP2
		1.288		Station E

Calculate the plan length, bearing and average gradient of the line AE.   (LU Ans.   70.45 m; 145° 11'; 1 in 22.75)

10 The following are the notes of an underground theodolite traverse.

Line	Azimuth	Distance (m)	Vertical angle
AB	180° 00'	—	
BC	119° 01'	55.35	+ 15° 25'
CD	160° 35'	95.10	+ 12° 45'
DE	207° 38'	97.54	− 19° 30'
EF	333° 26'	60.96	− 14° 12'

It is proposed to drive a cross-measures drift dipping from station B at a gradient of 1 in 10 on the line of AB produced to intersect at a point X, a level cross-measures drift to be driven from station F.

Calculate the azimuth and length of the proposed drift FX.
(Ans.   340° 34'; 25.33 m)

11 The following are the notes of an underground theodolite traverse:

Line	Azimuth	Distance (m)	Vertical angle
AB	089° 54'	106.68	—
BC	150° 12'	57.91	—
CD	180° 00'	182.88	—
DE	140° 18'	47.24	+ 28° 00'
EF	228° 36'	243.84	− 12° 00'

It is proposed to drive a cross-measures drift to connect stations A and F. Calculate the gradient and length of the cross-measures drift, and the azimuth of the line FA.
(MQB/M Ans.   1 in 14.8 (3° 52'); 424.07 m (incl); 182° 17')

12 The data for a link traverse are as follows:

Fixed points	E (m)	N (m)
A	5000.000	5000.000
E	5970.010	4870.280

Observed data:

Line	Bearing	Lengths
AB	060° 33′ 00″	212.120
BC	130° 12′ 00″	320.070
CD	064° 03′ 00″	315.820
DE	122° 45′ 00″	304.650

Adjust the traverse by the Bowditch rule.

	(Ans.	E	N
	B	5184.686	5104.261
	C	5429.881	4896.978
	D	5713.822	5035.132)

Bibliography

ALLAN, A. L., HOLLWEY, J. R., and MAYNES, J. H. B., *Practical Field Surveying and Computations*. Heinemann (1968)

CLARK, D., *Plane and Geodetic Surveying for Engineers*, 6th edition, **1** (1972), **2** (1973). Constable

IRVINE, W., *Surveying for Construction*, 2nd edition. McGraw-Hill (1980)

JAMESON, A. H., *Advanced Surveying*. Pitman (1948)

RAINSFORD, H. F., *Survey Adjustments and Least Squares*. Constable (1957)

SCHOFIELD, W., *Engineering Surveying*, 3rd edition. Butterworth (1978)

9

Tacheometry

The word tacheometry is derived from the Greek ταχνς swift, μετρον a measure. This form of surveying is usually confined to the optical measurement of distance.

In all forms of tacheometry there are two alternatives:

(a) a fixed angle with a variable length observed;
(b) a fixed length with a variable angle observed.

In each case the standard instrument is the theodolite, modified to suit the conditions.

The alternatives are classified as follows.

(a) Fixed angle:
    (1) stadia systems;
    (2) optical wedge systems.
(b) Variable angle:
    (1) tangential system—vertical staff;
    (2) subtense system—horizontal staff.

There are two forms of stadia: fixed stadia, found in all theodolites and levels; variable stadia, used in self reducing tacheometers.

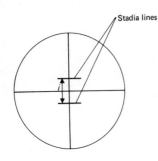

Stadia lines

Fig. 9.1 Diaphragm

9.1 Stadia systems—fixed stadia

The stadia lines are fine lines cut on glass diaphragms placed close to the eyepiece of the telescope, Fig. 9.1.

From Chapter 5, the basic formulae are:

$$D = ms + K \qquad \text{(Equation 5.31)}$$

$$= \frac{f}{i}s + (f+d) \qquad \text{(Equation 5.30)}$$

where $m = f/i =$ the multiplying constant
      $f =$ the focal length of the object lens
      $i =$ the spacing of the stadia lines on the diaphragm
      $d =$ the distance from the object lens to the vertical axis

9.2 Determination of the tacheometric constants m and K

Two methods are available: (a) by physical measurement of the instrument itself; (b) by reference to linear base lines.

9.2.1 By physical measurement of the instrument

From the general equation,

$$D = ms + K$$

where $m = f/i$ and $K = f+d$.

In the equation

$$\frac{1}{f} = \frac{1}{u} + \frac{1}{v} \qquad \text{(Equation 5.19)}$$

where $u =$ the distance from the objective to the staff is very large compared with f and v and thus $1/u$ is negligible compared with $1/v$ and $1/f$.

$$1/f \simeq 1/v$$

i.e. $\qquad f \simeq v$

i.e. $f \simeq$ the length from the objective to the diaphragm with the focus at ∞.

With the external focussing telescope, this distance can be changed to correspond to the value of u in one of two ways:

(1) by moving the objective forward;
(2) by moving the eyepiece backwards.

In the former case the value of K varies with u, whilst the latter gives a constant value.

The physical value i cannot easily be measured, so that a linear value D is required for the substitution of the value of f to give the factors i and K, i.e.

$$D = s\frac{f}{i} + (f+d)$$

Thus a vertical staff is observed at a distance D, the readings on the staff giving the value of s.

Example 9.1 A vertical staff is observed with a horizontal external focussing telescope at a distance of 112.489 m.

Measurements of the telescope are recorded as:

Objective to diaphragm   230 mm
Objective to vertical axis   150 mm

If the readings taken to the staff were 1.073, 1.629 and 2.185 calculate

(a) the distance apart of the stadia lines (i);
(b) the multiplying constant (m);
(c) the additive constant (K).

From equations 5.30 and 5.31,

$$D = ms + K$$

$$= \frac{f}{i}s + (f+d)$$

therefore

$$i = \frac{fs}{D - (f + d)}$$

$$= \frac{230 \times (2.185 - 1.073)}{112.489 - (0.230 + 0.150)}$$

$$= \frac{230 \times 1.100}{112.489 - 0.380} = 2.3 \, \text{mm}$$

therefore

$$m = \frac{f}{i} = \frac{230}{2.3} = 100$$

$$K = f + d = 230 + 150 = 380 \, \text{mm}$$

9.2.2   By field measurement

The more usual approach is to set out on a level site a base line of say 120 m with pegs at 30 m intervals.

The instrument is then set up at one end of the line and stadia readings are taken successively on to a staff held vertically at the pegs.

By substitution into the formula for selected pairs of observations, the solution of simultaneous equations will give the factors m and K, i.e.

$$D_1 = ms_1 + K \quad \text{and} \quad D_2 = ms_2 + K$$

Example 9.2   The following readings were taken with a vernier theodolite on to a vertical staff:

Stadia readings	Vertical angle	Horizontal distance
0.796 1.024 1.251	$0°$	45.736 m
1.873 2.179 2.485	$5° 00'$	61.013 m

Calculate the tacheometric constants.

$$D_1 = m(1.251 - 0.796) + K = 45.736$$
$$= 0.455 \, m + K \qquad = 45.736$$
$$D_2 = m(2.485 - 1.873) \cos^2 5° 00' + K \cos 5° 00' = 61.013$$
$$= 0.607 \, 35 \, m + 0.996 \, 19 \, K \qquad = 61.013$$

Solving these two equations simultaneously gives

$$m = 100.07 \, (\text{say } 100)$$

$$K = 0.2 \, \text{m}$$

NB: The three readings at each staff station should produce a check, i.e. middle − upper = lower − middle to within say 2 mm.

$$1.024 - 0.796 = 0.228 \quad 1.251 - 1.024 = 0.227$$
$$2.179 - 1.873 = 0.306 \quad 2.485 - 2.179 = 0.306$$

9.3 Inclined sights

The staff may be held (a) normal to the line of sight or (b) vertical.

9.3.1 Staff normal to the line of sight (Fig. 9.2)

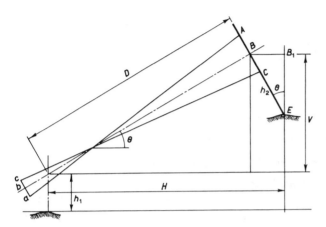

Fig. 9.2 Inclined sights with staff normal

As before,

$$D = ms + K$$

but
$$H = D \cos \theta + BB_1$$
$$= D \cos \theta + BE \sin \theta \tag{9.1}$$

i.e.
$$H = (ms + K) \cos \theta + BE \sin \theta \tag{9.2}$$

NB: $BE = h_2 = $ staff reading of middle line of diaphragm. BB_1 is $-ve$ when θ is a depression.

vertical difference
$$V = D \sin \theta \tag{9.3}$$

i.e.
$$V = (ms + K) \sin \theta \tag{9.4}$$

As the factor K may be neglected generally,

$$H = ms \cos \theta + BE \sin \theta \tag{9.5}$$
$$V = ms \sin \theta \tag{9.6}$$

If the height of the instrument to the trunnion axis is h_1 and the middle staff reading h_2, then the *difference in elevation*

$$= h_1 \pm V - h_2 \cos \theta \tag{9.7}$$

Setting the staff normal to the line of sight is not easy in practice and it is more common to use the vertical staff.

9.3.2 Staff vertical (Fig. 9.3)

As before,

$$D = ms_1 + K$$

i.e.
$$= m(A_1 C_1) + K$$

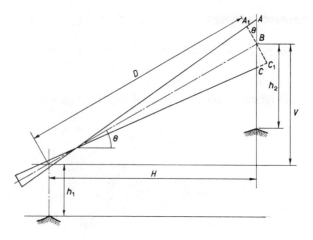

Fig. 9.3 Inclined sights with staff vertical

but A and C are the staff readings, thus

$$D = m(AC \cos \theta) + K$$

(assuming $B\hat{A}_1 A = B\hat{C}_1 C = 90°$)

$$= ms \cos \theta + K \qquad (9.8)$$

therefore

$$H = D \cos \theta$$

$$= ms \cos^2 \theta + K \cos \theta \qquad (9.9)$$

Also $\quad V = D \sin \theta$

$$= ms \sin \theta \cos \theta + K \sin \theta \qquad (9.10)$$

Also $\quad V = H \tan \theta \qquad (9.11)$

It can be readily seen that the constant $K = 0$ simplifies the equations.

Therefore the equations are generally modified to

$$H = ms \cos^2 \theta \qquad (9.12)$$

$$V = \tfrac{1}{2} ms \sin 2\theta \qquad (9.13)$$

If it is felt that the additive factor is required, then the following approximations are justified:

$$H = (ms + K) \cos^2 \theta \qquad (9.14)$$

$$V = \tfrac{1}{2}(ms + K) \sin 2\theta \qquad (9.15)$$

The difference in elevation now becomes

$$= h_1 \pm V - h_2 \qquad (9.16)$$

Example 9.3 A line of third order levelling is run by theodolite, using tacheometric methods with a staff held vertically. The usual three staff readings, of centre and both stadia hairs, are recorded together with the vertical angle (VA). A second value of height difference is found by altering the telescope elevation and recording the new readings by the vertical circle and centre hair only.

The two values of the height differences are then meaned. Compute the difference in height between the points A and B from the following data:

The stadia constants are: multiplying constant = 100; additive constant = 0.

Backsights		Foresights		Remarks (all measure-
VA	Staff	VA	Staff	ments in m)
$+0°\,02'\,00''$	1.890			
	1.417			Point A
	0.945			
$+0°\,20'\,00''$	1.908			
		$-0°\,18'\,00''$	3.109	Point B
			2.012	
			0.914	
		$0°\,00'\,00''$	3.161	

(*Aide memoire*: Height difference between the two ends of the theodolite ray $= 100\,s \sin\theta\cos\theta$, where $s =$ stadia intercept and $\theta =$ VA.)   (RICS)

$$V = 100\,s \sin\theta\cos\theta$$
$$= 50\,s \sin 2\theta$$

To A, $V = 50(1.890 - 0.945)\sin 0°\,04'\,00'' = 0.055$ m

Difference in level from instrument axis $= 0.055 - 1.417 = -1.362$.

Check reading

$$V = 50(0.945)\sin 0°\,40'\,00'' = 0.550 \text{ m}$$

Difference in level from instrument axis $= 0.550 - 1.908 = -1.358$.

mean $= 1.360$ m

To B,

$$V = 50(3.109 - 0.914)\sin -0°\,36'\,00'' = -1.149$$

Difference in level from instrument axis $= -1.149 - 2.012 = -3.161$.

Check level $= -3.161$.

mean $= -3.161$ m

Difference in level $AB = -3.161 + 1.360 = -1.801$ m.

Example 9.4 The readings below were obtained from an instrument station B using a theodolite with a stadia multiplying constant of 100.

Instrument at	Height of instrument	To bearing	Vertical angle	Stadia readings
B	1.503 m	A $69°\,30'\,00''$	$+5°\,00'\,00''$	0.658/1.055/1.451
		C $159°\,30'\,00''$	$0°\,00'\,00''$	2.231/2.847/3.463

Remarks: Staff held vertical for both observations.

Boreholes were sunk at A, B and C to expose a plane bed of rock, the ground surface being respectively 11.918 m, 10.266 m and 5.624 m above the rock plane. Given that the reduced level of B was 36.582 m, determine the line of steepest rock slope relative to the direction AB.

At station B:

To A, $H = 100 \times 0.793 \cos^2 5° 00' \quad = \quad 78.7\,\text{m}$

                $V = H \tan 5° 00' \qquad\qquad = \quad 6.89\,\text{m}$

therefore

$\quad$ level of $A = 36.58 + 6.89 + 1.50 - 1.06 = \quad 43.91\,\text{m}$

To C, $H = 100 \times 1.232 \qquad\qquad = 123.2\,\text{m}$

                $V \qquad\qquad\qquad\qquad\qquad\quad = \quad 0$

therefore

$\quad$ level of $C = 36.58 + 0 + 1.50 - 2.85 = 35.23$

$\quad\quad$ gradient $AB = (31.99 - 26.31)$ in 78.7

i.e. $= 5.68$ in 78.7

At point X in Fig. 9.4, i.e. on line AB where the bed level is that of C,

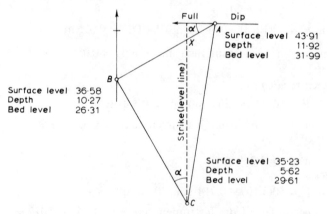

Fig. 9.4

$\quad\quad$ difference in level $AC = 31.99 - 29.61 = 2.38\,\text{m}$

therefore

$$\text{length } AX = 2.38 \times \frac{78.7}{5.68} = 32.9\,(8)\,\text{m}$$

$$BX = 78.7 - 33.0 = 45.7\,\text{m}$$

$$\text{angle } B = 159° 30' - 69° 30' = 90° 00'$$

In triangle BXC,

$$\text{angle } BCX\,(\alpha) = \tan^{-1} BX/BC$$

$$= \tan^{-1} 45.7/123.2 = 20° 21'$$

Therefore the bearing of full dip is perpendicular to the level line CX, i.e.

$$= \text{bearing } AB + \alpha$$
$$= 69° \ 30' + 180° \ 00' + 20° \ 21'$$
$$= 269° \ 51'$$

9.4   The effect of errors in stadia tacheometry

9.4.1   Staff tilted from the normal (Fig. 9.5)

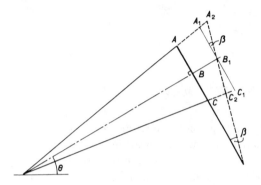

Fig. 9.5 Staff tilted from the normal

If the angle of tilt β is small, then

$$A_1 C_1 \simeq AC = s$$
$$A_1 C_1 = A_2 C_2 \cos \beta$$

i.e. $s = s_1 \cos \beta$

Thus the ratio of error

$$e = \frac{ms_1 - ms}{ms_1}$$

$$= 1 - \frac{s}{s_1}$$

$$= 1 - \cos \beta \qquad (9.17)$$

Thus the error e is independent of the inclination θ.

9.4.2   Error in the angle of elevation θ with the staff normal

$$H = D \cos \theta + BE \sin \theta$$

Differentiating gives

$$\frac{\delta H}{\delta \theta} = -D \sin \theta + BE \cos \theta$$

$$\delta H = (-D \sin \theta + BE \cos \theta) \delta \theta \qquad (9.18)$$

9.4.3 Staff tilted from the vertical (Fig. 9.6)

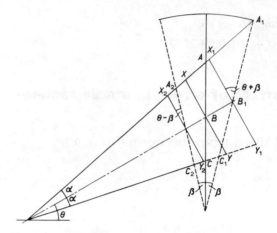

Fig. 9.6 Staff tilted from the vertical

Consider the staff readings on the vertical staff at A, B and C, Fig. 9.6. If the staff is inclined at an angle β away from the observer, the position of the staff normal to the line of collimation will be at XY when vertical and $X_1 Y_1$ when normal to the collimation at the intersection with the inclined staff.

Assuming that

$$BXX_1 = BYC = B_1 X_1 A_1 = B_1 Y_1 Y \simeq 90°$$

then, with angle θ an *elevation*,

$$XY = AC \cos \theta = s \cos \theta$$

$$X_1 Y_1 = A_1 C_1 \cos (\theta + \beta) = s_1 \cos (\theta + \beta)$$

Assuming that $XY \simeq X_1 Y_1$, then

$$s \cos \theta \simeq s_1 \cos (\theta + \beta)$$

therefore

$$s = \frac{s_1 \cos (\theta + \beta)}{\cos \theta} \tag{9.19}$$

i.e. the reading s on the staff if it had been held vertically compared with the actual reading s_1 taken on to the inclined staff.

Similarly, if the staff is inclined towards the observer,

$$s = \frac{s_1 \cos (\theta - \beta)}{\cos \theta} \tag{9.20}$$

If the angle θ is a *depression* the equations have the opposite sense, i.e.

away from the observer

$$s = \frac{s_1 \cos (\theta - \beta)}{\cos \theta} \tag{9.21}$$

towards the observer

$$s = \frac{s_1 \cos(\theta + \beta)}{\cos \theta} \tag{9.22}$$

Thus the general expression may be written as

$$s = \frac{s_1 \cos(\theta \pm \beta)}{\cos \theta} \tag{9.23}$$

The error e in the horizontal length due to reading s_1 instead of s is thus shown as

$$\text{true length} = H_T = ms \cos^2 \theta$$

$$= \frac{ms_1 \cos(\theta \pm \beta)}{\cos \theta} \cos^2 \theta \tag{9.24}$$

$$\text{apparent length} = H_A = ms_1 \cos^2 \theta$$

$$\text{error } e = H_T - H_A = ms_1 \cos^2 \theta \left(\frac{\cos(\theta \pm \beta)}{\cos \theta} - 1 \right) \tag{9.25}$$

The error expressed as a ratio

$$= \frac{H_T - H_A}{H_A}$$

$$= \frac{ms_1 \cos^2 \theta \left(\dfrac{\cos(\theta \pm \beta)}{\cos \theta} - 1 \right)}{ms_1 \cos^2 \theta}$$

$$= \frac{\cos(\theta \pm \beta)}{\cos \theta} - 1 \tag{9.26}$$

$$= \frac{\cos \theta \cos \beta \mp \sin \theta \sin \beta - \cos \theta}{\cos \theta}$$

$$= \cos \beta \pm \tan \theta \sin \beta - 1$$

If β is small, $< 5°$, then $e = \beta \tan \theta$.

Example 9.5   In a tacheometric survey, an intercept of 0.753 m was recorded on a staff which was believed to be vertical and the vertical angle measured on the theodolite was 15°. Actually the staff which was 3.5 m long was 122 mm out of plumb and leaning backwards away from the instrument position.

Assuming it was an anallatic instrument with a multiplying constant of 100, what would have been the error in the computed horizontal distance?

In what conditions will the effect of not holding the staff vertical but at the same time assuming it to be vertical be most serious? What alternative procedure can be adopted in such conditions?

Consider Fig. 9.7. By equation 9.19,

$$s = \frac{s_1 \cos (\theta + \beta)}{\cos \theta}$$

$$\beta = \tan^{-1} 0.122/3.5$$
$$= 1° 59' 47''$$

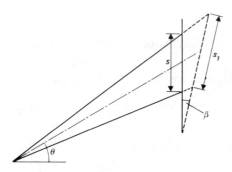

Fig. 9.7

Thus $\quad s = \dfrac{0.753 \cos (15° 00' 00'' + 1° 59' 47'')}{\cos 15° 00' 00''}$

$$= 0.745(51)$$

By equation 9.12,

$$H = ms \cos^2 \theta$$
$$\delta H = m \cos^2 \theta \, \delta s$$
$$= 100 \times \cos^2 15° 00' \times (0.753 - 0.746)$$
$$= 0.699, \text{ say } 0.70 \text{ m}$$

Alternatively,

$$\delta H = ms_1 \cos^2 \theta \left(\frac{\cos (\theta + \beta)}{\cos \theta} - 1 \right)$$

By equation 9.25,

$$\delta H = 75.3 \cos^2 15° 00' 00'' \left(\frac{\cos 16° 59' 47''}{\cos 15° 00' 00''} - 1 \right)$$

$$= -0.698, \text{ say } -0.70 \text{ m}$$

9.4.4   Accuracy of the vertical angle θ to conform to the overall accuracy

(Assuming an accuracy of 1/1000.) From

$$H = ms \cos^2 \theta$$

differentiation gives

$$\delta H = -2ms \cos \theta \sin \theta \, \delta \theta$$

For the ratio

$$\frac{\delta H}{H} = \frac{1}{1000} = \frac{2ms\cos\theta\sin\theta\,\delta\theta}{ms\cos^2\theta}$$

$$\delta\theta = \frac{\cos\theta}{2\sin\theta \times 1000}$$

$$= \frac{1}{2000}\cot\theta$$

If $\theta = 30°$,

$$\delta\theta = \frac{206\,265\cot 30°}{2000}\text{ seconds}$$

$$= 178\text{ seconds; i.e. } \simeq 3\text{ minutes}$$

NB: 1 in 1000 represents 100 mm in 100 m. The staff is graduated to 10 mm capable of estimation to ± 1 mm but as the multiplying factor is usually 100 this would represent ± 100 mm.

  As both stadia lines need to be read, the error in the stadia intercept would be $\sqrt{2}$ mm, i.e. 1.4 mm.

9.4.5   Errors in distance and level due to reading errors

By equation 9.12,

$$H = ms\cos^2\theta$$
$$\delta H = m\cos^2\theta\,\delta s \tag{9.27}$$

By equation 9.13,

$$V = 1/2\,ms\sin 2\theta$$
$$\delta V = 1/2\,m\sin 2\theta\,\delta s \tag{9.28}$$

then $\delta H = 0.14\cos^2\theta$

and $\delta V = 0.07\sin 2\theta$

Assuming $m = 100$, $\delta s = 0.0014$ m.

For	θ	δH	δV
	0	0.140	0.000
	1	0.140	0.002
	2	0.140	0.005
	3	0.140	0.007
	5	0.139	0.012
	10	0.136	0.024

From the above, it can be seen that the horizontal distances should not be quoted to better than 0.1 m whilst the levels should seldom be given better than 0.01 m unless the vertical angle is less than 3°.

9.4.6 The effect of the stadia intercept assumption

(Assuming $B\hat{A}_1 A = B\hat{C}_1 C = 90°$; Fig. 9.8.)

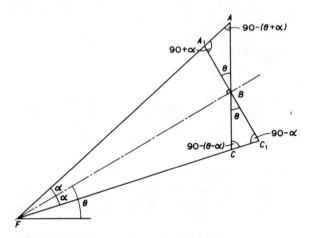

Fig. 9.8

Let the multiplying factor $m = 100$. Then

$$\alpha = \tan^{-1}\frac{1}{200} = \frac{206\,265}{200}\,\text{sec}$$

$$= 0°\,17'\,11.35''$$

$$2\alpha = 0°\,34'\,23''$$

In triangle BA_1A,

$$A_1B = \frac{s_1 \sin[90-(\theta+\alpha)]}{\sin(90+\alpha)}$$

$$= \frac{s_1 \cos(\theta+\alpha)}{\cos\alpha} = \frac{s_1(\cos\theta\cos\alpha - \sin\theta\sin\alpha)}{\cos\alpha}$$

In triangle BC_1C,

$$BC_1 = \frac{s_2 \sin[90-(\theta-\alpha)]}{\sin(90-\alpha)}$$

$$= \frac{s_2 \cos(\theta-\alpha)}{\cos\alpha} = \frac{s_2(\cos\theta\cos\alpha + \sin\theta\sin\alpha)}{\cos\alpha}$$

$$A_1B + BC_1 = \frac{s_1(\cos\theta\cos\alpha - \sin\theta\sin\alpha)}{\cos\alpha} + \frac{s_2(\cos\theta\cos\alpha + \sin\theta\sin\alpha)}{\cos\alpha}$$

$$= s_1(\cos\theta - \sin\theta\tan\alpha) + s_2(\cos\theta + \sin\theta\tan\alpha)$$

$$A_1C_1 = (s_1 + s_2)\cos\theta + (s_2 - s_1)(\sin\theta\tan\alpha) \qquad (9.29)$$

Thus the accuracy of assuming $A_1C_1 = AC\cos\theta$ depends on the second term $(s_2 - s_1)(\sin\theta\tan\alpha)$.

Example 9.6 (See Fig. 9.8.) If $\theta = 30°$, $FB = 100\,\text{m}$, $m = 100$ and $K = 0$,

$$A_1 B = BC_1 = \frac{100}{200} = 0.500\,\text{m}$$

$$s_1 = \frac{A_1 B}{\cos \theta - \sin \theta \tan \alpha}$$

$$= \frac{0.500}{0.8660 - 0.5 \times 0.005} = 0.5790\,\text{m}$$

Similarly,

$$s_2 = \frac{BC_1}{\cos \theta + \sin \theta \tan \alpha}$$

$$= \frac{0.500}{0.8660 + 0.5 \times 0.005} = 0.5757\,\text{m}$$

Therefore the effect of ignoring the second term

$$(s_2 - s_1)(\sin \theta \tan \alpha) = (0.5757 - 0.5790)(0.0025)$$
$$= -8.25 \times 10^{-6}$$

The inaccuracy in the measurement FB is thus -8.25×10^4 and the effect is negligible.

Thus the relative accuracy is very dependent on the ability to estimate the stadia readings. For all distances the staff must be read to 0.001 m, and as the distances increase this will become increasingly difficult to attain and the accuracy will diminish, although the error ratio may be maintained.

Example 9.7 A theodolite had a tacheometric multiplying constant of 100 and an additive constant of zero. The centre reading on a vertical staff held at a point B was 2.292 m when sighted from A. If the vertical angle was $+25°$ and the horizontal distance AB 109.326 m, calculate the other staff readings and show that the two intercept intervals are not equal.

Using these values, calculate the level of B if A is 37.950 m AOD and the height of the instrument 1.35 m.   (LU)

Consider Fig. 9.9.

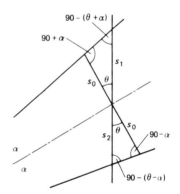

Fig. 9.9

$$\text{horizontal distance} = ms \cos^2 \theta$$

therefore

$$s = \frac{HD}{m \cos^2 \theta} = \frac{190.326}{100 \cos^2 25°}$$

$$= 2.317\,\text{m}$$

$$\text{inclined distance} = HD \sec \theta$$

$$= 190.326 \sec 25°$$

$$= 210.002\,\text{m}$$

therefore $\qquad s_0 = 210.002/200 = 1.050\,\text{m}$

$$s_1 = \frac{s_0 \cos \alpha}{\cos (\theta + \alpha)} \quad \text{(sine rule)}$$

$$= \frac{1.050 \cos 0° \, 17' \, 11''}{\cos 25° \, 17' \, 11''}$$

$$= 1.161\,\text{m}$$

Similarly, $\qquad s_2 = \dfrac{s_0 \cos \alpha}{\cos (\theta - \alpha)} \quad \text{(sine rule)}$

$$= \frac{1.050 \cos 0° \, 17' \, 11''}{\cos 24° \, 42' \, 49''}$$

$$= 1.156\,\text{m}$$

Check: $\quad 1.161 + 1.156 = 2.317\,\text{m}$

Staff readings are: $\quad$ upper $\;2.292 + 1.161 = 3.453$
$\qquad\qquad\qquad\qquad$ lower $\;2.292 - 1.156 = \underline{1.136}$
$\qquad\qquad\qquad\qquad\qquad$ *Check*: $s = \overline{2.317}$

$\qquad$ vertical difference $= HD \tan \theta$

$$= 190.326 \tan 25°$$

$$= 88.75\,\text{m}$$

$\qquad\qquad$ level at $B = 37.95 + 88.75 + 1.35 - 2.29$

$$= 125.76\,\text{m}$$

Alternatively,

$$s_1 = HD \left[\tan (\theta + \alpha) - \tan \theta \right]$$

$$= 190.326 \, (\tan 25° \, 17' \, 11'' - \tan 25°)$$

$$= 1.161\,\text{m}$$

$$s_2 = HD \left[\tan \theta - \tan (\theta - \alpha) \right]$$

$$= 190.326 \, (\tan 25° - \tan 24° \, 42' \, 49'')$$

$$= 1.156\,\text{m, as above}$$

Example 9.8 Two sets of tacheometric readings were taken from an instrument station A, the reduced level of which was 4.587 m, to a staff station B.
(a) Instrument P—multiplying constant 100, additive constant 0.366 m, staff held vertical.
(b) Instrument Q—multiplying constant 95, additive constant 0.381 m, staff held normal to line of sight.

Inst	At	To	Height of inst	Vertical angle	Stadia readings
P	A	B	1.378 m	30°	0.722/1.012/1.301
Q	A	B	1.362 m	30°	

What should be the stadia readings with instrument Q?

$\hfill$ (LU)

To find level of B (using instrument P)
By equation 9.10,

$$V = ms \sin \theta \cos \theta + K \sin \theta$$

therefore

$$V_1 = 100 \times (1.301 - 0.722) \sin 30° \cos 30° + 0.366 \sin 30°$$
$$= 57.900 \times 0.5 \times 0.866\,03 + 0.366 \times 0.5$$
$$= 25.071 + 0.183$$
$$= 25.254\,\text{m} \quad (25.25\,\text{m})$$

By equation 9.9,

$$H_1 = ms \cos^2 \theta + K \cos \theta$$
$$= 57.900 \times 0.866\,03^2 + 0.366 \times 0.866\,03$$
$$= 43.424 + 0.317$$
$$= 43.741\,\text{m} \quad (43.7\,\text{m})$$

Also, by equation 9.11,

$$V_1 = H_1 \tan \theta$$
$$= 43.741 \times 0.577\,35 = 25.254\,\text{m} \quad (\textit{Check})$$

level of *B*

$$= 4.587 + 1.378 + 25.254 - 1.012$$
$$= 30.207\,\text{m} \quad (30.21\,\text{m})$$

Using instrument Q
In Fig. 9.2,

$$V = (H - BE \sin \theta) \tan \theta$$

therefore

$$V_2 = (43.741 - BE \sin 30°) \tan 30°$$
$$= (43.741 \times 0.577\,35) - (BE \times 0.5 \times 0.577\,35)$$
$$= 25.254 - 0.288\,68\,BE$$

level of *B*

$$= 4.587 + 1.362 + V_2 - BE \cos \theta = 30.207\,\text{m}$$
$$= 5.959 + 25.254 - 0.288\,68\,BE - 0.866\,03\,BE = 30.207\,\text{m}$$
$$- 1.154\,71\,BE = -0.996\,00$$

$$BE = \text{middle reading} = 0.862\,\text{m}$$

By equation 9.5,

$$H_2 = ms \cos \theta + BE \sin \theta$$
$$= 95 \times 0.866\,03\,s + 0.862 \times 0.5 = 43.741$$
$$= 82.273\,s + 0.431 = 43.741$$

therefore

$$s = 0.526$$
$$\tfrac{1}{2}s = 0.263$$

Therefore readings are 0.862 ± 0.263, i.e. $1.125/0.862/0.599$.

Example 9.9 Three points A, B and C lie on the centre line of an existing mine roadway. A theodolite is set up at B and the following observations were taken on to a vertical staff.

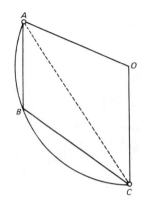

Fig. 9.10

Staff at	Horizontal circle	Vertical circle	Staff readings Stadia	Staff readings Collimation
A	002° 10′ 20″	+2° 10′	2.082/1.350	1.716
C	135° 24′ 40″	−1° 24′	2.274/1.256	1.765

If the multiplying constant is 100 and the additive constant zero calculate:
(a) the radius of the circular curve which will pass through A, B and C;
(b) the gradient of the track laid from A to C if the instrument height is 1.573 m. (RICS)

Consider Fig. 9.10.

$$\text{Assumed bearing } BA = 002° 10′ 20″$$
$$BC = 135° 24′ 40″$$
$$\text{angle } ABC = 133° 14′ 20″$$
$$\text{angle } AOC = 360 - 2(133° 14′ 20″)$$
$$= 93° 31′ 20″$$

Line AB
Horizontal length

$$H = ms \cos^2 \theta$$
$$= 100(2.082 - 1.350)\cos^2 2° 10′$$
$$= 73.095 \text{ m} \quad (73.1 \text{ m})$$

Vertical difference

$$V = H \tan \theta$$
$$= 73.095 \tan 2° 10′$$
$$= 2.765 \text{ m} \quad (2.77 \text{ m})$$

Line BC

$$H = 100(2.274 - 1.256) \cos^2 1° 24′$$
$$= 101.739 \text{ m} \quad (101.7 \text{ m})$$
$$V = 101.739 \tan - 1° 24′$$
$$= -2.486 \text{ m} \quad (-2.49 \text{ m})$$

In triangle ABC,

$$\tan \frac{A - C}{2} = \frac{a - c}{a + c} \tan \frac{A + C}{2}$$

$$= \frac{101.739 - 73.095}{101.739 + 73.095} \tan \frac{180° - 133° 14′ 20″}{2}$$

$$= \frac{28.644}{174.834} \tan 23° 22′ 50″$$

$$\frac{A-C}{2} = 4° 03' 06''$$

$$\frac{A+C}{2} = 23° 22' 50''$$

therefore

angle $A = 27° 25' 56''$

angle $C = 19° 19' 44''$

$$\frac{AB}{\sin C} = 2R \quad \text{(sine rule)}$$

therefore

$$R = \frac{73.095}{2 \sin 19° 19' 44''}$$

$$= 110.419 \, \text{m} \quad (110.4 \, \text{m})$$

Differences in level

$$BA = 1.573 + 2.765 - 1.716 = \quad 2.622 \, \text{m} \quad (2.62 \, \text{m})$$

$$BC = 1.573 - 2.486 - 1.765 = -2.678 \, \text{m} \quad (-2.68 \, \text{m})$$

$$AC \qquad\qquad\qquad\qquad = \quad 5.300 \, \text{m} \quad (5.30 \, \text{m})$$

Length of arc

$$AC = 110.419 \times 93° 31' 20''$$

$$= 110.419 \times 1.632\,271 \, \text{(rad)} = 180.234 \, \text{m} \quad (180.2 \, \text{m})$$

gradient $= 5.3$ in 180.234

$$= 1 \text{ in } 34$$

Example 9.10 The following observations were taken during a tacheometric survey using the stadia lines of a theodolite (multiplying constant 100, no additive constant.)

Station set at	Station observed	Staff readings U	M	L	Vertical angle	Bearing
B	A	1.713	2.109	2.505	+5° 32'	026° 36'
	C	0.957	1.356	1.756	−6° 46'	174° 18'

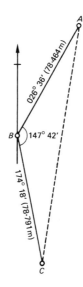

Fig. 9.11

Calculate:

(a) the horizontal lengths AB and BC;
(b) the difference in level between A and C;
(c) the horizontal length AC.

Consider Fig. 9.11.

Line BA

$$s = 2.505 - 1.713 = 0.792$$

horizontal length $= 100 \, s \cos^2 \theta$

$$= 100 \times 0.792 \cos^2 5° 32'$$

$$= 78.464 \, \text{m} \quad (78.5 \, \text{m})$$

$$\text{vertical difference} = H \tan \theta$$
$$= 78.464 \tan 5° 32'$$
$$= 7.601 \text{ m} \quad (7.60 \text{ m})$$

Line BC

$$s = 1.756 - 0.957 = 0.799$$
$$H = 100 \times 0.799 \cos^2 6° 46'$$
$$= 78.791 \text{ m} \quad (78.8 \text{ m})$$
$$V = 78.791 \tan - 6° 46'$$
$$= -9.349 \text{ m} \quad (-9.35 \text{ m})$$

Relative levels

$$A = (7.601 - 2.109) = 5.492 \text{ above } B, \text{ say } +5.49 \text{ m}$$
$$C = (-9.349 - 1.356) = -10.705 \text{ m below } B, \text{ say } -10.71 \text{ m}$$

Difference in level $AC = 16.20$ m
In triangle ABC,

$$\tan \frac{A-C}{2} = \frac{78.791 - 78.464}{78.791 + 78.464} \tan \frac{180° - 147° 42'}{2}$$

$$= \frac{0.327}{157.255} \tan 16° 09'$$

$$\frac{A-C}{2} = 0° 02'$$

$$\frac{A+C}{2} = 16° 09'$$

$$A = 16° 11'$$

$$\text{length } AC = \frac{78.791 \sin 147° 42'}{\sin 16° 11'}$$

$$= 151.060 \text{ m}, \quad \text{say } 151.1 \text{ m}$$

9.5 Fieldwork

Stadia tacheometry is mainly used in surveying detail in selected areas. Adequate horizontal and vertical control, supplied by traversing and levelling, is required to orientate the survey and to provide station levels. It is best suited to open ground where few hard levels are required. The theodolite is set up over the control point, oriented on to other control points and then directed as required:

  (i) linear detail: hedges, roads, streams, etc.;
 (ii) isolated detail: trees, buildings, pylons etc.;
(iii) spot heights.

9.5.1 Recording of observations

Stadia tacheometry is best booked in tabulated form:

Station .... 3 ....   Back station ..... 4 ..   bearing ...... 0° 00′
Station level . 35.87 .. Forward station .... 5 ... bearing .. 92° 57′
Inst ht .... 1.35 .
Axis level .... 37.22 ...

Pt	U	M	L	U − L	VA	φ	H	V	V − M	RL	Remarks
1	2.891	2.660	2.446	0.445	93° 00′	142° 40′	44.4	−2.33	−4.99	32.32	Hedge
12	1.092	1.005	0.918	0.174	99° 20′	180° 00′	16.9	−2.78	−3.78	33.44	Fence
13	1.258	1.088	0.915	0.343	100° 40′	210° 00′	33.1	−6.23	−7.32	29.90	Road
19	1.165	0.917	0.665	0.500	93° 00′	300° 00′	49.9	−2.62	−3.54	33.68	Tree
22	1.355	1.192	1.029	0.326	88° 40′	330° 00′	32.6	0.76	−0.43	36.79	Corner of house
25	1.222	1.118	1.014	0.208	84° 40′	359° 20′	20.6	1.92	0.80	38.02	Pylon

NB: The difference between $U − M$ and $M − L$ should not exceed say 0.004 m because values for Pt 1 are suspect.

Explanation of the booking

Col. 1   Point number or letter shown on diagram.
Cols 2, 3, 4   Upper, middle and lower stadia readings.
Col. 5 $U − L$, i.e. stadia intercept s.
Col. 6   VA, the vertical angle (θ) or the zenith angle $Z = (\theta = 90 − Z)$. For example, for pt 1 $Z = 93° 00′$, then $\theta = −3° 00′$.
Col. 7 ϕ, the bearing of the ray oriented on to the control points.
Col. 8 H, the horizontal length $= m . s \cos^2 \theta$ (usually $100 . s . \cos^2 \theta) = m . s \sin^2 Z$.
Col. 9 V, the vertical difference $= H \tan \theta$ or $m . s \sin 2\theta/2$ (usually $50 \sin 2\theta$) $= H/\tan Z$.
Col. 10 $V − M$, i.e. col. 9 − middle reading.
Col. 11   RL, the reduced level of the point $=$ axis level $+ (V − M)$.
Col. 12   Remarks—amplification of diagram.

9.6 Self reducing tacheometers

There are two types of self reducing instruments: (1) variable stadia tacheometers and (2) double image tacheometers.

9.6.1 Variable stadia type instruments

(a) Diagram circle tacheometers,
(b) Variable stadia instruments involving mechanical/optical mechanisms e.g. Kern K1 RA.

(a) *Diagram circle tacheometers* (based upon the Hammer–Fennel principle). In stadia tacheometry the basic equations are:

$$H = m.s.\cos^2\theta = (f/i).s.\cos^2\theta \quad \text{for horizontal lengths}$$

$$V = \frac{m.s}{2}\sin 2\theta = H\tan\theta \qquad \text{for vertical differences}$$

If the stadia intervals can be changed such that direct reading of the horizontal length occurs then the equation must be of the form:

$$H = m_H s_H$$

where m_H is the multiplying factor and s_H the intercept).

Consider Fig. 9.12.

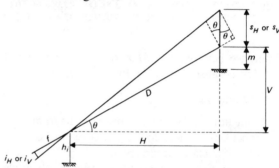

Fig. 9.12

$$\frac{i_H}{f} = \frac{s_H.\cos\theta}{D + s_H.\sin\theta}$$

$$i_H = \frac{f.s_H.\cos\theta}{H/\cos\theta + s_H.\sin\theta}$$

and if $\qquad H = m_H s_H$

then $\qquad i_H = \dfrac{f.\cos^2\theta}{m_H + (\sin 2\theta)/2}$

Similarly

$$i_V = \frac{f.s_V.\cos\theta}{H/\cos\theta + s_V.\sin\theta}$$

and if $\qquad V = H\tan\theta = m_V s_V$

then $\qquad i_V = \dfrac{f.s_V.\cos\theta}{V/\sin\theta + s_V.\sin\theta}$

$$= \frac{f.\sin 2\theta}{2(m_V + \sin^2\theta)}$$

As generally $m_H = 100$ and m_V varies from 10 to 100 whilst θ is less than $15°$, then the approximate formulae are:

$$i_H = f.\cos^2\theta/m_H$$

and $\quad i_V = f.\sin 2\theta/2m_V$

and it can be seen that the intercept between the datum line and the distance line will vary as $\cos^2 \theta$ whilst the intercept between the datum line and the vertical height line will vary as $\sin 2\theta/2$.

The diagram lines are obtained by plotting the equations radially from the centre of rotation.

In order that the height curves may be spread over the whole range of vertical angles involved the functions are changed to provide for various multiplying factors.

Angular range	Function	Zeiss (m)	Wild (m/100)
$\pm 0°$ to $\pm 4° 45'$	$5 \sin 2\theta$	± 10	± 0.1
$\pm 3° 45'$ to $\pm 9° 45'$	$2.5 \sin 2\theta$	± 20	± 0.2
$\pm 8° 45'$ to $\pm 22° 45'$	$\sin 2\theta$	± 50	$\pm 1/2$
$\pm 21° 45'$ to $\pm 43° 45'$	$0.5 \sin 2\theta$	± 100	± 1

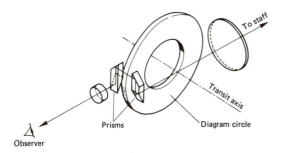

Fig. 9.13 Path of light rays from staff through diagram circle to observer. (Courtesy of Carl Zeiss Jena Ltd.)

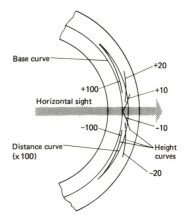

Fig. 9.14 Reduction curves on the diagram circle of the Zeiss 'Dahlta' tacheometer. (Courtesy of Carl Zeiss Jena Ltd.)

In the Zeiss (Jena) Dahlta instrument (Fig. 9.13) the curves appear as in Figs 9.14 to 9.16. In the Wild RDS the curves are exaggerated four times and to compensate for this a set of gears are introduced as in Fig. 9.17 such that the diagrams are moved through 4θ. This has the effect of flattening the curves for observing.

NB: In each case the base curve is set to the zero of the staff as indicated by the zero symbol (Figs 9.18 and 9.19).

(b) *The Kern K1 RA*

This instrument removes the necessity to measure the vertical angle by reducing the spacing of the stadia using a combination of mechanical and optical devices (Fig. 9.20). The upper horizontal line is fixed and the lower one is moveable by the action of the lever on the cam system. The shape of the active parts of the cam approximate to the functions $\cos^2 \theta$ and $\sin 2\theta$ as before. Here (Fig. 9.21)

$$s_H = s(1 + \cos 2\theta)/2 = s \cos^2 \theta \quad \text{indicated by the change-over knob marked 'D'}$$

$$s_V = s(\sin 2\theta)/2 \quad \text{indicated by the symbol } \Delta H$$

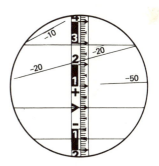

Fig. 9.15 Field of view showing staff image.
Horizontal distance $= 0.292 \times 100$ $= 29.2$ m or $0.146 \times 200 = 29.2$ m. Difference in height between instrument trunnion axis and staff index mark $= 0.218 \times -20 = -4.36$ m. (Courtesy of Carl Zeiss Jena Ltd.)

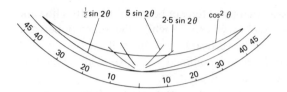

Fig. 9.16 (Courtesy of Carl Zeiss Jena Ltd.)

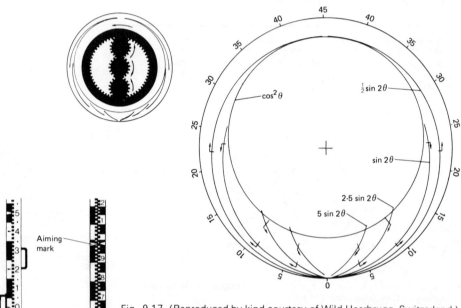

Fig. 9.17 (Reproduced by kind courtesy of Wild Heerbrugg, Switzerland.)

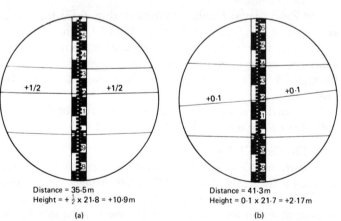

Distance = 35·5 m
Height = $+\frac{1}{2}$ × 21·8 = +10·9 m

(a)

Distance = 41·3 m
Height = 0·1 × 21·7 = +2·17 m

(b)

Fig. 9.18 Views through eyepiece RDS. (Reproduced by kind courtesy of Wild Heerbrugg, Switzerland.)

Fig. 9.19 Tacheometric staves. (Reproduced by kind courtesy of Wild Heerbrugg, Switzerland.)

(a)   (b)

thus $H = 50 \cdot s(1 + \cos 2\theta) = 100 \, s_H$
$V = 50 \cdot s \cdot \sin 2\theta \quad = 100 \, s_V$

The stadia lines are horizontal and for horizontal lengths this is probably the best of the self reducing instruments capable of

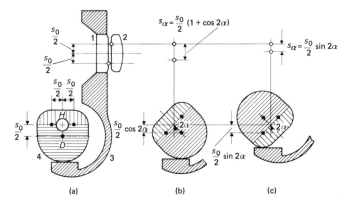

Fig. 9.20(a) Basic arrangement of the reticule mechanism with the cam in the horizontal distance position and with the vertical angle $\alpha = 0°$. The reticule interval S_α is, in this case, equal to the stadia interval S_0. (b) The telescope is inclined at an angle α. The corresponding rotation of the cam is 2α and the reticule interval is reduced to S_α. The rod interval between the reticule lines defines the horizontal distance between instrument and rod. (c) For the same telescope inclination as in (b), the cam position for measuring difference of elevation. The cam positions differ by 180° in the two figures. The reticule interval S_α now intercepts on the rod difference of elevation between the instrument and the sighted rod point. (Courtesy of Survey & General Instrument Co. Ltd.)

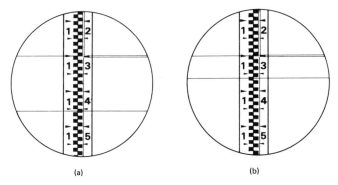

Fig. 9.21(a) Horizontal distance reading 15.6 m. The switching ring is set at D. (b) Height difference reading 6.4 m. The switching ring is set at ΔH. Sighting height on the rod 1.30 m. (Courtesy of Survey & General Instrument Co. Ltd.)

being read to ± 0.1 m but as the multiplying factor for the vertical differences is also 100 then the values of V are also ± 0.1 m and thus inferior to other tacheometers. To counteract this deficiency the vertical circle is replaced by a scale reading $\tan \theta$ and thus a more precise value for V is obtained by using $V = H \tan \theta$.

Self-reduction booking (Fig. 9.22)
These are similar to the stadia tacheometry bookings but much simpler as the vertical angle is not required and the values of H and V are automatically reduced.

TRENT ▮
POLYTECHNIC
NOTTINGHAM ▮

DEPARTMENT OF CIVIL AND STRUCTURAL ENGINEERING

TACHEOMETRIC SURVEY SHEET No. *1*

Location .... *HOME FARM*

of Station Occupied *3*

Instrument .... *ZEISS/WILD*

Staff .... *ZEISS (1·40)/WILD(1·0)*

Observer .... *I.C.WELL*   Date .... *25·3·82*

Booker .... *I. WRIGHTWELL*

Back Station *4*

Fore Station *5*

Back Bearing .... *0°00'*

Forward Brg. .... *92°57'*

Reduced Level .... *35·87*

Inst. Height .... *1·35*

Axis Red' Level .... *37·22*

Pt.	Ht.K	Ht. R'g	Δh	m	Δh-m	R.Level	Dist.	Brg. φ	Remarks
1	·10	0·359	-3·59	1·40	-4·99	32·32	44·4	142° 40'	Hedge
12	-20	0·119	-2·38		-3·78	33·44	16·9	180°00	Fence
13	-50	0·118	-5·90		-7·30	29·92	33·1	210°00	Road CL
	Instrument changed to Wild RDS								
				Axis level=37·22				Wild Staff (1·0)	
1	·0·1	39·9	-3·99	1·00	-4·99	32·32	44·4	142 40	Hedge
19	-0·1	25·4	-2·54		-3·54	33·68	49·9	300 00	Tree
22	0·2	2·8	0·56		-0·44	36·78	32·6	330 10	House corner
25	0·1	18·0	1·80		0·80	38·02	20·6	359 20	Pylon

Diagram

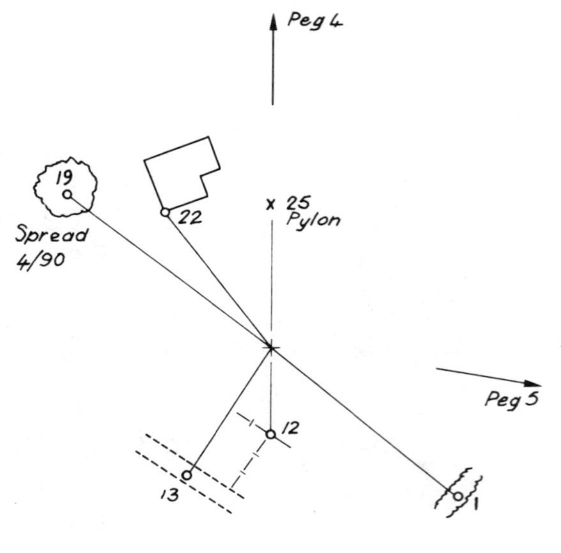

Fig. 9.22

Explanation of the bookings

Col. 2   These represent the multiplying factors for vertical differences.

Col. 3   For Zeiss—the readings are read normally.
          For Wild—the cm readings are taken as metres.

Col. 4 $\Delta h = V = m_V \times$ reading.

Col. 7   RL, the reduced level = the axis level $+ (\Delta h - m)$ where m is the height of the zero mark on the staff.

Col. 8 H is the horizontal length—cm read as metres.

9.6.2   Double image tacheometers

Wedge type tacheometers use graduated horizontal bars and in their integrated form incorporated the Bosshardt–Zeiss rotating wedge principle. These have now become obsolete with the expansion of the electromagnetic distance (EDM) measuring instruments.

Range finder type tacheometers in the form of the Zeiss (Jena) BRT 006 are still a very viable instrument, particularly in traffic environments. The basic principles are shown in Fig. 9.23 which shows the pentaprism A fixed whilst B is free to move along a graduated bar until coincidence occurs between the two images.

The angle δ is fixed at $\cot^{-1} 200$, i.e. $17'\,11.3''$.

In order to obtain horizontal lengths the deviation angle δ is varied as the function $\cos\theta$, i.e. $\delta_\theta = \delta_0 \cos\theta$.

(For more precise details see *Optical distance measurement* by J. R. Smith.)

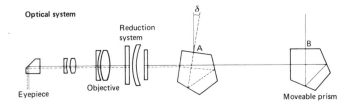

Fig. 9.23. Optical system for Zeiss (Jena) BRT 006 Reducing Telemeter. (Courtesy of Carl Zeiss Jena Ltd.)

Exercises 9.1

1   *P* and *Q* are two points on opposite banks of a river about 100 m wide. A level with an anallatic telescope and a constant of 100 is set up at *A* on the line *QP* produced, then at *B* on the line *PQ* produced and the following readings taken on to a graduated staff held vertically at *P* and *Q*.

What is the true difference in level between *P* and *Q* and what is the collimation error of the level expressed in seconds of arc, there being 206 265 seconds in a radian.

| | | Staff readings in metres | | |
From	To	Upper stadia	Collimation	Lower stadia
A	P	1.567	1.423	1.280
	Q	0.997	0.369	below ground
B	P	3.240	2.594	1.948
	Q	1.603	1.442	1.280

(ICE Ans. 1.103 m; 105″ above horizontal)

2 Readings taken with a tacheometer that has a multiplying constant of 100 and an additive constant of 0.610 m were recorded as follows:

Instrument at	Staff at	Vertical angle	Stadia readings	Remarks
P	Q	30° 00′ elevation	1.747, 2.027, 2.307	Vertical staff

Although the calculations were made on the assumption that the staff was vertical, it was in fact made at right angles to the collimation.

Compute the errors, caused by the mistake, in the calculation of horizontal and vertical distances from the instrument to the foot of the staff. Give the sign of each error.

If the collimation is not horizontal, is it preferable to have the staff vertical or at right angles to the collimation? Give reasons for your preference.

(ICE Ans. Horizontal error − 7.5 m, Vertical error − 4.01 m)

3 The following readings were taken with an anallatic tacheometer set up at each station in turn and a staff held vertically on the forward station, the forward station from D being A.

Station	Height of instrument	Stadia readings	Inclination (elevation +ve)
A	1.35 m	1.503 1.079 0.655	+0° 54′
B	1.41 m	1.817 1.448 1.079	−2° 48′
C	1.44 m	1.570 1.134 0.698	+2° 48′
D	1.40 m	1.850 1.414 0.978	−1° 48′

The reduced level of A is 52.43 m and the constant of the tacheometer is 100.

Determine the reduced levels of B, C and D, adjusted to close on A, indicating and justifying your method of adjustment.

(ICE Ans. 54.12 m, 50.45 m, 55.10 m)

4 With a tacheometer stationed at X sights were taken on three points A, B, and C as follows:

Instrument at	To	Vertical	Stadia readings	Remarks
X	A	− 4° 30′	2.417/2.115/1.814	RL of A = 108.84 m (Staff normal to line of sight)
	B	0° 00′	1.387/1.079/0.774	RL of B = 114.59 m (Staff vertical)
	C	+ 2° 30′	2.697/1.713/0.728	Staff vertical

The telescope was of the draw-tube type and the focal length of the object glass was 250 mm. For the sights to A and B, which were of equal length, the distance of the object glass from the vertical axis was 120 mm.

Derive any formulae you use. Calculate (a) the spacing of the cross hairs in the diaphragm and (b) the reduced level of C.

(LU Ans.   2.5 mm, 122.54 m)

5   Tacheometric readings were taken from a survey station S to a staff held vertically at two pegs A and B, and the following readings were recorded:

Point	Horizontal circle	Vertical circle	Stadia readings
A	62° 00′	+ 4° 10′ 30″	1.250/1.881/2.512
B	152° 00′	− 5° 05′ 00″	0.881/1.881/2.880

The multiplying constant of the instrument was 100 and the additive constant zero. Calculate the horizontal distance from A to B and the height of peg A above the axis level of the instrument.

(ICE Ans.   234.7 m, 7.28 m)

6   In a tacheometric survey made with an instrument whose constants were $f/i = 100$, $(f+d) = 0.457$ m, the staff was held inclined so as to be normal to the line of sight for each reading. How is the correct inclination assured in the field?

Two sets of readings were as given below. Calculate the gradient between the staff stations C and D and the reduced level of each. The reduced level of station A was 38.222 m.

Instrument at	Staff at	Height of instrument	Azimuth	Vertical angle	Stadia readings
A	C	1.463 m	44°	+ 4° 30′	0.914/1.295/1.676
	D		97°	− 4° 00′	0.914/1.515/2.115

(LU Ans.   1 in 6.57)

7 (a)   A telescope with tacheometric constants m and c is set up at A and sighted on a staff held vertically at B. Assuming the usual relationship $D = ms + c$ derive expressions for the horizontal and vertical distances between A and B.

(b)   An instrument at A, sighted on to a vertical staff held at B and C in turn, gave the following readings:

Sight	Horizontal circle	Vertical circle	Staff readings (m)
B	05° 20′	+ 4° 29′ 00″	0.442/0.744/1.045
C	95° 20′	− 0° 11′ 40″	0.655/0.960/1.265

If the instrument constants are $m = 100$, $c = 0$, calculate the gradient of the straight line BC.   (NU Ans.   1 in 16.7)

8   A self reducing tacheometric instrument is set up over a control station of reduced level 36.47 m AOD. If the height of the instrument axis is 1.06 m and the zero on the staff is 1.40 m, describe how the contours are derived from the following recorded data.

Contour	Multiplying constant
35.00 m	− 10
40.00 m	+ 20
10.00 m	− 50

9   Using an EDM as a tacheometer, the following observations to three points (X, Y, Z) on a concrete slab are recorded from a station A (100.00 m E, 100.00 m N) relative to a reference station B (100.00 m E, 158.36 m N).

Line	Slope length (m)	Zenith angle	Horizontal circle reading
AB			25° 30′ 25″
AX	212.375	76° 47′ 00″	249° 42′ 25″
AY	311.505	80° 12′ 50″	292° 41′ 05″
AZ	225.115	92° 35′ 30″	327° 39′ 25″

Assume that the line of sight is parallel to the EDM axis.
The level of station $A = 142.63$ m AOD.
Instrument and reflector heights $= 1.25$ m.
Thickness of slab at $X = 150$ mm; at $Y = 250$ mm; at $Z = 225$ mm.
   Calculate:
 (i)   the rate and direction of full dip of the slab;
 (ii)   the volume of the slab.
                (RICS Ans.   1 in 2.82, 222° 42′ 20″; 3892.1 m^3)

Bibliography

ALLAN, A. L., HOLLWEY, J. R., and MAYNES, J. H. B., *Practical Field Surveying and Computations.* Heinemann (1968)

BANNISTER, A., and RAYMOND, S., *Surveying*, 4th edition. Pitman (1977)

GREENWOOD, J. B., and HODGES, D. J., Quarry surveying with vertical staff self-reduction tacheometer. *The Quarry Managers' Journal* 53 (1969)

HODGES, D. J., and GREENWOOD, J. B., *Optical Distance Measurement.* Butterworth (1971)

IRVINE, W., *Surveying for Construction*, 2nd edition. McGraw-Hill (1980)

SCHOFIELD, W., *Engineering Surveying*, 3rd edition. Butterworth (1978)

SMITH, J. R., *Optical Distance Measurement.* Crosby Lockwood Staples (1970)

10

Areas and volumes

It is frequently necessary as part of engineering surveying projects to determine the area enclosed by the boundaries of a site or the volume of earthwork required to be moved.

Many of the figures involve accepted mensuration formulae (see the Appendix) but it is more common to meet irregular shapes and these require special attention.

10.1 Plan areas

The basic unit of area in SI units is the square metre (m^2) but for large areas the hectare is a derived unit.

$$1 \text{ hectare (ha)} = 10\,000 \text{ m}^2 = 2.471\,05 \text{ acres}$$

10.1.1 Conversion of planimetric areas into actual values

Let the scale of the plan be 1 in H (or as a representative fraction $1/H$). Then 1 mm is equivalent to H mm and 1 mm^2 is equivalent to H^2 mm^2, i.e.

$$H^2 \times 10^{-6} \text{ m}^2$$

10.2 Areas enclosed by straight lines

These may be considered as exact values as they conform to specified geometrical figures (see the Appendix).

10.2.1 Areas made up of triangles (Fig. 10.1)

There are two fundamental equations:

(a) $A = (a.b.\sin C)/2$ (10.1)

where a and b are the known sides and C is the value of the enclosed angle.

(b) $A = \sqrt{[s(s-a)(s-b)(s-c)]}$ (10.2)

where $s = (a+b+c)/2$.

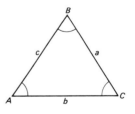

Fig. 10.1

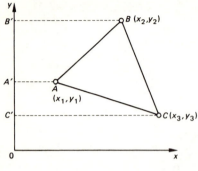

Fig. 10.2

10.2.2 Areas from coordinates

Consider the triangle ABC (Fig. 10.2).

$$A = \text{area } B'BCC' - \text{area } B'BAA' - \text{area } A'ACC'$$
$$= (x_2 + x_3)(y_2 - y_3)/2 - (x_2 + x_1)(y_2 - y_1)/2$$
$$\quad - (x_1 + x_3)(y_1 - y_3)/2$$
$$2A = x_2 y_2 - x_3 y_3 + x_3 y_2 - x_2 y_3 - x_2 y_2 + x_1 y_1 - x_1 y_2 + x_2 y_1$$
$$\quad - x_1 y_1 + x_3 y_3 - x_3 y_1 + x_1 y_3$$
$$= x_2 y_1 + x_3 y_2 + x_1 y_3 - x_1 y_2 - x_2 y_3 - x_3 y_1$$

This may be written in general terms as:

$$A = \tfrac{1}{2}\Sigma\left[y_n(x_{n+1} - x_{n-1})\right] \quad \text{or} \quad \tfrac{1}{2}\Sigma\left[x_n(y_{n+1} - y_{n-1})\right]$$
$$(10.3)$$

Applied to rectangular coordinates this becomes:

$$A = \Sigma\left[N_n(E_{n+1} - E_{n-1})\right] \tag{10.4}$$

This equation when applied to hand held calculators becomes:

Station	E	N
1	E_1	N_1
2	E_2	N_2
3	E_3	N_3
1	E_1	N_1

This is interpreted as 'the area equals half the sum of the product of the coordinates joined by the solid lines minus half the sum of the product of the coordinates joined by the dotted lines'. NB: The first station coordinates are repeated at the end of the list.

Example 10.1 The coordinates of the corners of an area of ground are given in metres as follows:

	E	N
A	1000	1000
B	1200	840
C	1630	795
D	2000	1070
E	1720	1400
F	1310	1540
G	905	1135

Calculate
(a) the area in hectares,
(b) the coordinates of the far end of a straight fence from A which cuts the area in half. (TP)

(a) From the coordinates, the area of the enclosed figure is 508 225 m^2, i.e. 50.8225 ha.
NB: The negative sign if present is ignored.

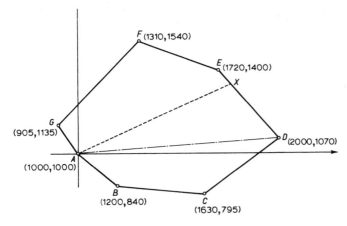

Fig. 10.3

(b)   By visual inspection of Fig. 10.3 it is apparent that the bisector of the area AX will cut the line ED.

The area of the figure $ABCD$ is then found as $154\,450\,\text{m}^2$ and the area of the triangle AXD must equal

$$508\,225/2 - 154\,450 = 99\,662.5\,\text{m}^2$$

$$= (AD.DX.\sin D)/2$$

Then

$$DX = (2 \times 99\,662.5)/(AD.\sin D)$$

The line DA has

$$\Delta E = -1000; \quad \Delta N = -70$$

then

$$\phi_{DA} = 265.996 \quad \text{and} \quad S_{DA} = 1002.45$$

The line DE has

$$\Delta E = -280; \quad \Delta N = 330$$

then

$$\phi_{DE} = 319.686$$

$$\text{angle } D = \phi_{DE} - \phi_{DA} = 53.690$$

thus $DX = 199\,325/(1002.45 \sin 53.690) = 246.75$

The line DX has

$$\phi = 319.686 \quad \text{and} \quad S = 246.75$$
$$\Delta E = -159.64 \qquad E_X = 1840.36, \quad \text{say} \quad 1840.4$$
$$\Delta N = 188.15 \qquad N_X = 1258.15, \quad \text{say} \quad 1258.2$$

10.2.3   Equalisation of a boundary to give straight lines (Fig. 10.4)

The irregular boundary is to be equalised by a line from a point on YY to F. Construction:

(1)   Join A to C; draw a line Bb parallel to AC cutting YY at b. $\triangle AbC$ is then equal to $\triangle ABC$.

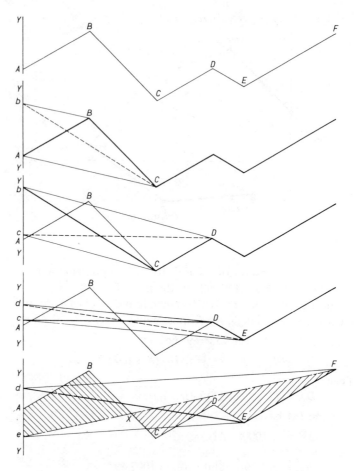

Fig. 10.4 Equalisation of an irregular boundary

(2) Repeat this procedure: join *bD*, parallel through *C*, to give *c* on *YY*.

(3) This process is now repeated as shown until the final line *eF* equalises the boundary so that the area *eABX* = area *XFEDC*.

NB: With practice there is little need to draw the construction lines, but merely with the aid of a pair of set squares or a parallel rule, to record the position of the points *b*, *c*, *d*, *e*, etc. on the line *YY*.

10.3   Irregular areas

Where irregular boundaries exist the figure may be plotted and the area derived:

(1) by equalisation to derive a regular figure as above; or
(2) by an ordinate method; or
(3) by the use of a mechanical integrator (i.e. a planimeter).

10.3.1 Ordinate methods

There are two important methods: the trapezoidal rule and Simpson's rule.

The trapezoidal rule assumes that the boundary between the extremities of the ordinates are straight lines (Fig. 10.5).

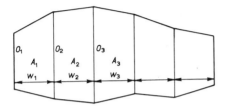

Fig. 10.5 The trapezoidal rule

the area of trapezium $A_1 \quad = w_1(o_1 + o_2)/2$

the area of trapezium $A_2 \quad = w_2(o_2 + o_3)/2$

the area of trapezium $A_{n-1} = w_{n-1}(o_{n-1} + o_n)/2 \qquad (10.5)$

If $w_1 = w_2 = w_{n-1} = w$ then

the total area $= w[o_1 + 2(o_2 + o_3 + \cdots + o_{n-1}) + o_n]/2$
$\qquad\qquad (10.6)$

Simpson's rule assumes that the boundaries are curved lines and are considered as portions of parabolic arcs of the form $y = ax^2 + bx + c$.

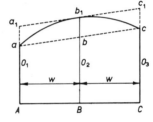

Fig. 10.6 Simpson's rule

In Fig. 10.6 the area of Aab_1cCB is made up of two parts: the trapezium $AabcC$ + the curved portion above the line abc.

the area of $ab_1cb = 2/3$ of the parallelogram $aa_1b_1c_1cb$

$$= 4w[o_2 - (o_1 + o_3)/2]/3$$

the total area $= 2w(o_1 + o_3)/2 + 4w[o_2 - (o_1 + o_3)/2]/3$

$$= w(o_1 + 4o_2 + o_3)/3 \qquad (10.7)$$

NB: This is the same form as the prismoidal formula (10.17, p. 332) with the linear values of the ordinates replacing the cross-sectional areas.

If the figure is divided into an even number of parts giving an odd number of ordinates, the total area of the figure is given as:

$$A = w[(o_1 + 4o_2 + o_3) + (o_3 + 4o_4 + o_5) + \cdots$$
$$+ (o_{n-2} + 4o_{n-1} + o_n)]/3$$
$$= w[o_1 + 4(o_2 + o_4 + \ldots + o_{n-1}) + 2(o_3 + o_5 + o_{n-2}) + o_n]/3$$
$$\qquad\qquad (10.8)$$

$$= w[o_1 + 4(\text{even numbered ordinates})$$
$$+ 2(\text{odd numbered ordinates}) + o_n]/3$$

Example 10.2 A plot of land has two straight boundaries AB and BC and the third boundary is irregular (Fig. 10.7). The

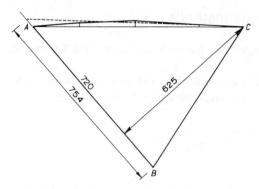

Fig. 10.7

dimensions, in metres, are $AB = 720$, $BC = 650$ and the straight line $CA = 828$. Offsets from CA on the side away from B are 0, 16, 25, 9, and 0 at chainages 0, 186, 402, 652, and 828 respectively from A.

(a) Describe briefly three methods of obtaining the area of such a plot.

(b) Obtain, by any method, the area of the above plot in square metres. (RICS)

The area of the figure can be found by:

(1) the use of the planimeter (see p. 327), or
(2) the equalisation of the irregular boundary to form a straight line and thus a triangle of equal area, or
(3) the solution of a triangle ABC + the area of the irregular boundary above the line by one of the ordinate methods.

By (2), $A = 754 \times 625/2 = \underline{235\,625\,\text{m}^2}$ (scaled values)

By (3), area of $\triangle ABC = [s(s-a)(s-b)(s-c)]^{1/2}$

(Equation 10.2)

$$
\begin{array}{llll}
a = & 650 & s-a = & 449 \\
b = & 828 & s-b = & 271 \\
c = & 720 & s-c = & 379 \\
& 2)\overline{2198} & & \\
s = & \overline{1099} & s = & \overline{1099}
\end{array}
$$

$$A = (1099 \times 449 \times 271 \times 379)^{1/2} = \underline{225\,126\,\text{m}^2}$$

Area of irregular boundary
By the trapezoidal rule (Equation 10.6):

$$A = [186(0+16) + (402-186)(16+25)$$
$$+ (652-402)(25+9) + (828-652)(9+0)]/2$$
$$= (2976 + 8856 + 8500 + 1584)/2 = \underline{10\,958\,\text{m}^2}$$

Therefore the total area is $\underline{236\,084\,\text{m}^2}$.

NB: As the distance apart of the offsets is irregular neither the full trapezoidal rule nor Simpson's rule is applicable.

10.3.2 The planimeter (Fig. 10.8)

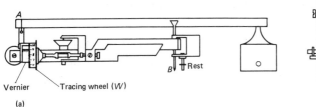
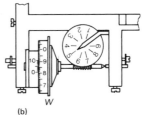

Vernier Tracing wheel (W)

(a)

W

(b)

Fig. 10.8

This is a mechanical integrator used for measuring the area of irregular figures.

It consists essentially of two bars OA and AB, with O fixed as a fulcrum and A forming a freely moving joint between the bars. Thus A is allowed to rotate along the circumference of a circle of radius OA whilst B can move in any direction with a limiting circle OB.

Theory of the planimeter (Fig. 10.9)

Let the joint at A move to A_1 and the tracing point B move first to B_1 and then through a small angle $\delta\alpha$ to B_2.

If the whole motion is very small, the area traced out by the tracing bar AB is $ABB_1B_2A_1$, i.e.

$$AB \times \delta h + \tfrac{1}{2}AB^2\delta\alpha$$

or $\qquad \delta A = l\delta h + \tfrac{1}{2}l^2\delta\alpha \qquad\qquad (10.9)$

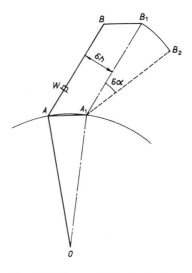

Fig. 10.9 Theory of the planimeter

where δA = the small increment of area
$\qquad l$ = the length of the tracing bar AB
$\qquad \delta h$ = the perpendicular height between the parallel lines
$\qquad\qquad AB$ and A_1B_1
$\qquad \delta\alpha$ = the small angle of rotation

A small wheel is now introduced at W on the tracing bar which will rotate when moved at right angles to the bar AB, and slide when moved in a direction parallel to its axis, i.e. the bar AB.

Let the length $AW = k.AB$. Then the recorded value on the wheel will be

$$\delta w = \delta h + kAB\delta\alpha$$

i.e. $\qquad \delta h = \delta w - kAB\delta\alpha = \delta w - kl\delta\alpha$

which when substituted in equation 10.9 gives

$$\delta A = l(\delta w - kl\delta\alpha) + \tfrac{1}{2}l^2\delta\alpha$$
$$= l\delta w + l^2(\tfrac{1}{2} - k)\delta\alpha \qquad\qquad (10.10)$$

To obtain the total area with respect to the recorded value on the wheel and the total rotation of the arm, by integrating,

$$A = lw + l^2(\tfrac{1}{2} - k)\alpha \qquad\qquad (10.11)$$

where A = the area traced by the bar

  w = the total displacement recorded on the wheel

  α = the total angle of rotation of the bar

Two cases are now considered:

(1) when the fulcrum O is *outside* the figure being traced.

(2) when the fulcrum O is *inside* the figure being traced.

(1) *When the fulcrum O is outside the figure* (Fig. 10.10)
Commencing at a, the joint is at A.

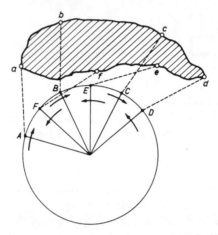

Fig. 10.10 Planimeter fulcrum outside the figure

Moving to the right, the line *abcd* is traced by the pointer
whilst the bar traces out the positive area (A_1) *abcdDCBA*.

Moving to the left, the line *defa* is traced out by the pointer
whilst the negative area (A_2) *defaAFED* is traced out by the bar.

The difference between these two areas is the area of the figure
abcdefa, i.e.

$$A = A_1 - A_2 = lw + l^2(\tfrac{1}{2} - k)\alpha \tag{10.12}$$

but $\alpha = 0$

therefore

$$A = lw \tag{10.13}$$

NB: The joint has moved along the arc *ABCD* to the right,
then along *DEFA* to the left.

In measuring such an area the following procedure should be
followed:

(1) With the pole and tracing arms approximately at right
     angles, place the tracing point in the centre of the area to be
     measured.

(2) Approximately circumscribe the area, to judge the size of
     the area compared with the capacity of the instrument. If
     not possible the pole should be placed elsewhere, or if the

area is too large it can be divided into sections, each being measured separately.

(3) Note the position on the figure where the drum does not record—this is a good starting point (*A*).

(4) Record the reading of the vernier whilst the pointer is at *A*.

(5) Circumscribe the area carefully in a *clockwise* direction and again read the vernier on returning to *A*.

(6) The difference between the first and second readings will be the required area. (This process should be repeated for accurate results.)

(7) Some instruments have a variable scale on the tracing arm to give conversion for scale factors.

(2) *When the fulcrum O is inside the figure* (Fig. 10.11)

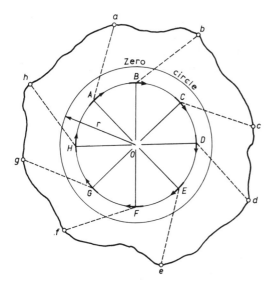

Fig. 10.11 Planimeter fulcrum inside the figure

In this case the bar traces out the figure (A_T) *abcdefgha* – the area of the circle (A_c) $ABCDEFGHA$; it has rotated through a full circle, i.e. $a = 2\pi$.

$$A_T - A_c = lw + l^2(\tfrac{1}{2} - k)2\pi$$

therefore

$$A_T = lw + l^2(\tfrac{1}{2} - k)2\pi + A_c$$

$$= lw + l^2(\tfrac{1}{2} - k)2\pi + \pi b^2 \quad \text{(where } b = OA)$$

$$= lw + \pi[b^2 + l^2(1 - 2k)] \tag{10.14}$$

This is explained as follows (Fig. 10.12).

If the pointer *B* were to rotate without the wheel *W* moving, the angle *OWP* would be 90°.

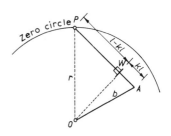

Fig. 10.12 Theory of the zero circle

The figure thus described is known as the *zero circle* of radius r; i.e.

$$OW^2 = b^2 - (kl)^2$$
$$r^2 = OW^2 + (l - kl)^2$$
$$= b^2 - (kl)^2 + l^2 - 2kl^2 + (kl)^2$$
$$= b^2 + l^2(1 - 2k) \tag{10.15}$$

Therefore in equation 10.14,

$$A_T = lw + \pi r^2$$
$$= lw + \text{the area of the zero circle} \tag{10.16}$$

The value of the zero circle is quoted by the manufacturer.

Notes

(1) $A_T - A_c = lw$. If $A_c > A_T$ then lw will be negative, i.e. the second reading will be less than the first, the wheel having a resultant negative recording.

(2) The area of the zero circle is converted by the manufacturer into revolutions on the measuring wheel and this *constant* is normally added to the recorded number of revolutions.

Example 10.3

$A_T > A_c$			or	$A_T < A_c$		
	1st reading	3.597			1st reading	6.424
	2nd reading	12.642			2nd reading	3.165
	Difference	9.045			Difference	−3.259
	Constant	23.515			Constant	23.515
	Total value	32.560			Total value	20.256

Procedure for using the planimeter

(1) Set up the instrument with the pivot (O) in such a position that the tracing point B can cover the periphery of the area.

(2) Choose a well-defined point on the boundary as a starting point. Note the scale reading (main scale + vernier).

(3) Moving clockwise, carefully trace out the boundary to finish at the starting point. Note the second scale reading.

(4) The difference between the second and the first readings represents the area. (If the pivot lies inside the area, add the zero circle constant.)

(5) Repeat the process several times and take the mean value of the area.

Notes

(1) If the area is too large, divide it into measurable portions.

(2) If the planimeter is the moveable arm type, the index on the arm may be set for the scale of the plan, using the manufacturer's value.

(3) If the planimeter is the fixed arm type, 1 division on the scale (i.e. 1 rev) usually equals $100 \, \text{cm}^2$ and thus the equivalent area, based upon the plan scale (1 in X) is given as $A = X^2 \times 10^{-2} \text{m}^2$.

If 1 cm $= X$ cm, then

$$1 \text{ cm}^2 = X^2 \text{ cm}^2 = X^2 \times 10^{-4} \text{ m}^2$$

therefore

$$100 \text{ cm}^2 = 100 X^2 \text{ cm}^2 = X^2 \times 10^{-2} \text{ m}^2$$

As $10\,000 \text{ m}^2 = 1$ hectare (1 ha),

$$1 \text{ division} = X^2 \times 10^{-6} \text{ ha} = X^2 \times 10^{-2} \text{ m}^2$$

Exercises 10.1 (Areas)

1 Plot to a scale of 1/2500, a square representing 2.5 ha. By construction, draw an equilateral triangle of the area and check the plotting by calculation.
(Ans.   side of the square 63 mm; side of triangle 96 mm)

2 The following offsets 15 m apart were measured from a chain line to an irregular boundary: 23.8, 18.6, 14.2, 16.0, 21.4, 30.4, 29.6, 24.2, 0 metres. Calculate the area in square metres using (a) the mean ordinate rule; (b) the mid-ordinate rule; and (c) the trapezoidal rule.   (Ans.   2376 m², 2676 m², 2555 m²)

3 In a quadrilateral $ABCD$, the coordinates of the points are as follows:

Point	E(m)	N(m)
A	1000.0	1000.0
B	1000.0	106.2
C	1634.8	271.2
D	2068.4	1699.3

(a) Find the area of the figure.
(b) If E is the midpoint of AB find, graphically or by calculation, the coordinates of F, on the line CD, such that the area $AEFD =$ area $EBCF$.
(Ans.   area $= 894\,974.9$ m², $E_F = 1892.5$ m, $N_F = 1119.8$ m)

4 If the bearing of the dividing line in Exercise 3 is 057° 35′ 10″, calculate the coordinates of E and F.
(Ans. E, 1000.0 m E, 553.1 m N; F, 1892.5 m E, 1119.8 m N)

5 If the dividing line in Exercise 3 is to pass through the point whose coordinates are 1703.8 m E, 1000.0 m N, find the coordinates of E.   (Ans.   1000.0 m E, 553.1 m N)

6 Using the data given in the traverse table below, compute the area of the figure $ABCDEA$.

Side	ΔE (m)	ΔN (m)
AB	−231.04	404.47
BC	245.36	233.48
CD	300.84	−290.17
DE	163.68	−594.06
EA	−478.84	246.28

(ICE Ans.   29.670 ha)

7   State what is meant by the term 'zero circle' when used in connection with the planimeter.

A planimeter reading 1 rev = 100 cm^2 is handed to you to measure certain areas on plans drawn to scales of (a) 1/360, (b) 1/1584 (c) 1/2500 (d) 1/10560. State the multiplying factor that you would use in each instance to convert the instrumental readings to hectares.   (Ans.   0.1296, 2.509, 6.250, 111.51)

8   The following data relate to a closed traverse:

Line	Azimuth	Length (m)
AB	241° 30′ 00″	301.5
BC	149° 27′ 00″	145.2
CD	134° 20′ 30″	415.7
DE	079° 18′ 00″	800.9

Calculate:
(a)   the length and bearing of the line *EA* to the nearest 30″;
(b)   the area of the figure *ABCDEA*;
(c)   the length and bearing of the line *BX* which will divide the area into two equal parts;
(d)   the length of a line *XY* of bearing 068° 50′ which will divide the area into two equal parts.

(Ans.   (a)   983.1 m, 294° 42′ 00″   (b)   317 652.72 m^2
(c)   1036.5 m, 106° 43′ 00″ (d)   550.2 m)

10.3.3   Volume of regular solids

Most volumetric formulae can be derived by the use of the *prismoidal* formula

$$V = h(A + 4M + B)/6 \tag{10.17}$$

where h = the perpendicular distance between the end areas
    A, B = the two parallel end areas
    M = the area of the mid-section between the end areas
    The *prismoid* is defined as: 'a solid having two parallel end areas A and B, which may be of any shape, provided that the surfaces joining their perimeters are capable of being generated by straight lines' (Fig. 10.13). Newton's proof of the formula is to take any point X on the mid-section and to join the vertices of the three sections. The total volume then becomes the sum of the volumes of the 10 pyramids so formed.

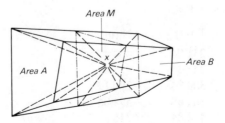

Fig.10.13

The value of this formula is shown by applying it to simple solids:

The cone   (Fig. 10.14)

$$V = h[(\pi r^2) + 4\pi(r/2)^2 + 0]/6$$
$$= \pi r^2 h/3 \qquad (10.18)$$

The sphere   (Fig. 10.15)

$$V = 2r[0 + 4\pi r^2 + 0]/6$$
$$= 4\pi r^3/3 \qquad (10.19)$$

The frustum of a cone   (Fig. 10.16)

$$V = h[\pi R^2 + 4\pi(R+r)^2/4 + \pi r^2]/6$$
$$= \pi h(R^2 + r^2 + Rr)/3 \qquad (10.20)$$

The wedge   (Fig. 10.17)

$$V = h[w(x+y)/2 + 4\{(x+z) + (z+y)\}w/8 + 0]/6$$
$$= wh(x+y+z)/6 \qquad (10.21)$$

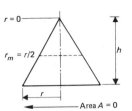

Fig. 10.14

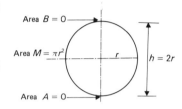

Fig. 10.15

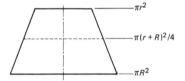

Fig. 10.16

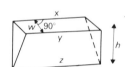

Fig. 10.17

10.4   Earthworks

One of the major considerations in civil engineering projects is the volume of earthwork to be excavated, moved around and perhaps used as fill at some other part of the site.

There are three general methods of calculating volumes using:

(1)   cross-sectional areas;
(2)   contours;
(3)   spot heights.

10.4.1   The use of cross-sections

Volumetric calculations involve cross-sections taken at right angles to a convenient base line, e.g. the centre line of the construction such as a motorway, railway or waterway, which generally runs longitudinally through the earthworks.

In the sections the following terms are used (Fig. 10.18):

Formation width (w)
Formation height (h_0) measured on the centre line
Side width (W) for the fixing of formation side width pegs, measured from the centre line
Side slope or batter, 1 in m, i.e. 1 vertical to m horizontal
Crossfall gradient 1 in k

Three types of configurations are possible:

(1)   two level—the crossfall is level;
(2)   three level—the crossfall is uniform;
(3)   multi-level—the ground is irregular in shape.

(1)   *Two level section* (Fig. 10.18)

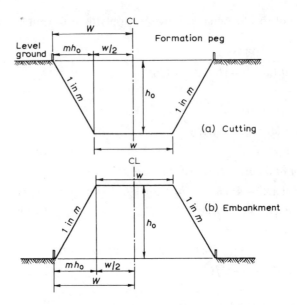

Fig. 10.18 Sections without crossfall

Here $W = w/2 + mh_0$ (10.22)

and $A = h_0(w + 2W)/2$

$\qquad = h_0(w + mh_0)$ (10.23)

(2)   *Three level section* (Fig. 10.19)

Here the levels involved are the formation level and the levels at the side width pegs A and E, giving the respective heights h_0, h_a and h_e. The area (A) then consists of two triangles and a trapezium.

In triangle AHB,

$\qquad A_1 = h_1 d_1/2$ (10.24)

In trapezium $BHFD$,

$\qquad A_2 = h_0 w$ (10.25)

In triangle DFE,

$\qquad A_3 = h_2 d_2/2$ (10.26)

$\qquad h_1 = h_0 - w/2k$ (10.27)

$\qquad h_2 = h_0 + w/2k$ (10.28)

Using the rate of approach method,

$\qquad d_1 = h_1/(1/m + 1/k)$ (10.29)

$\qquad d_2 = h_2/(1/m - 1/k)$ (10.30)

The side widths are then given as:

$\qquad W_1 = w/2 + d_1$ (10.31)

$\qquad W_2 = w/2 + d_2$ (10.32)

and the area of the cross-section is then

$\qquad A = (h_1 d_1 + 2h_0 w + h_2 d_2)/2$ (10.33)

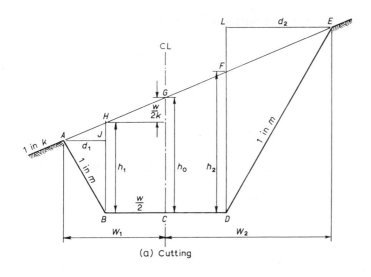

(a) Cutting

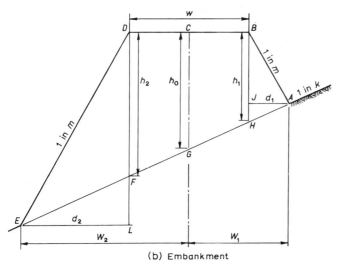

(b) Embankment

Fig. 10.19 Sections with crossfall

Applying all the data into a common formula,

$$A = \frac{m(h_0^2 k^2 + (w/2)^2 + wh_0 m)}{k^2 - m^2} + wh_0 \qquad (10.34)$$

It can be seen that this is an unwieldy formula and generally it is preferable for students to work from first principles.

If the corners of the figure are converted into cartesian coordinates the area may be computed using these figures (Fig. 10.19(a)):

Assuming the level of the formation as L_f, then

the level at A $(L_a) = L_f + h_0 - w/2k - d_1/k \qquad (10.35)$

at E $(L_e) = L_f + h_0 + w/2k + d_2/k \qquad (10.36)$

The cartesian coordinates then become:

	x	y
A	$-W_1$	L_a
E	$+W_2$	L_e
D	$+w/2$	L_f
B	$-w/2$	L_f

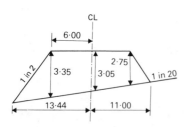

Fig. 10.20

Example 10.4 The ground slopes at 1 in 20 at right angles to the centre line of a proposed embankment which is 12.00 m wide at a formation height of 3.05 m above the ground. If the batter of the sides is 1 in 2, calculate (a) the side widths and (b) the area of the cross-section.

In Fig. 10.20,

$$w = 12.00, \ h_0 = 3.05, \ m = 2 \text{ and } k = 20$$

Then

$$h_1 = h_0 - w/2k = 3.05 - 12.00/40 \qquad = 2.75 \text{ m}$$
$$h_2 = h_0 + w/2k = 3.05 + 0.30 \qquad = 3.35 \text{ m}$$
$$d_1 = h_1 mk/(k+m) = 2.75 \times 2 \times 20/(20+2) = 5.00 \text{ m}$$
$$d_2 = h_2 mk/(k-m) = 3.35 \times 40/18 \qquad = 7.44 \text{ m}$$
$$W_1 = w/2 + d_1 = 6.00 + 5.00 \qquad = \underline{11.00 \text{ m}}$$
$$W_2 = w/2 + d_2 = 6.00 + 7.44 \qquad = \underline{13.44 \text{ m}}$$

By equation 10.33,

$$\text{area } A = (h_1 d_1 + h_2 d_2 + 2wh_0)/2$$

$$= \frac{(2.75 \times 5.00) + (3.35 \times 7.44) + (24.00 \times 3.05)}{2}$$

$$= \underline{55.94 \text{ m}^2}$$

By equation 10.34,

$$A = \frac{2[(3.05 \times 20)^2 + (12.00/2)^2 + (12.00 \times 3.05 \times 2)]}{20^2 - 2^2}$$

$$+ (12.00 \times 3.05)$$

$$= \underline{55.94 \text{ m}^2}$$

Sections with part cut and part fill (Fig. 10.21)
As before, the formation width $BD = w$
the formation height $FG = h_0$
the ground slope = 1 in k
but here the batter on the cut and fill may differ, so let the batter of fill be 1 in n, and the batter of cut 1 in m.

The total area is made up of only 2 parts:
(1) Triangle ABC,

$$\text{area} = h_1 d_1/2$$

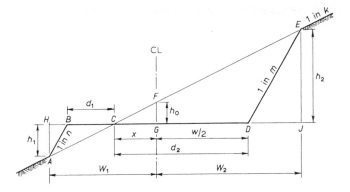

Fig. 10.21 Section part cut/part fill

(2) Triangle CED,

$$\text{area} = h_2 d_2/2$$
$$d_1 = w/2 - x = w/2 - kh_0$$
$$d_2 = w/2 + x = w/2 + kh_0$$

By the rate of approach method and noting that h_1 and h_2 are now required, it will be seen that to conform to the basic figure of the method, the gradients must be transformed into (n in 1), (m in 1) and (k in 1). Therefore

$$h_1 = d_1/(k - n) \tag{10.37}$$
and $\quad h_2 = d_2/(k - m) \tag{10.38}$

The side widths

$$W_1 = w/2 + HB = w/2 + nh_1 \tag{10.39}$$
and $\quad W_2 = w/2 + DJ = w/2 + mh_2 \tag{10.40}$

The area of fill

$$A_f = h_1 d_1/2 = d_1^2/2(k - n)$$
$$= (w/2 - kh_0)^2/2(k - n) \tag{10.41}$$

The area of cut

$$A_c = h_2 d_2/2 = d_2^2/2(k - m)$$
$$= (w/2 + kh_0)^2/2(k - m) \tag{10.42}$$

In the above h_0 has been treated as $-$ve, occurring in the cut. If it is $+$ve and the centre line is in fill, then

$$A_f = (w/2 + kh_0)^2/2(k - n) \tag{10.43}$$
and $\quad A_c = (w/2 - kh_0)^2/2(k - m) \tag{10.44}$

NB: If $h_0 = 0$ and $m = n$ then

$$A_f = A_c = w^2/8(k - m) \tag{10.45}$$

Example 10.5 A proposed road is to have a formation width of 12.00 m with side slopes of 1 in 1 in cut and 1 in 2 in fill. The ground falls at 1 in 3 at right angles to the centre line which has a reduced level of 79.34 m. If the reduced level of the road is to be

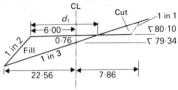

Fig. 10.22

80.10 m, calculate (a) the side widths, (b) the area of cut and (c) the area of fill.

In Fig. 10.22,

$$
\begin{aligned}
d_1 &= w/2 - (-h_0 k) = 6.00 + [(80.10 - 79.34) \times 3] &&= 8.28 \text{ m} \\
d_2 &= w/2 - h_0 k &&= 6.00 - 2.28 &&= 3.72 \text{ m} \\
h_1 &= d_1/(k-n) &&= 8.28/(3-2) &&= 8.28 \text{ m} \\
h_2 &= d_2/(k-m) &&= 3.72/(3-1) &&= 1.86 \text{ m} \\
W_1 &= w/2 + nh_1 &&= 6.00 + (2 \times 8.28) &&= \underline{22.56 \text{ m}} \\
W_2 &= w/2 + mh_2 &&= 6.00 + (1 \times 1.86) &&= \underline{7.86 \text{ m}} \\
A_c &= d_2 h_2/2 &&= 0.50 \times 3.72 \times 1.86 &&= \underline{3.46 \text{ m}^2} \\
A_f &= d_1 h_1/2 &&= 0.50 \times 8.28 \times 8.28 &&= \underline{34.28 \text{ m}^2}
\end{aligned}
$$

By equation 10.41,

$$
\begin{aligned}
A_c &= (w/2 - kh_0)^2/[2(k-m)] = [6.00 - (3 \times 0.76)]^2/[2(3-1)] \\
&= \underline{3.46 \text{ m}^2}
\end{aligned}
$$

By equation 10.42,

$$
\begin{aligned}
A_f &= (w/2 + kh_0)^2/[2(k-n)] = (6.00 + 2.28)^2/[2(3-2)] \\
&= \underline{34.28 \text{ m}^2}
\end{aligned}
$$

Example 10.6 An access road to a small mine is to be constructed to rise at 1 in 20 across a hillside having a maximum slope of 1 in 10. The road is to have a formation width of 7.30 m, and the volumes of cut and fill are to be equalised. Find the width of the cutting, and the volume of excavation in 100 m of road. Side slopes are to batter at 1 in 1 in cut and 1 in 2 in fill. (TP)

To find the transverse slope:

In Fig. 10.23, let AB be the proposed road dipping 1 in 20 (20 units), AC the full dip 1 in 10 (10 units) and AD the transverse slope 1 in t (t units).

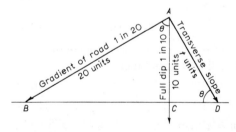

Fig. 10.23

In triangle ABC,
$$\theta = \cos^{-1} 10/20 = 60°$$
In triangle ADC,
$$AD = t = 10/\sin 60° = 11.55 \text{ (gradient value)}$$

If the area of cut $(A_c) =$ the area of fill (A_f) (Fig. 10.24), for $h_0 +$ ve:

$$(w/2 - kh_0)^2 / [2(k - m)] = (w/2 + kh_0)^2 / [2(k - n)]$$

i.e.

$$\frac{(3.65 - 11.55\,h_0)^2}{2(11.55 - 1)} = \frac{(3.65 + 11.55\,h_0)^2}{2(11.55 - 2)}$$

$$3.65 - 11.55\,h_0 = (10.55/9.55)^{1/2}\,(3.65 + 11.55\,h_0)$$
$$= 1.051\,(3.65 + 11.55\,h_0)$$
$$h_0 = -0.186/23.689$$
$$= \underline{-0.0079\ \text{m}},\ \text{i.e. in cut}$$

$$x = kh_0 = 11.55 \times (-0.0079) = -0.091\ \text{m}$$

width of cutting $= 3.65 + 0.09 = \underline{3.74\ \text{m}}$

area of cutting $= (3.65 + 0.09)^2 / [2(11.55 - 1)] = \underline{0.66\ \text{m}^2}$

volume of 100 m of cutting $= 100 \times 0.66 = \underline{66\ \text{m}^3}$

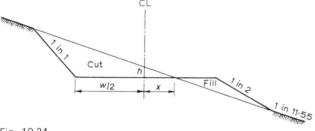

Fig. 10.24

(3)   Multi-level crossfall

It must be assumed that a cross-sectional levelling has been carried out and thus at each change in gradient the level is known together with linear offsets from the centre line. The cartesian coordinates are thus known and a cross-section can be drawn from which the area can be computed by:

(1)   an equalised crossfall to give a regular figure;
(2)   the area from the derived coordinates; or
(3)   the use of the planimeter.

The second method is preferred as the cartesian coordinates are readily computed and the derived values for the offsets and the levels are useful in setting out the figure.

Example 10.7   Given the following sectional levelling: the formation is to be 20 m wide; formation level 2.00 and the sides batter at 1 in $1\frac{1}{2}$, calculate the area of the cross-section.

In Fig. 10.25,

Level of $L = H = 2.00 + 10/1.5 = 8.66\dot{6}$

Gradient $AB = (16.01 - 13.41)$ in $10 = 1$ in $10/2.60$

$L(BL) = 16.01 - 8.66\dot{6} = 7.344$

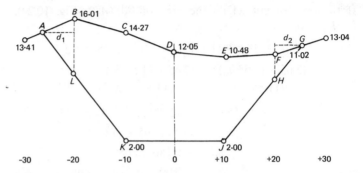

Fig. 10.25

$$d_1 = 7.344 \bigg/ \left(\frac{1}{1.5} + \frac{2.60}{10} \right) = 7.925$$

$$x_A = -27.925$$

$$y_A = 8.66\dot{6} + 7.925/1.5 = 13.950$$

Gradient $EG = (13.04 - 11.02)$ in $10 = 1$ in $10/2.02$

$$L(FH) = 11.02 - 8.66\dot{6} = 2.353$$

$$d_2 = 2.353 \bigg/ \left(\frac{1}{1.5} - \frac{2.02}{10} \right) = 5.064$$

$$x_G = 25.064$$

$$y_G = 8.66\dot{6} + 5.064/1.5 = 12.043$$

Area	x	y
A	-27.93	13.95
B	-20.00	16.01
C	-10.00	14.27
D	0.00	12.05
E	10.00	10.48
F	20.00	11.02
G	25.06	12.04
J	10.00	2.00
K	-10.00	2.00
A	-27.93	13.95

Area $= 391.57\,\text{m}^2$.

Example 10.8   The access to a tunnel has a level formation width of 10 m and runs into a plane hillside, whose natural ground slope is 1 in 10. The intersection line of this formation and the natural ground is perpendicular to the centre line of the tunnel. The level formation is to run a distance of 360 m into the hillside, terminating at the base of a cutting of slope 1 vertical to 1 horizontal. Calculate the amount of excavation in cubic metres. Marks will be deducted if calculations are not clearly related to diagrams.   (LU)

Volume of the solid which forms a wedge (equation 10.21)

$$= wh(x+y+z)/6$$

In Fig. 10.26,

$$x = y = 10.00\,\text{m}$$

$$h = 360/(10-1) \quad \text{(by the rate of approach method)}$$

$$= 40\,\text{m}$$

As the side slopes are 1 in 1.5 then for 40 m in height the horizontal value $d = 60$ m. The value of z is then

$$10 + (2 \times 60) = 130\,\text{m}$$

The volume is then

$$= 40 \times 360 \times (10 + 10 + 130)/6 = \underline{360\,000\,\text{m}^3}$$

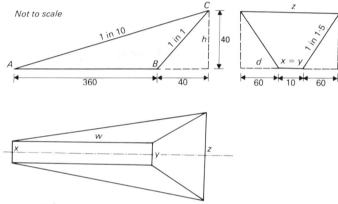

Fig. 10.26

Exercises 10.2 (Cross-sectional areas)

1   At a point A on the surface of ground dipping uniformly due South, 1 in 3, excavation is about to commence to form a short cutting for a branch railway bearing 030° 00′ and rising at 1 in 60 from A. The width at formation level is 6.00 m and the sides batter at 1 vertical to 1 horizontal. Plot two cross-sections at points B and C, 30 m and 50 m respectively from A and calculate the cross-sectional area at B.   (TP Ans.   119.1 m²)

2   Calculate the side widths and cross-sectional area of an embankment to a road with a formation width of 12.00 m. The sides slope 1 in 2, when the centre height is 3.00 m and the existing ground has a crossfall of 1 in 12 at right angles to the centre line of the embankment.

(TP Ans.   10.28 m, 14.40 m, 56.05 m²)

3   A road is to be constructed on the side of a hill having a crossfall of 1 vertical to 8 horizontal at right angles to the centre line of the road; the side slopes are to be similarly 1 in 2 in cut and 1 in 3 in fill; the formation is 15.00 m wide and level. Find the distance of the centre line of the road from the point of intersection of the formation with the natural ground to give equality of cut and fill, ignoring any consideration of 'bulking'.

(LU Ans.   0.34 m on the fill side)

4   The earth embankment for a new road is to have a top width of 12.00 m and the sides slope at 1 vertically to 2 horizontally, the reduced level of the top surface being 30.00 m OD.

At a certain cross-section, the chainages and reduced levels of the natural ground are as follows, the chainage of the centre line being zero, those on the left and right being treated as negative and positive respectively:

Chainage (m) -15.00 -9.00 -4.50 0 $+3.00$ $+13.20$
Reduced level (m)   25.98   26.58   26.76 27.00 27.21   27.72

Find the area of the cross-section of the filling to the nearest square metre, by calculation.   (LU Ans.   51 m^2)

10.5 Volumes derived from the cross-sections

The ordinate formulae applied to irregular areas are equally applicable to irregular solids by changing the ordinate variable into areas. The trapezoidal rule then becomes:

$$V = w[A_1 + 2(A_2 + A_3 + \ldots + A_{n-1}) + A_n]/2 \quad (10.46)$$

The prismoidal rule is then:

$$V = w[A_1 + 4(A_2 + A_4 + \ldots) + 2(A_3 + A_5 + \ldots) + A_n]/3$$
$$(10.47)$$

which is the application of Simson's rule to volumes.

The cross-sections are all parallel and the distance apart can be made equal, the alternate sections being considered as the mid-section.

The formula assumes that the mid-section is derived from the mean of all linear dimensions of the end areas. This is difficult to apply in practice but the above application is considered justified particularly if the distance apart of the sections is kept small.

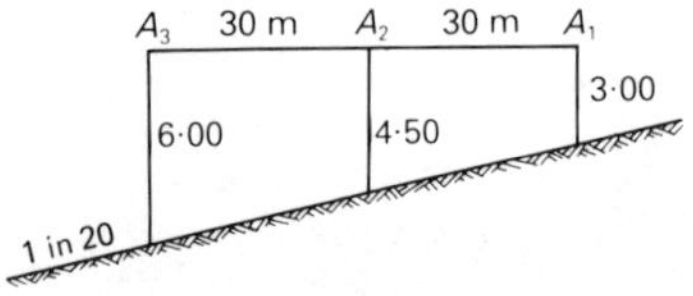

Example 10.9   An embankment is to be formed with its centre line on the surface (in the form of a plane) on full dip of 1 in 20. If the formation width is 12.00 m and the formation heights are 3.00, 4.50 and 6.00 m at intervals of 30.00 m, with side slopes 1 in 2, calculate the volume between the end sections.

Fig. 10.27

In Fig. 10.27,

Area (1) $= h_1(w + mh_1)$
$= 3.00[12.00 + (2 \times 3.00)] = \ 54.00 \, \text{m}^2$
Area (2) $= 4.50[12.00 + (2 \times 4.50)] = \ 94.50 \, \text{m}^2$
Area (3) $= 6.00[12.00 + (2 \times 6.00)] = 144.00 \, \text{m}^2$

Volume:

(a) By mean areas

$$V = W(A/n) = 60.00(54.00 + 94.50 + 144.00)/3$$

$$= 5850.0\,\text{m}^3$$

(b) By trapezoidal rule (end areas)

$$V = w(A_1 + 2A_2 + A_3)/2$$

$$= 30.00(54.00 + 189.00 + 144.00)/2$$

$$= 5805.0\,\text{m}^3$$

(c) By prismoidal rule

$$V = w(A_1 + 4A_2 + A_3)/3$$

$$= 30.00(54.00 + 378.00 + 144.00)/3$$

$$= 5760.0\,\text{m}^3$$

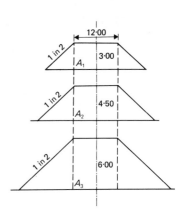

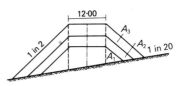

Example 10.10 Given the previous example but with the centre line turned through 90°.

In Fig. 10.28, by equation 10.34,

Fig. 10.28

$$A = \frac{m(h_0^2 k^2 + w^2/4 + wh_0 m)}{(k^2 - m^2)} + wh_0$$

Cross-sectional areas

$$A_1 = \frac{2[(3.00^2 \times 20^2) + (0.25 \times 12.00^2) + (12.00 \times 3.00 \times 2)]}{(20^2 - 2^2)} + (12 \times 3.00)$$

$$= [(3600.00 + 36.00 + 72.00)/198] + 36.00 \quad = \quad 54.73\,\text{m}^2$$

$$A_2 = [(8100.00 + 36.00 + 108.00)/198] + 54.00 \quad = \quad 95.64\,\text{m}^2$$

$$A_3 = [(14400.00 + 36.00 + 144.00)/198] + 72.00 = 145.64\,\text{m}^2$$

Volume

(a) By mean areas

$$V = 60.00(54.73 + 95.64 + 145.64)/3 \quad = 5920.2\,\text{m}^3$$

(b) By end areas

$$V = 30.00(54.73 + 191.28 + 145.64)/2 = 5874.8\,\text{m}^3$$

(c) By prismoidal rule

$$V = 30.00(54.73 + 382.56 + 145.64)/3 = 5829.3\,\text{m}^3$$

10.6 Curvature correction

When the centre line of the construction is curved, the cross-sectional areas will be no longer parallel but radial to the curve (Fig. 10.29).

Volumes of this form are obtained by using the *Theorem of Pappus* which states that 'a volume swept out by a constant area revolving about a fixed axis is given by the product of the area and the length of the path of the centroid of the area'.

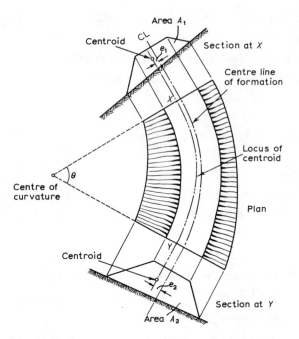

Fig. 10.29 Curvature correction

The volume of earthworks involved in cuttings and embankments as part of transport systems following circular curves may thus be determined by considering cross-sectional areas revolving about the centre of such circular curves.

If the cross-sectional area is constant, then the volume will equal the product of this area and the length of the arc traced by the centroid.

If the sections are not uniform, an approximate volume can be derived by considering a mean eccentric distance $(e) = (e_1 + e_2)/2$ relative to the centre line of the formation.

This will give a mean radius for the path of the centroid $(R \pm e)$, the negative sign being taken as on the same side as the centre of curvature.

Length of path of centroid

$$XY = (R \pm e)\theta_{\text{rad}}.$$

but $\theta_{\text{rad}} = S/R$

where S = length of arc on the centre line, therefore

$$XY = \frac{S}{R}(R \pm e) = S\left(1 \pm \frac{e}{R}\right)$$

Therefore volume is given approximately as,

$$V = \frac{S}{2}(A_1 + A_2)\left(1 \pm \frac{e}{R}\right) \tag{10.48}$$

Alternatively each area may be corrected for the eccentricity of its centroid.

If e_1 is the eccentricity of the centroid of an area A_1 then the volume swept out through a small arc $\delta\theta$ is $\delta V = A_1(R \pm e_1)\delta\theta$.

If the eccentricity had been neglected then

$$\delta V = A_1 R \delta\theta$$

with a resulting error

$$= A_1 e_1 \delta\theta$$

$$= A_1 e_1 / R \quad \text{per unit length} \tag{10.49}$$

Thus, if each area is corrected by an amount $\pm Ae/R$, these new equivalent areas can be used in the volume formula adopted.

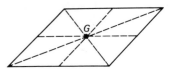

Fig. 10.30

10.6.1   Derivation of the eccentricity e of the centroid G

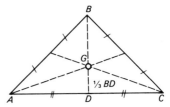

Fig. 10.31

Centroids of simple shapes

Parallelogram (Fig. 10.30)
G lies on the intersection of the diagonals or the intersection of lines joining the midpoints of their opposite sides.   (10.50)

Triangle (Fig. 10.31)
G lies at the intersection of the medians and is $\frac{2}{3}$ of their length from each apex.   (10.51)

Trapezium (Fig. 10.32)

$$y = \frac{h(2a+b)}{3(a+b)} \tag{10.52}$$

$$x = \frac{b}{2} + \frac{(2a+b)(2t+a-b)}{6(a+b)} \tag{10.53}$$

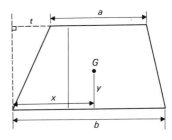

Fig. 10.32

A compound body (Fig. 10.33)
If the areas of the separate parts are A_1 and A_2 and their centroids G_1 and G_2, with the compounded centroid G,

$$G_1 G = \frac{A_2 \times G_1 G_2}{A_1 + A_2} \tag{10.54}$$

or $\quad G_2 G = \dfrac{A_1 \times G_1 G_2}{A_1 + A_2} \tag{10.55}$

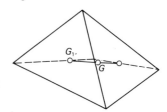

Fig. 10.33

Thus for typical cross-sectional areas met with in earthwork calculations, the figures can be divided into triangles and the centre of gravity derived from the compounding of the separate centroids of the triangles or trapezium, Fig. 10.34.

Alternatively, Fig. 10.35, let the diagonals of $ABCD$ intersect at E. Then

$$BO = OD \text{ on line } BD$$

$$AE = FC \text{ on line } AC$$

then $2 . OG = GF \tag{10.56}$

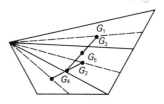

Fig. 10.34

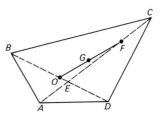

Fig. 10.35

To find the eccentricity e of the centroid G

Case 1　*Where the surface has no crossfall*, the area is symmetrical and the centroid lies on the centre line, i.e. $e = 0$.

Case 2　*Where the surface has a crossfall 1 in k* (Fig. 10.36): let total area of $ABDE = A_T$.

$$\text{area of triangle } AEF = \tfrac{1}{2}AF(H_2 - H_1)$$

$$= \frac{W_1(W_1 + W_2)}{k} \tag{10.57}$$

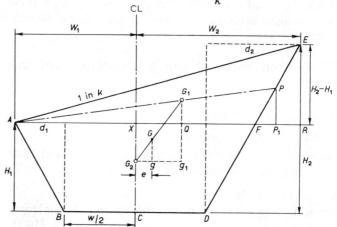

Fig. 10.36 Section with crossfall

Let G_1 and G_2 be the centroids of areas AEF and $AFDB$ respectively.

$$\text{length } AQ = \text{horizontal projection of } AG_1$$
$$= \tfrac{2}{3}(AP_1)$$
$$= \tfrac{2}{3}\left(\frac{AR + AF}{2}\right) = \tfrac{1}{3}(W_1 + W_2 + 2W_1)$$
$$= \tfrac{1}{3}(3W_1 + W_2) = W_1 + \frac{W_2}{3}$$

Distance of Q from centre line:

$$XQ = G_2g_1 = W_1 + \frac{W_2}{3} - W_1$$
$$= W_2/3 \tag{10.58}$$

Distance of centroid G for the whole figure (from the centre line)

$$e = \frac{\text{area } \triangle AEF \times XQ}{\text{total area } A_T} = \frac{W_1 W_2(W_1 + W_2)}{3kA_T} \tag{10.59}$$

conversion area $A_c = \pm A_T e/R$

i.e.
$$= \pm \frac{A_T[W_1 W_2(W_1 + W_2)]}{3k\, A_T R}$$
$$= \frac{W_1 W_2(W_1 + W_2)}{3kR}$$

therefore

$$\text{corrected area} = A_T \pm \frac{W_1 W_2 (W_1 + W_2)}{3kR} \qquad (10.60)$$

Case 3   *Sections with part cut and part fill* (Fig. 10.37): for section in cut, i.e. triangle *CED*, *G* lies on the median *EQ*.

$$JQ = \tfrac{1}{2}\left(\frac{w}{2} + x\right) - x = \tfrac{1}{2}\left(\frac{w}{2} - x\right)$$

$$= \tfrac{1}{2}\left(\frac{w}{2} - kh_0\right)$$

$$e_2 = JQ + \tfrac{1}{3}(W_2 - JQ) = \tfrac{1}{3}(W_2 + 2JQ)$$

$$= \tfrac{1}{3}\left(W_2 + \frac{w}{2} - kh_0\right) \qquad (10.61)$$

Similarly for fill

$$e_1 = \tfrac{1}{3}\left(W_1 + \frac{w}{2} + kh_0\right) \qquad (10.62)$$

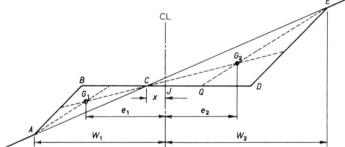

Fig. 10.37 Section part cut/part fill

Example 10.10   Using the information in Example 10.9, viz. embankment with surface crossfall of 1 in 20, side slopes 1 in 2, formation width 12.00 m and formation heights of 3.00, 4.50 and 6.00 m at 30.00 m centres, if the formation lies with its centre line on the arc of a circle of radius 150 m, calculate:

(a)   the side widths of each section;
(b)   the eccentricity of their centroids;
(c)   the volume of the embankment over this length if the centre of curvature is (i) uphill and (ii) downhill.

(a)   *Side widths*

(Section 1): $W_1 = w/2 + (h_0 - w/2k)mk/(k + m)$

$$= 6.00 + (3.00 - 12.00/40)(2 \times 20)/22$$

$$= 6.00 + (3.00 - 0.30) \times 1.818 = 10.91 \text{ m}$$

$$W_2 = w/2 + (h_0 + w/2k)mk/(k - m)$$

$$= 6.00 + (3.00 + 0.30) \times 2.222 = 13.33 \text{ m}$$

(Section 2): $W_1 = 6.00 + (4.50 - 0.30) \times 1.818 = 13.64\,\text{m}$
$\ \ W_2 = 6.00 + (4.50 + 0.30) \times 2.222 = 16.67\,\text{m}$

(Section 3): $W_1 = 6.00 + (6.00 - 0.30) \times 1.818 = 16.36\,\text{m}$
$\ \ W_3 = 6.00 + (6.00 + 0.30) \times 2.222 = 20.00\,\text{m}$

(b)   *Eccentricity (e)*

By equation 10.60,

$$e = [W_1 W_2 (W_1 + W_2)]/3kA$$
$$e_1 = (10.91 \times 13.33)(10.91 + 13.33)/(3 \times 20 \times 54.73)$$

(Area from Example 10.9)

$$= 1.07\,\text{m}$$
$$e_2 = (13.64 \times 16.67)(13.64 + 16.67)/(60 \times 95.64)\ \ = 1.20\,\text{m}$$
$$e_3 = (16.36 \times 20.00)(16.36 + 20.00)/(60 \times 145.64) = 1.36\,\text{m}$$

(c)   *Volumes*

Using the above values of eccentricity in the prismoidal formula the volume correction is:

$$V_c = \pm 30.00[(54.73 \times 1.07/150) + 4(95.64 \times 1.20/150) + (145.64 \times 1.36/150)]/3$$
$$= \pm 47.71\,\text{m}^3$$

The correction is $+$ ve if the centre line lies on the uphill side and $-$ ve if on the downhill side.

$$V_1 = 5829.3 \pm 47.71 = 5877.0\,\text{m}^3 \quad \text{or} \quad 5781.6\,\text{m}^3$$

Applying the area correction (equation 10.61),

$$A_1' = A_1 + A_{c1} = \ \ 54.73 \pm 0.39 = \ \ 55.12\,\text{m}^2 \quad \text{or} \quad 54.34\,\text{m}^2$$
$$A_2' = A_2 + A_{c2} = \ \ 95.64 \pm 0.77 = \ \ 96.41\,\text{m}^2 \quad \text{or} \quad 94.87\,\text{m}^2$$
$$A_3' = A_3 + A_{c3} = 145.64 \pm 1.32 = 146.96\,\text{m}^2 \quad \text{or} \quad 144.32\,\text{m}^2$$

The corrected volumes are then:

(uphill) $V_1' = 10.00(55.12 + 385.64 + 146.96) = 5877.2\,\text{m}^3$
(downhill) $V_2' = 10.00(54.34 + 379.48 + 144.32) = 5781.4\,\text{m}^3$

10.7   Calculation of volumes from contour maps

Here the volume is derived from the areas contained in the plane of the contour. For accurate determinations the contour interval must be kept to a minimum and this value will be the width (w) in the formulae previously discussed.

The areas will generally be obtained by means of a planimeter, the latter tracing out the enclosing line of the contour.

For most practical purposes the prismoidal formula is satisfactory, with alternate areas as 'mid-areas' or, if the contour interval is large, an interpolated mid-contour giving the required 'mid-area' may be used.

10.8 Calculation of volumes from spot-heights

This method uses grid levels from which the depth of construction is derived.

The volume is computed from the mean depth of construction in each section forming a truncated prism, the end area of which may be rectangular but preferably triangular (Fig. 10.38).

$$V = \text{plan area} \times \text{mean height} \qquad (10.63)$$

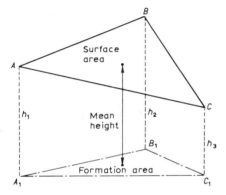

Fig. 10.38 Volume from spot-heights

If a grid is used, the triangular prisms are formed by drawing diagonals, and then each prism is considered in turn (Fig. 10.39).

The total volume is then derived (each triangle is of the same area) as one third of the area of the triangle multiplied by the sum of each height in turn multiplied by the number of applications of that height, i.e.

$$V = \frac{\Delta}{3}\left(\Sigma nh\right) \qquad (10.64)$$

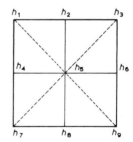

Fig. 10.39

e.g.

$$V = \frac{\Delta}{3}\left(2h_1 + 2h_2 + 2h_3 + 2h_4 + 8h_5 + 2h_6 + 2h_7 + 2h_8 + 2h_9\right)$$

10.9 Digital ground models (DGMs)

10.9.1 These are mathematical models used by the computer to represent land surfaces, whether they are natural topographical features or man-made structures.

The computer is used to store the coordinates (x, y and z, i.e. E, N and levels) of selected points, together with parameters such as chainages, offsets, radii of curves and polar coordinates.

The mathematical models, derived from direct observations, computed data or photogrammetric interpolations, consist of types of 'strings' in the form of connected points.

The system, retrieving the stored data, may be used as a survey tool, e.g. computerised automated plotting data for making plans, revising existing OS sheets, constructing sections, computing areas and volumes, optimising route designs and for computing setting out data. The original input may thus be in the form of original ground fieldwork, or as vertical or terrestrial photogrammetry. As a design tool for highway engineers, the DGM forms the basis for horizontal and vertical alignment.

10.9.2　Types of 'strings'

(1)　Two dimensional (x, y) for a constant level (z), i.e. a contour.
(2)　Three dimensional (x, y, z), i.e. a feature line.
(3)　Section strings relative to a baseline.
(4)　Alignment strings—chainage, bearing, radius of curvature, etc.
(5)　Point strings—successive points assumed to be joined by a series of straight lines.

10.9.3　The DGM coordinates are stored in matrix form (Fig. 10.40), describing the topography in the form of a square, rectangular or triangular grid, with the intersection points referenced, e.g. a_{11}, a_{12}, a_{1n} etc., assigned values at the corresponding point on the grid, and known by the location of the element in the matrix. The column number represents the Easting and the row number the Northing.

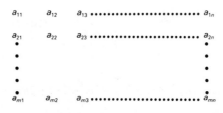

Fig. 10.40

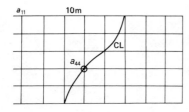

Fig. 10.41

The grid is formed adjacent to the proposed scheme, as in Fig. 10.41. If the coordinates of an element (a_{11}) are say 1960 E, 2140 N and the grid has a 10 m size, then the coordinates of the element (a_{44}) will be:

$$E: \ 1960 + (4-1) \times 10 = 1990$$

$$N: \ 2140 - (4-1) \times 10 = 2110$$

Hence any set of coordinates may be converted to matrix reference numbers or *vice versa*.

10.9.4 Square grid interpolation

To find the level (z_p) of a point $P(x_p, y_p)$ assuming that the grid is sufficiently small that the ground surface is considered to be a uniform plane: in Fig. 10.42,

$$z_5 = x_p(z_3 - z_4)/L + z_4$$

$$z_6 = x_p(z_2 - z_1)/L + z_1$$

then $z_p = z_5 + y_p(z_6 - z_5)/L$

$$= z_4 + x_p(z_3 - z_4)/L + y_p\{[x_p(z_2 - z_1)/L + z_1]$$
$$- [x_p(z_3 - z_4)/L + z_4)]/L\}$$

$$= z_4 + x_p(z_3 - z_4)/L + y_p[(z_1 - z_4)$$
$$+ x_p(z_2 + z_4 - z_1 - z_3)/L]/L \qquad (10.65)$$

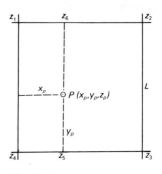

Fig. 10.42

10.9.5 Triangular grid interpolation (Fig. 10.43)

Using the 'cut' formulae (p. 115),

$$x_q = \frac{x_1 \cot \phi_{1,3} - x_2 \cot \phi_{2,p} + \Delta y_{1,2}}{\cot \phi_{1,3} - \cot \phi_{2,p}} \qquad (10.66)$$

$$x_q = \frac{x_1(\Delta y_{1,3}/\Delta x_{1,3}) - x_2(\Delta y_{2,p}/\Delta x_{2,p}) + \Delta y_{1,2}}{(\Delta y_{1,3}/\Delta x_{1,3}) - (\Delta y_{2,p}/\Delta x_{2,p})} \qquad (10.67)$$

Then $z_q = (x_q - x_1)(z_3 - z_1)/(x_3 - x_1) + z_1$

$$= z_1 + (\Delta x_{1,q} \cdot \Delta z_{1,3})/\Delta x_{1,3} \qquad (10.68)$$

$$z_p = z_2 + (x_p - x_2)(z_q - z_2)/(x_q - x_2)$$

$$= z_2 + \Delta x_{2,p} \cdot \Delta z_{2,q}/\Delta x_{2,q} \qquad (10.69)$$

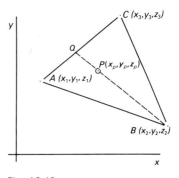

Fig. 10.43

10.9.6 Interpolation from point strings

Here random points having x, y and z coordinates are stored in the computer's data bank. Interpolation is then obtained by the use of an equation of the form $z_p = ax_i^2 + bx_i y_i + cy_i^2 + dx_i + ey_i + f$ using a number of points (n) surrounding the unfixed point (n must be equal to or greater than 6; $i = 1$ to n).

A 'best fit' value for z is obtained by solving the simultaneous equations if $n = 6$ or using the principle of least squares to provide normal equations when n is greater than 6. Weights are attributed to levels of the points surrounding the interpolated point depending upon their distances from the point, i.e. $w \propto 1/S^2$ (see Shepherd, *Advanced Engineering Surveying*, Chapter 1).

Example 10.11 In a DGM square grid of 20 m sides, the levels of the corners (1 to 4) are given as: 10.00, 11.50, 11.50, 13.00 m. If the point P has coordinates $x = 5.00$, $y = 15.00$ m find the level of P.

By equation 10.65,

$$z_p = z_4 + x_p(z_3 - z_4)/L + y_p[(z_1 - z_4) + x_p(z_2 + z_4 - z_1 - z_3)/L]/L$$
$$= 13.00 + 0.25(11.50 - 13.00) + 0.75[(10.00 - 13.00) + 0.25(11.50 + 13.00 - 10.00 - 11.50)]$$
$$= 13.00 - 0.38 - 1.69 = 10.93 \text{ m}$$

Example 10.12 Given a triangle ABC with coordinates $A(10.00, 10.00, 10.00)$, $B(20.00, 5.00, 8.25)$ and $C(15.00, 15.00, 12.75)$, compute the level of the point P whose coordinates are $(15.50, 12.25)$.

$$\cot \phi_{1,3} = (15.00 - 10.00)/(15.00 - 10.00) = 1.000$$
$$\cot \phi_{2,p} = (12.25 - 5.00)/(15.50 - 20.00) = -1.611$$

By equation 10.66,

$$x_q = \frac{(10.00 \times 1.000) - (20.00 \times -1.611)45.00}{(1.000 + 1.611)} = 14.26$$

By equation 10.68,

$$z_q = 10.00 + (4.26 \times 2.75)/5.00 \qquad = 12.34$$

By equation 10.69,

$$z_p = 8.25 + (-4.50 \times 4.09)/5.74 \qquad = 11.46$$

10.9.7 Calculation of volumes

Using the DGM volumes may be calculated as follows:

(1) After the route location the coordinates of the horizontal alignment is determined.
(2) Vertical alignment design gives formation levels on the centre line.
(3) Existing ground levels on the cross-sections are calculated by interpolation as above.
(4) Using the specified parameters such as the side slopes and the centre line levels as in (2) the area of the cross-sections are obtained.
(5) Earthwork volumes are calculated from the cross-sectional areas with allowance for 'bulking' where necessary.
(6) Alternative vertical alignments may be tried to optimise the earth movement using 'mass-haul' diagrams.

10.10 Mass-haul diagrams

These are used in planning the haulage of large volumes of earthwork for construction works in railway and trunk road projects.

10.10.1 Definitions

Bulking An increase in volume of earthwork after excavation.
Shrinkage A decrease in volume of earthwork after deposition and compaction.

Haul distance (d) The distance from the working face of the excavation to the tipping point.

Average haul distance (D) The distance from the centre of gravity of the cutting to that of the filling.

Freehaul distance The distance, given in the Bill of Quantities, included in the price of excavation per cubic metre.

Overhaul distance The extra distance of transport of earthwork volumes beyond the freehaul distance.

Haul The sum of the product of each load by its haul distance. This must equal the total volume of excavation multiplied by the average haul distance, i.e. $\Sigma vd = VD$.

Overhaul The products of volumes by their respective overhaul distance. Excess payment will depend upon overhaul.

Station metre A unit of overhaul, viz. $1\,m^3 \times 100\,m$.

Borrow The volume of material brought into a section due to a deficiency.

Waste The volume of material taken from a section due to excess.

10.10.2 Construction of the mass-haul diagram
(Fig. 10.44)

(1) Calculate the cross-sectional areas at given intervals along the project.
(2) Calculate the volumes of cut and fill between the given areas relative to the proposed formation.

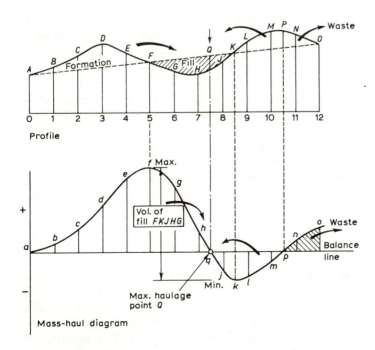

Fig. 10.44 Mass-haul curves

Notes
(a)   Volumes of cut are considered positive.
(b)   Volumes of fill are considered negative.
(3)   Calculate the aggregated algebraic volume for each section.
(4)   Plot the profile of the existing ground and the formation.
(5)   Using the same scale for the horizontal base line, plot the mass-haul curve with the aggregated volumes as ordinates.

10.10.3   Characteristics of the mass-haul diagram

(1)   A rising curve indicates cutting as the aggregate volume is increasing ($a–f$ is seen to agree with AF on the profile).
(2)   A maximum point on the curve agrees with the end of the cut, i.e. $f–F$.
(3)   A falling curve indicates filling as the aggregate volume is decreasing ($f–k$ is seen to agree with $F–K$ on the profile).
(4)   The vertical difference between a maximum point and the next minimum point represents the volume of the embankment, i.e. $ff_1 + k_1 k$ (the vertical difference between any two points not having a minimum or maximum between them represents the volume of earthwork between them).
(5)   If any horizontal line is drawn cutting the mass-haul curve, e.g. aqp, the volume of cut equals the volume of fill between these points. In each case the algebraic sum of the quantities must equal zero.
(6)   When the horizontal balancing line cuts the curve, the area above the line indicates that the earthwork volume must be moved forward. When the area cut off lies below the balancing line, then the earthwork must be moved backwards.
(7)   The length of the balancing line between intersection points, e.g. aq, qp, represents the maximum haul distance in that section (q is the maximum haulage point both forward, aq, and backwards, pq).
(8)   The area cut off by the balancing line represents the haul in that section.
NB: As the vertical and horizontal scales are different, i.e. 1 in V (vertical) and 1 in H (horizontal) then the area in mm^2 is given as

$$A = aVH \times 10^{-6}\,m^3\,m^{-1}$$

$$= aVH \times 10^{-8}\ \text{station metres}$$

10.10.4   Freehaul and overhaul (Fig. 10.45)

The mass-haul diagram is used for finding the overhaul charge as follows: freehaul distance is marked off parallel to the balance line on any haul area, e.g. bd. The ordinate cc_2 represents the volume dealt with as illustrated in the profile.

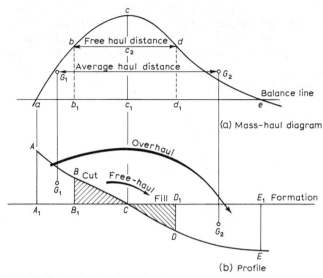

(a) Mass-haul diagram

(b) Profile

Fig. 10.45 Freehaul and overhaul

Any cut within the section ABB_1A_1 has to be transported through the freehaul length to be deposited in the section D_1E_1ED. This represents the 'overhaul' of volume (ordinate bb_1) which is moved from the centroid G_1 of the cut to the centroid G_2 of the fill.

The overhaul distance is given as the distance between the centroids less the freehaul distance, i.e.

$$(G_1G_2) - bd$$

The amount of overhaul is given as the volume (ordinate $bb_1 = dd_1$) × the overhaul distance.

Where long haulage distances are involved, it may be more economical to waste material from the excavation and to borrow from a location within the freehaul limit.

If l is the overhaul distance, c the cost of overhaul and e the cost of excavation, then to move $1\,m^{-3}$ from cut to fill the cost is given as

$$e + lc$$

whereas the cost to cut, waste the material, borrow and tip without overhaul will equal $2e$. Economically

$$e + lc = 2e$$

therefore

$$l = e/c \quad \text{(assuming no cost for wasting)}$$

Thus if the cost of excavation is $100\,p\,m^{-3}$ and the cost of overhaul is $10\,p$ per station metre, then the total economic overhaul distance

$$= 100 \times 100/10 = 1000\,m$$

If the freehaul is given as $100\,m$ the maximum economic haul

$$= 1000 + 100 = 1100\,m$$

The overhaul distance is found from the mass-haul diagram by determining the distance from the centroid of the mass of the excavation to the centroid of the mass of the embankment.

The centroid of the excavation and of the embankment can be determined (1) graphically or (2) planimetrically. These methods are illustrated in the following example.

Example 10.12   Volumes of cut and fill along a length of proposed road are as follows:

	Volume (m³)	
Chainage	Cut	Fill
0		
100	290	
200	760	
300	1680	
400	620	
480	120	
500		20
600		110
700		350
800		600
900		780
1000		690
1100		400
1200		120

Draw a mass-haul diagram, and excluding the surplus excavated material along this length determine the overhaul if the freehaul distance is 300 m.   (ICE)

	Volume (m³)		Aggregate
Chainage	Cut	Fill	volume (m³)
0			
100	290		+ 290
200	760		+ 1050
300	1680		+ 2730
400	620		+ 3350
480	120		+ 3470
500		20	+ 3450
600		110	+ 3340
700		350	+ 2990
800		600	+ 2390
900		780	+ 1610
1000		690	+ 920
1100		400	+ 520
1200		120	+ 400
	3470	3070	
	3070		
Check	400		

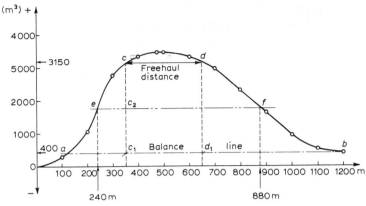

Fig. 10.46 Mass-haul diagram (Example 10.12)

(1) *Graphical method* (Fig. 10.46)

(i) As the surplus of $400 \, m^3$ is to be neglected, the balancing line is drawn from the end of the mass-haul curve, parallel to the base line, to form a new balancing line ab.

(ii) As the freehaul distance is 300 m, this is drawn as a balancing line cd.

(iii) From c and d, draw ordinates cutting the new base line at $c_1 d_1$.

(iv) To find the overhaul:

(a) Bisect cc_1 to give c_2 and draw a line through c_2 parallel to the base line and cutting the curve at e and f, which now represent the centroids of the masses acc_1 and dbd_1.

(b) The average haul distance from acc_1 in excavation to make up the embankment $dbd_1 = ef$.

(c) The overhaul distance = the haul distance – the free-haul distance, i.e.

$$ef - cd$$

i.e. scaled value $= 640 - 300 = 340 \, m$.

(d) The overhaul of material at acc_1

$$= \text{volume } (cc_1) \times \text{overhaul distance } (ef - cd)$$
$$= 2750 \, m^3 \times 340 \, m$$
$$= 9350 \text{ station metres}$$

(2) *Planimetric method*

distance to centroid = haul/volume

$$= \frac{\text{area} \times \text{horizontal scale} \times \text{vertical scale}}{\text{volume ordinate}}$$

From area acc_1

area scaled from mass-haul curve $= 0.9375 \, cm^2$

horizontal scale $= 1 \, cm = 200 \, m$

vertical scale $= 1 \, cm = 1600 \, m^3$

Therefore
$$\text{haul} = 0.9375 \times 200 \times 1600 = 300\,000$$
$$\text{volume (ordinate } cc_1) = \quad 2750$$
$$\text{distance to centroid} = 300\,000/2750$$
$$= 109.1 \text{ m}$$
$$\text{chainage of centroid} = 350 - 109.1$$
$$= \underline{240.9 \text{ m}}$$

For area dbd_1
$$\text{area scaled} = 1.9688 \text{ cm}^2$$

Therefore
$$\text{haul} = 1.9688 \times 320\,000 = 630\,016$$
$$\text{volume (ordinate } dd_1) = 2750$$
$$\text{distance to centroid} = 229.1 \text{ m}$$
$$\text{chainage of centroid} = 650 + 229.1$$
$$= \underline{879.1 \text{ m}}$$
$$\text{average haul distance} = 879.1 - 240.9$$
$$= 638.2 \text{ m}$$
$$\text{overhaul distance} = 638.2 - 300$$
$$= \underline{338.2 \text{ m}}$$

Therefore
$$\text{overhaul} = 338.2 \times 2750$$
$$= \underline{9300} \text{ station metres}$$

NB: Instead of the above calculation the overhaul can be obtained direct as the sum of the two mass-haul curve areas acc_1 and dbd_1.

$$\text{area } acc_1 = \frac{301\,950}{100} \text{ station metres}$$

$$\text{area } dbd_1 = \frac{634\,950}{100} \text{ station metres}$$

$$\text{total area} = \text{overhaul} = \frac{936\,900}{100} = 9369 \text{ station metres}$$

Proof: Take any area cut off by a balancing line, Fig. 10.47. Let a small increment of area $\delta A = $ (say) 1 m³ and length of haul be L. Then

$$\delta A = 1 \text{ m}^3 \times L/100 \text{ station metres}$$

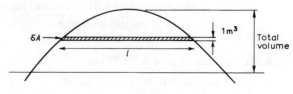

Fig. 10.47

therefore

$$A = n \times 1\,\mathrm{m}^3 \times \frac{\Sigma L}{n}$$

$\qquad$ = total volume $\times$ average haul distance

Therefore

$\qquad$ area = total haul

Exercises 10.3 (Earthwork volumes)

1 The following cross-sectional areas were derived at 20 m intervals along an embankment. Calculate the volume contained between the first and last cross-section.

Section $\quad$ 1 $\quad$ 2 $\quad$ 3 $\quad$ 4 $\quad$ 5 $\quad$ 6 $\quad$ 7
Area (m²) $\;$ 28.4 $\;$ 57.6 $\;$ 68.9 $\;$ 109.8 $\;$ 98.6 $\;$ 112.6 $\;$ 87.6
$\qquad\qquad$ (TP Ans. $\;$ (a) $\;$ By end areas $10\,110.0\,\mathrm{m}^3$
$\qquad\qquad\qquad\qquad$ (b) $\;$ By prismoidal rule $10\,473.3\,\mathrm{m}^3$)

2 In a cutting, formation width 10.00 m, the following centre line heights with corresponding ground slopes are measured:

	Centre line height	Slopes
(1)	2.40 m	constant ground slope 1 : 8 falling left.
(2)	3.60 m	slope left; 1 : 12 fall, right; 1 : 20 rise.
(3)	4.40 m	slope left; 1 : 16 rise, right; 1 : 10 fall.

Calculate the side widths, the cross-sectional areas and the volume of the sections are 20 m apart. (NB: The sides slope at 1.5 vertical to 1 horizontal.)

$\qquad\qquad$ (TP Ans. $\;$ (1) $\quad$ 7.24, 10.58, 34.43 m²
$\qquad\qquad\qquad\qquad$ (2) $\quad$ 9.24, 11.24, 54.34 m²
$\qquad\qquad\qquad\qquad$ (3) $\quad$ 12.79, 10.08, 71.79 m²
$\qquad\qquad\qquad\qquad\qquad$ Volume 2157.2 m³)

3 Calculate the cubic contents, using the prismoidal formula, of the length of embankment of which the cross-sectional areas at 50 m intervals are as follows:

Distance (m) $\quad$ 0 $\quad$ 50 $\quad$ 100 $\quad$ 150 $\quad$ 200 $\quad$ 250 $\quad$ 300
Area (m²) $\quad$ 110 $\quad$ 425 $\quad$ 640 $\quad$ 726 $\quad$ 1590 $\quad$ 1790 $\quad$ 2600

Make a similar calculation using the trapezoidal method and explain why the results differ.
$\qquad\qquad$ (ICE Ans. $\;$ $315\,566.7\,\mathrm{m}^3$, $326\,300.0\,\mathrm{m}^3$)

4 A level cutting is made on ground having a uniform cross-slope of 1 in 8. The formation width is 9.70 m and the sides slope at 1 vertical to 1.75 horizontal. At 3 sections, spaced 20 m apart, the depths to the centre line are 10.30 m, 8.48 m and 6.06 m respectively. Calculate (a) the side widths of each section; (b) the volume of the cutting.
$\qquad\qquad$ (TP Ans. $\;$ (a) 18.76 m, 29.29 m, (b) 8717.33 m³)

5 Calculate the volume, in cubic metres, contained between three successive sections of a railway cutting 15 m apart. The width of formation is 3.00 m, the side slope 1 vertical to 2 horizontal and the heights at the top of the slopes above the formation level are as follows:

	Left	Centre	Right
1st cross-section	4.53	4.00	4.67
2nd cross-section	5.33	5.17	5.93
3rd cross-section	6.10	5.33	5.33

(TP Ans.   2124.2 m^3)

6 A 100 m length of earthwork volume for a proposed road has a constant cross-section of cut and fill, in which the cut area equals the fill area. The level formation is 30 m wide, the transverse ground slope is 1 in 20 and the side slopes in cut and fill are 0.5 horizontal to 1 vertical and 1 horizontal to 1 vertical respectively. Calculate the volume of excavation in 100 m length.   (LU Ans.   5646 m^3)

7 The centre line of a highway cutting is on a curve of 80 m radius, the original surface of the ground being approximately level. The cutting is to be widened by increasing the formation width from 4.000 m to 6.000 m, the excavation to be entirely on the inside of the curve and to retain the existing side slopes of 1.5 horizontal to 1 vertical. If the depth of formation increases uniformly from 1.600 m at chainage 600 to 3.400 m at chainage 660 m, calculate the volume of earth to be removed in this 60 m length.   (Ans.   281.41 m^3)

8 The central heights of the ground above formation at three sections 20 m apart are 2.000 m, 2.400 m and 3.000 m and the crossfalls at these sections are 1 in 30, 1 in 40 and 1 in 20. If the formation width is 8.000 m and the sides slope at 1 in 2, calculate the volume of excavation in the 40 m length: (a) if the centre line is straight; and (b) if the centre line is an arc of 80 m radius.
(Ans.   (a) 1266.2 m^3, (b) 1274.1 or 1258.3 m^3)

9 The following notes refer to 1200 m section of a proposed railway and the earthwork distribution in this section is to be planned without regard to the adjoining sections. The table shows the stations and the surface levels along the centre line, the formation level being at an elevation above datum of 43.50 m at chainage 70 and thence rising uniformly on a gradient of 1.2 %. The volumes are recorded in m^3, the cuts are plus and the fills minus.
(i) Plot the longitudinal section using a horizontal scale of 1/1200 and a vertical scale of 1/240.
(ii) Assuming a correction factor of 0.8 applicable to fill, plot the mass-haul diagram to a vertical scale of 20 mm to 1000 m^3.
(iii) Calculate the total haul in station metres and indicate the haul limits on the curve and section.

(iv) State which of the following estimates you would recommend:

(a) No freehaul at 35p per m³ for excavating, hauling and filling.

(b) A freehaul distance of 300 m at 30p per m³ plus 2p per station metre for overhaul.

Chainage (m)	Surface level (m)	Volume (m³)	Chainage (m)	Surface level (m)	Volume (m³)	Chainage (m)	Surface level (m)	Volume (m³)
70	52.8		74	44.7		78	49.5	
		+1860			−1080			−237
71	57.3		75	39.7		79	54.3	
		+1525			−2025			+362
72	53.4		76	37.5		80	60.9	
		+ 547			−2110			+724
73	47.1		77	41.5		81	62.1	
		− 238			−1120			+430
74	44.7		78	49.5		82	78.5	

(LU Ans.   22 790 station m. (a) £1932; (b) £1793)

10 The volumes between sections along a 1200 m length of proposed road are shown below, positive volumes denoting cut and negative volumes denoting fill:

Chainage (m)	0	100	200	300	400	500	600	700	800	900	1000	1100	1200
Volume between sections (m³ × 10³)		+2.1	+2.8	+1.6	−0.9	−2.0	−4.6	−2.5	−2.2	−2.4	+1.1	+3.9	+2.8

Plot a mass-haul diagram for this length of road to a suitable scale and determine suitable positions of balancing lines so that there is:

(a) a surplus at chainage 1200 but none at chainage 0;
(b) a surplus at chainage 0 but none at chainage 1200;
(c) an equal surplus at chainage 0 and chainage 1200.

Hence determine the cost of earth removal for each of the above conditions based upon the following prices and a freehaul limit of 400 m.

Excavate, cart and fill (freehaul)	60 p/m³
Excavate, cart and fill (overhaul)	85 p/m³
Removal of surplus to tip from chainage 0,	125 p/m³
Removal of surplus to tip from chainage 1200,	150 p/m³

(ICE Ans.   £14 800, £14 100, £14 400)

11 The volumes in m³ between successive sections 100 m apart on a 900 m length of a proposed road are given below: excavation is positive and fill negative.

Section	0	1	2	3	4	5	6	7	8	9
Volume		+1700	−100	−3200	−3400	−1400	+100	+2600	+4600	+1100

Determine the maximum haul distance when earth may be wasted only at the 900 m end. Show and evaluate on your diagram the overhaul if the freehaul limit is 300 m.

(LU Ans.   558 m, 5500 station m)

12   The areas of ground within contour lines at the site of a reservoir are as follows:

Contour in m above datum	Area (m^2)
40.0	5056.02
39.5	4421.04
39.0	3016.35
38.5	2322.03
38.0	940.56
37.5	568.21
37.0	341.07
36.5	158.34
36.0	4.72

Taking 36.0 m AOD as the level of the bottom of the reservoir and 40.0 m AOD as the water level, estimate the quantity of water in m^3 contained in the reservoir. (Use Simpson's rule.)

(Ans.   7255.86 m^3)

Bibliography

BANNISTER, A., and RAYMOND, S., *Surveying*, 4th edition. Pitman (1977)

MURCHISON, D. E., *Surveying and Photogrammetry: Computation for Civil Engineers*. Newnes-Butterworth (1977)

RITTER, L. J., and PAQUETTE, R. J., *Highway Engineering*, 4th edition. Wiley (1979)

SHEPHERD, F. A., *Surveying Problems and Solutions*. Edward Arnold (1968)

UREN, J., and PRICE, W. F., *Surveying for Engineers*. Macmillan (1978)

Appendix

Areas of regular figures

Triangle

  (a) $A = b \cdot h/2$

  (b) $A = (a \cdot b \cdot \sin C)/2$

  (c) $A = \sqrt{[s(s-a)(s-b)(s-c)]}$

             $s = (a+b+c)/2$

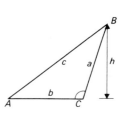

Quadrilateral

  (a)   Rectangle $A = b \cdot l$

  (b)   Parallelogram $A = a \cdot h$

                    $= a \cdot b \cdot \sin \alpha$

  (c)   Trapezium $A = (a+b)h/2$

  (d)   Irregular $A = AC(Bb + Dd)/2$

                    $= (AC \cdot BD)(\sin \theta)/2$

Circle

  (a) $A = \pi r^2 = \pi d^2/4$

  (b)   Sector $A = \pi r^2 \theta/360$

                    $= r^2 \theta_{\text{rad}}/2$

  (c)   Segment   (i) $A = r^2[(\theta/360) - (\sin \theta)/2]$

                (ii) $A = r^2(\theta - \sin \theta)/2$

                (iii) $A = \dfrac{4h}{3}\left(\dfrac{25h^2}{64} + \dfrac{c^2}{4}\right)^{\frac{1}{2}}$

                    $\simeq h^3/2c + 2ch/3$

  (d)   Annulus $A = \pi(R^2 - r^2)$

Ellipse $A = \pi a \cdot b/4$

Parabola $A = 2 \cdot b \cdot h/3$

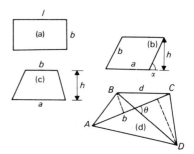

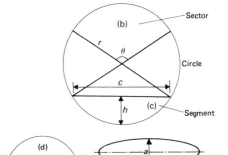

Surface areas

Cylinder $SA = 2\pi r \cdot h$

Cone $SA = \pi r \cdot l$

Sphere

  (a) $SA = 4\pi r^2$

  (b)   Segment $SA = 2\pi r \cdot h$

Index

365